IEE CONTROL ENGINEERING SERIES 56

Series Editors: Professor D. P. Atherton
Professor G. W. Irwin

SYMBOLIC METHODS *in*
control system analysis and design

Other volumes in print in this series:

SYMBOLIC METHODS *in*
control system analysis and design

Edited by Neil Munro

The Institution of Electrical Engineers

Published by: The Institution of Electrical Engineers, London,
United Kingdom

British Library Cataloguing in Publication Data

A CIP catalogue record for this book
is available from the British Library

ISBN 0 85296 943 0

Printed in England by Bookcraft, Bath

Contents

Foreword

Advances in computer technology have had a major effect on control engineering, not only in the implementation of control systems but also for control system analysis and design. They have allowed new algorithms and techniques to be implemented and allowed techniques to be accomplished much more efficiently and faster, also graphical facilities have greatly improved the way in which data can be presented. This contrasts remarkably with when I started my PhD around forty years ago at the University of Manchester where I was lucky enough to have access to one of the most powerful computers then available. Its vacuum tubes, magnetic drums and cathode ray tubes filled a large room but its capabilities were but a fraction of today's personal computers. Its input was paper tape and its output a slow alpha-numeric printer. Today's personal computers have powerful software for control engineering calculations and simulation studies which, incidentally, were done on vacuum tube analogue computers some forty years ago. Such software has control-specific tools and user interfaces are provided with excellent graphical display facilities both for input and output. The majority, however, of this control analysis and design software which has been developed has been concerned with numerical work. Only recently have symbolic methods for control system analysis and design started to appear and I am sure these techniques will continue to develop rapidly in the future.

I was delighted when Neil Munro agreed to contribute and edit this book and I am pleased to be able to provide a few comments in this foreword. I am sure that many of the ideas and techniques presented in this book will provide the basis for further research and development in this important area of control engineering. The chapters are gathered together in four parts, under the sub-headings System modelling, System analysis, Design and synthesis methods and Nonlinear aspects. This is a logical hierarchy in control engineering but, as each chapter was written to stand alone, a reader with interests in certain topics may choose to take the chapters in a different sequence.

The book contains 14 chapters, many of which include recent results on the topic addressed and extensive references to enable the interested reader to explore the topic further. Although control engineers have always recognised uncertainty in the parameters of the models they use, it has only been in recent years, since the introduction of Kharitonov's theorem to the control community, that a large body of knowledge has been published on the robustness of systems with parametric uncertainty, a theme which is developed in several chapters. Chapter 4 presents a collection of the recent results in this field which form a 'toolkit' for the control engineer interested in systems with uncertain parameters. Apart from Kharitonov's theorem and its generalisation, more basic control-related results on stability margins and frequency domain bounds are presented. The results are extended in Chapter 7 to MIMO systems with parametric uncertainty, where it is shown how the classical methods used in multivariable systems, such as the Nyquist array, can be extended to such systems. These two chapters are primarily concerned with analysis but the material is extended for use in design and synthesis in Chapters 8 and 10. In the former, it is shown how classical controllers can be designed to achieve desired stability margins for these systems. Pole placement is a common procedure used in control system design but changes in plant parameters result in poles moving from their desired locations. Chapter 10 examines this problem and in particular considers the situation in multivariable systems where there is not a unique solution to the pole assignment problem. Chapter 14 looks at an approach to the solution of systems of inequalities for multivariate polynomials in real variables. These types of problems can arise in assessing the stability of control systems with variable parameters and therefore fit the theme of methods for systems with parametric uncertainty.

The first chapter in the book is concerned with the modelling of nonlinear systems and shows how some of the early ideas in this field developed at MIT in the 1960s can be exploited using symbolic languages. Routines are developed for manipulating multidimensional Laplace transforms and showing how this modelling approach can be used in the analysis of simple nonlinear systems. The second chapter deals with the use of symbolic computations for the manipulation of bond graphs, again a topic first introduced at MIT in the sixties. Bond graphs have been used for modelling for many years but the advent of symbolic computation gives appreciable extra power providing, for instance, excellent possibilities for the re-use of models, even with different parameters, and the building of system libraries. Chapter 3 presents a survey of software which has been, and is being, developed for the dynamic modelling of multibody systems. The extensive references provided to relevant work should prove of value to the reader.

The two chapters in the Analysis section not already mentioned are Chapters 5 and 6. The former returns to the theme of nonlinear systems and in particular systems represented by differential equations with polynomial nonlinear terms. It is shown how symbolic computer-based methods using differential and real algebra can be used to explore properties of nonlinear differential equations. The final chapter in the Analysis section is also concerned with algebraic computations but this time for linear systems. The use of symbolic tools and computation algorithms associated with systems described by polynomial matrices is discussed. This is another theme which is expounded further in the synthesis part of the book where it comprises the material of Chapter 11.

One aspect of providing computer software to support control design techniques is that it enables design methods to be used by people, particularly in industry, who are not completely knowledgeable on the theoretical background of the method. An approach which is now starting to be accepted in industrial applications is sliding mode control, which can be particularly effective for systems with uncertain or changing parameters. Chapter 9 shows how symbolic methods can be used in controller design for these systems.

The last part of the book contains four chapters, one of which has already been mentioned. The first, Chapter 12, is a continuation of the nonlinear system theory theme of Chapters 1 and 5. Here the focus is also on design and synthesis, as well as modelling, and it is shown how symbolic methods can be used to investigate the zero dynamics and exact-linearisation methods for nonlinear systems. Chapter 13 addresses optimisation, a topic of considerable importance in all areas of engineering. In particular the spatial branch-and-bound approach is discussed and it is shown how it can be used to test for diagonal dominance in a multivariable system. Finally, Chapter 15 describes powerful software which has recently been developed to support integrated design of controlled structural mechanical systems. The software contains a large number of routines which are clearly explained, often by examples, both for linear and nonlinear design, the latter including both feedback linearisation and sliding mode control. The description of the control law design for a conical magnetic bearing system provides an excellent practical application of the use of the software.

Professor Derek Atherton
Series Editor
November 1998

Contributors

Professor H A Barker
Department of Electrical and
    Electronic Engineering
University of Wales
Swansea
SA2 8PP
UK

Professor S P Bhattacharyya
Department of Electrical Engineering
Texas A & M University
College Station
TX 77843-3128
USA

Professor G Blankenship
Department of Electrical Engineering
University of Maryland
College Park
Maryland
MD 20742
USA

Dr A G de Jager
Faculty of Mechanical Engineering
Eindhoven University of Technology
PO Box 513
5600 MB Eindhoven
The Netherlands

Professor J Garloff and B Graf
Fachbereich Informatik
Fachhochschule Konstanz
Postfach 10 05 43
D-78405
Germany

**Professor P J Gawthrop and
    Dr D J Ballance**
Department of Mechanical
    Engineering
James Watt Building
University of Glasgow
Glasgow
G12 8QQ
UK

**Professor T Glad and
    Dr M Jirstrand**
Division of Automatic Control
Department of Electrical Engineering
Linköping University
S-581 83 Linköping
Sweden

**Professor N Karcanias,
    D Vafiadis and J Leventides**
Department of Electrical and
    Electronic Information Engineering
Control Engineering Centre
City University

Northampton Square
London
EC1V 0HB
UK

Professor L H Keel
Center of Excellence in Information
    Systems
Tennessee State University
Nashville
Tennessee 37203-3401
USA

Dr P J Larcombe and I C Brown
School of Mathematics and Computing
University of Derby
Kedleston Road
Derby
DE22 1GB
UK

M Mitrouli
Department of Mathematics
University of Athens
Panepistimiopolis 15784
Athens
Greece

**Professor N Munro, M T Soylemez
    and E Kontogiannis**
Control Systems Centre
Department of Electrical and
    Electronic Engineering
UMIST
PO Box 88
Manchester
M60 1QD
UK

**Professor C C Pantelides and
    Dr E M B Smith**
Centre for Process Systems Engineering
Imperial College of Science,
    Technology and Medicine
London
SW7 2BY
UK

Dr S K Spurgeon and Dr X Y Lu
Control Systems Research
Department of Engineering
University of Leicester
University Road
Leicester
LE1 7RH
UK

Part I
System modelling

Chapter 1

Symbolic modelling and analysis of nonlinear systems

H. A. Barker

1.1 Introduction

Functional methods for the study of nonlinear systems have developed steadily for the 70 years since the publication of the seminal paper by Volterra [1]. For engineers, the most important contributions were made 40 years ago, by a group at the Massachusetts Institute of Technology. In particular, Brilliant [2] showed that the Volterra series representation of a nonlinear system converges when the system is composed entirely of subsystems that are either linear with memory or nonlinear without memory. This allowed functional methods to be applied to systems described by block-oriented models which are not only favoured by many engineers but also form the basis of much computer software for dynamic system simulation [3]. Later, George developed an algebra for this class of systems using multidimensional Laplace transforms [4]. This allowed the concepts of transform methods, also favoured by many engineers, to be applied to these systems. Despite this considerable advantage, the approach never found universal appeal, although it continued to have some adherents [5–7].

The reasons for this are not hard to find. George's nonlinear system algebra is neither easy to understand nor simple to apply, and the multidimensional transforms involved are extremely cumbersome algebraic expressions that are tedious and time-consuming to obtain, manipulate and simplify. Consequently the method is prone to error and is therefore unsuitable to become accepted engineering practice without significant refinement. A basis for such a refinement is provided by symbolic computation, which through computer algebra software overcomes the fundamental limitations of the approach. The way in which symbolic computation can be used to refine the method was first described ten years ago by Barker and Ko [8], who used the early computer algebra software *MACSYMA* [9] for the purpose. In a subsequent development [10], they used the more recent software *Mathematica* [11]. Further work has been reported by Barker and Zhuang [12–14].

1.2 Nonlinear systems and signals

1.2.1 *System elements and operations*

The nonlinear systems considered here are composed entirely of linear and nonlinear elements that satisfy Brilliant's criterion [2] for convergence of the Volterra series representation [1] of the input-output relationships. It is convenient to classify the elements through the operations that they perform, and to represent them as blocks.

The linear elements are those which appear in conventional block diagrams. They perform either a dynamic operation, defined by a transfer function $W(s)$, or an amplification operation, defined by a gain G, or a summation operation, as shown in Figure 1.1.

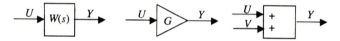

Figure 1.1   Representation of linear elements in a nonlinear system

Although the amplification operation could be regarded as a special case of the dynamic operation, the frequency of its occurrence justifies its separate treatment. The summation operation can be used with more than two inputs, and for both addition and subtraction by appropriate changes of sign in the block.

The nonlinear elements have no dynamics. They perform either a multiplication operation or a power operation, defined by a power $^\wedge n$ as shown in Figure 1.2.

Figure 1.2   Representation of nonlinear elements in a nonlinear system

Although the power operation could be regarded as a number of multiplication operations, the frequency of its occurrence justifies its separate treatment. The multiplication operation can be used with more than two inputs.

1.2.2 Signals and signal components

When a block diagram is constructed from these elemental blocks, the lines that interconnect the blocks represent the signals on which the elemental operations are performed. The composition of these signals provides the basis for an approach to nonlinear system modelling and analysis. Every signal is taken to be the sum of a series of components, each related to a specific order of nonlinearity. A signal v is therefore defined as

$$v = v_0 + v_1 + v_2 + \cdots + v_r + \cdots \tag{1.1}$$

where v_r is the component due to rth-order nonlinearities.

For the purposes of modelling and analysis, the component v_r must be taken to be an r-dimensional signal $v_r(t_1, t_2, \ldots, t_r)$ that equates to the time-domain signal $v_r(t)$ when the time variables $t_1, t_2, \ldots, t_r$ are associated and equated to t; that is,

$$v_r(t) = [v_r(t_1, t_2, \ldots, t_r)]_{t_1 = t_2 = \cdots = t_r = t} \tag{1.2}$$

The signal v and its components are therefore taken to be

$$v = v_0 + v_1(t) + v_2(t_1, t_2) + \cdots + v_r(t_1, t_2, \ldots, t_r) + \cdots \tag{1.3}$$

With the exception of v_0, which is a constant component on which all other components are superimposed, the components are zero when $t < 0$; that is,

$$v_r(t_1, t_2, \ldots, t_r) = 0 \qquad \text{for } r > 0 \quad \text{and} \quad t_1, t_2, \ldots, t_r < 0 \tag{1.4}$$

Each component of the signal v may be transformed by an appropriate multidimensional Laplace transform [4], the rth component transform being

$$V_r(s_1, s_2, \ldots, s_r) = \mathbf{L}^r v_r(t_1, t_2, \ldots, t_r)$$

$$= \int_0^\infty \int_0^\infty \cdots \int_0^\infty v_r(t_1, t_2, \ldots, t_r) e^{-(s_1 t_1 + s_2 t_2 + \cdots + s_r t_r)} \, \mathrm{d}t_1 \, \mathrm{d}t_2 \cdots \mathrm{d}t_r \tag{1.5}$$

The signal transform V and its components are therefore taken to be

$$V = V_0 + V_1(s) + V_2(s_1, s_2) + \cdots + V_r(s_1, s_2, \ldots, s_r) + \cdots \tag{1.6}$$

As the relationships between the signals in the block diagrams constructed from the elemental blocks in Figures 1.1 and 1.2 are developed in the transform domain, signal transforms are used to represent the signals in the diagrams.

1.3 Nonlinear system relationships

1.3.1 Elemental equations

The input-output relationships for each kind of element in the nonlinear system may be developed as equations relating the components of the input and output signal transforms. For the dynamic operation with transfer function $W(s)$ in Figure 1.1, the elemental equations are

$$Y_r(s_1, s_2, \ldots, s_r) = W(s_1 + s_2 + \cdots + s_r)U_r(s_1, s_2, \ldots, s_r) \qquad (1.7)$$

For the amplification operation with gain G in Figure 1.1, the elemental equations are

$$Y_r(s_1, s_2, \ldots, s_r) = GU_r(s_1, s_2, \ldots, s_r) \qquad (1.8)$$

For the summation operation in Figure 1.1, the elemental equations are

$$Y_r(s_1, s_2, \ldots, s_r) = U_r(s_1, s_2, \ldots, s_r) + V_r(s_1, s_2, \ldots, s_r) \qquad (1.9)$$

For the multiplication operation in Figure 1.2, the elemental equations are

$$Y_r(s_1, s_2, \ldots, s_r) = \sum_{q=0}^{r} U_q(s_1, s_2, \ldots, s_q)V_{r-q}(s_{q+1}, s_{q+2}, \ldots, s_r) \qquad (1.10)$$

For the power operation in Figure 1.2, the elemental equations are

$$Y_r(s_1, s_2, \ldots, s_r) = \sum_{q_1=0}^{q_2} \cdots \sum_{q_2=0}^{q_3} \sum_{q_{n-1}=0}^{r} U_{q_1}(s_1, s_2, \ldots, s_{q_1})$$
$$U_{q_2-q_1}(s_{q_1+1}, s_{q_1+2}, \ldots, s_{q_2}) \cdots U_{r-q_{n-1}}(s_{q_{n-1}+1}, s_{q_{n-1}+2}, \ldots, s_r) \qquad (1.11)$$

1.3.2 Signal transforms

When the elements are connected together to form a system, the signal transforms are not only defined by the relationships in Section 1.3.1, but also by the system topology which defines how the elements are interconnected by the signals. The component transforms may be synthesised by a manual bottom-up procedure that commences with the zero-order static components, for which $r = 0$, and continues with the linear components, for which $r = 1$, the quadratic components, for which $r = 2$, the cubic components, for which $r = 3$, and so on until the required order of approximation is obtained. At each stage, a set of equations, which may contain equations from previous stages, is obtained and solved.

1.3.3 Multidimensional transfer functions

If y is an output signal of a nonlinear system, then the rth-order component $Y_r(s_1, s_2, \ldots, s_r)$ of its transform Y, obtained by the procedure in Section 1.3.2, is typically related to the transform $U(s)$ of a system input signal u by an equation of the form

$$Y_r(s_1, s_2, \ldots, s_r) = W_r(s_1, s_2, \ldots, s_r)U(s_1)U(s_2) \cdots U(s_r) \qquad (1.12)$$

In eqn. 1.12, $W_r(s_1, s_2, \ldots s_r)$ is the multidimensional transfer function that relates the rth-order component $Y_r(s_1, s_2, \ldots s_r)$ of the output signal transform Y to the input signal transform $U(s)$. Multidimensional transfer functions are used to obtain the responses of nonlinear systems to input signals in both the time and frequency domains.

If a multidimensional transfer function $W_r^{\text{asym}}(s_1, s_2, \ldots, s_r)$ obtained in this way is asymmetric in $s_1, s_2, \ldots, s_r$, then it is easily converted into the equivalent multidimensional transfer function $W_r^{\text{sym}}(s_1, s_2, \ldots s_r)$ that is symmetric in $s_1, s_2, \ldots, s_r$, through

$$W_r^{\text{sym}}(s_1, s_2, \ldots, s_r) = \frac{1}{r!} \sum_{\substack{\text{all permutations} \\ \text{of } s_1, s_2, \ldots, s_r}} W_r^{\text{asym}}(s_1, s_2, \ldots, s_r) \qquad (1.13)$$

1.4 Application of symbolic computation

To obtain the multidimensional transfer functions of the system from the elemental equations in Section 1.3.1 by the procedure in Section 1.3.2 is tedious, time-consuming and error-prone, even when applied to quite simple systems. The use of symbolic computation overcomes these difficulties. Although *Mathematica* is used here, the principles are applicable to other kinds of computer algebra software. The approach is based mainly on the use of lists and procedures.

For the purposes of symbolic computation, the signal transform V in eqn. 1.6 is represented by the list

$$V = \{V_0, V_1[s_1], V_2[s_1, s_2], \ldots, V_r[s_1, s_2, \ldots, s_r], \ldots\} \qquad (1.14)$$

The operations performed by the system elements and defined by the elemental equations are implemented as procedures. For the dynamic operation with transfer function $W(s)$ in Figure 1.1, the elemental equations defined by eqn. 1.7 are implemented as the procedure

$$Y = \text{trans}[U, W] \qquad (1.15)$$

For the amplification operation with gain G in Figure 1.1, the elemental equations defined by eqn. 1.8 are implemented as the procedure

$$Y = \text{gain}[U, G] \qquad (1.16)$$

For the summation operation in Figure 1.1, the elemental equations defined by eqn. 1.9 are implemented as the procedure

$$Y = \text{sum}[U, V] \qquad (1.17)$$

For the multiplication operation in Figure 1.2, the elemental equations defined by eqn. 1.10 are implemented as the procedure

$$Y = \text{mult}[U, V] \qquad (1.18)$$

For the power operation with power $^\wedge n$ in Figure 1.2, the elemental equations defined by eqn. 1.11 are implemented as the procedure

$$Y = \text{power}[U, u^\wedge n] \qquad (1.19)$$

In order to implement the procedure in Section 1.3.2 for obtaining the component transforms and the corresponding multidimensional transfer functions, it is necessary to define the system topology. This is accomplished from the block diagram of the system. First, a number is allocated to every block and signal in the system. Then for every block in the system, the input signal(s), the output signal and the elemental operation performed by the block are defined by a list

blocknumber = {input signal number(s), output signal number, operation}

$$(1.20)$$

The system is then defined by a linked list that contains the list for every block:

$$\text{system} = \{\{\text{block 1}\}, \{\text{block 2}\}, \ldots, \{\text{block } n\}\} \qquad (1.21)$$

The block and system lists in eqns. 1.20 and 1.21 specify the component transforms to which the procedures in eqns. 1.15 to 1.19 are to be applied. The procedures define sets of equations that relate the component transforms of every order up to the highest order specified. The equations are solved symbolically to obtain the required component transforms and hence the corresponding multidimensional transfer functions. These are made symmetric by applying the procedure

$$Wsym = \text{symmetrize}[Wasym] \qquad (1.22)$$

which is an implementation of eqn. 1.13.

There is a stark contrast between the efficiency of the symbolic computation procedure and the complexity of the calculations required for a manual solution.

1.5 Example

As an example, the method will be applied to a Type 1 control system with input u, output y and error e. The forward path between the error and the output contains the saturation characteristic

$$\begin{aligned} v &= 3e - e^3 & \text{for } |e| < 1 \\ v &= 2 \text{ sgn } e & \text{for } |e| > 1 \end{aligned} \qquad (1.23)$$

as shown in Figure 1.3, followed by linear dynamics with transfer function

$$\frac{3}{s(s+10)}$$

It is subsequently assumed that, during operation of the system, the absolute error $|e|$ is always less than 1. The block diagram of the system is then as shown in Figure 1.4.

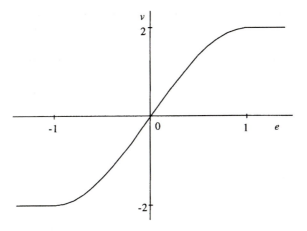

Figure 1.3   Nonlinear saturation characteristic

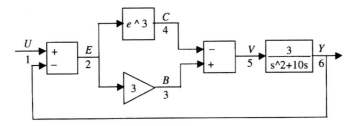

Figure 1.4   Block diagram of a nonlinear control system

1.5.1 *Manual procedure*

The set of equations for the zero-order components is

$$E_0 = U_0 - Y_0$$
$$B_0 = 3E_0$$
$$C_0 = E_0^3 \tag{1.24}$$
$$V_0 = B_0 - C_0$$
$$Y_0 = \frac{1}{0} V_0$$

Solving this set of equations gives $V_0 = 3E_0 - E_0^3 = 0$ so that $E_0 = 0$ or $\pm\sqrt{3}$. As the only solution in the range is $E_0 = 0$, the only possible nonzero component is

$$Y_0 = U_0 \tag{1.25}$$

The set of equations for the first-order components is

$$E_1(s_1) = U_1(s_1) - Y_1(s_1)$$
$$B_1(s_1) = 3E_1(s_1)$$
$$C_1(s_1) = 0 \tag{1.26}$$
$$V_1(s_1) = B_1(s_1) - C_1(s_1)$$
$$Y_1(s_1) = \frac{3}{s_1(s_1 + 10)} V_1(s_1)$$

Solving this set of equations gives, *inter alia*

$$Y_1(s_1) = \frac{9}{(s_1 + 1)(s_1 + 9)} U_1(s_1)$$
$$E_1(s_1) = \frac{s_1(s_1 + 10)}{(s_1 + 1)(s_1 + 9)} U_1(s_1) \tag{1.27}$$

The set of equations for the second-order components is

$$E_2(s_1, s_2) = -Y_2(s_1, s_2)$$
$$B_2(s_1, s_2) = 3E_2(s_1, s_2)$$
$$C_2(s_1, s_2) = 0 \tag{1.28}$$
$$V_2(s_1, s_2) = B_2(s_1, s_2) - C_2(s_1, s_2)$$
$$Y_2(s_1, s_2) = \frac{3}{(s_2 + s_2)(s_1 + s_2 + 10)} V_2(s_1, s_2)$$

Solving this set of equations gives

$$Y_2(s_1, s_2) = 0 \tag{1.29}$$

and all other second-order components are also zero.

The set of equations for the third-order components is

$$E_3(s_1, s_2, s_3) = -Y_3(s_1, s_2, s_3)$$
$$B_3(s_1, s_2, s_3) = 3E_3(s_1, s_2, s_3)$$
$$C_3(s_1, s_2, s_3) = E_1(s_1)E_1(s_2)E_1(s_3) \tag{1.30}$$
$$V_3(s_1, s_2, s_3) = B_3(s_1, s_2, s_3) - C_3(s_1, s_2, s_3)$$
$$Y_3(s_1, s_2, s_3) = \frac{3}{(s_1 + s_2 + s_3)(s_1 + s_2 + s_3 + 10)} V_3(s_1, s_2, s_3)$$

Solving this set of equations requires the definition of $E_1(s_1)$ from eqn. 1.27, and gives, *inter alia*,

$$Y_3(s_1, s_2, s_3) =$$

$$-\frac{3s_1(s_1 + 10)s_2(s_2 + 10)s_3(s_3 + 10)}{(s_1 + s_2 + s_3 + 1)(s_1 + s_2 + s_3 + 9)(s_1 + 1)(s_1 + 9)(s_2 + 1)(s_2 + 9)(s_3 + 1)(s_3 + 9)}$$

$$U(s_1)U(s_2)U(s_3) \qquad (1.31)$$

It is evident that continuing this manual procedure to obtain higher-order components is possible, but hardly practicable. The even-order components are zero, but there is an infinite number of odd-order components. The order of approximation for the manual procedure is therefore specified as four in this case, giving the multidimensional transfer functions relating the output to the input as

$$W_0 = 0$$

$$W_1(s_1) = \frac{9}{(s_1 + 1)(s_1 + 9)}$$

$$W_2(s_1, s_2) = 0$$

$$W_3(s_1, s_2, s_3) =$$

$$-\frac{3s_1(s_1 + 10)s_2(s_2 + 10)s_3(s_3 + 10)}{(s_1 + s_2 + s_3 + 1)(s_1 + s_2 + s_3 + 9)(s_1 + 1)(s_1 + 9)(s_2 + 1)(s_2 + 9)(s_3 + 1)(s_3 + 9)}$$

$$W_4(s_1, s_2, s_3, s_4) = 0 \qquad (1.32)$$

1.5.2 Symbolic procedure

In the symbolic procedure it is first necessary to represent the system shown in Figure 1.4 by the linked list

$$\text{system} = \{\{1, 6, 2, \text{``} + -\text{''}\}, \{2, 3, 3\}, \{2, 4, u\,^{\wedge}3\}\{3, 4, 5, \text{``} + -\text{''}\}\{5, 6, 3/(s\,^{\wedge}2 + 10s)\}\}$$

$$(1.33)$$

The procedures in eqns. 1.15 to 1.19 are then applied to obtain the sets of equations that relate the component transforms of every order up to the highest order specified. For a fourth-order approximation, these are equivalent to eqns. 1.24, 1.26, 1.28 and 1.30. The equations are then solved to obtain the required component transforms, as in eqns. 1.25, 1.27, 1.29 and 1.31, and these give the corresponding multidimensional transfer functions, as in eqn. 1.32. The procedure in eqn. 1.22 is then applied to the multidimensional transfer functions to ensure that they are symmetric.

1.6 Nonlinear system response

Nonlinear system responses in both the time and frequency domains may be obtained through the component transforms and multidimensional transfer functions of the system. In both cases, symbolic computation considerably reduces the time and effort required in comparison to the manual procedure.

1.6.1 Time-domain response

To obtain the time-domain response y of a nonlinear system to a specified input signal directly, the appropriate multidimensional inverse Laplace transform would first be applied to each component $Y_r(s_1, s_2, \ldots s_r)$ of the output transform Y to obtain the corresponding component $y_r(t_1, t_2, \ldots t_r)$ of the signal y, that is

$$y_r(t_1, t_2, \ldots, t_r) = \mathbf{L}^{-r} Y_r(s_1, s_2, \ldots, s_r) \qquad \text{for } r = 1, 2, \ldots \qquad (1.34)$$

and then the variables $t_1, t_2, \ldots, t_r$ in each component $y_r(t_1, t_2, \ldots, t_r)$ would be associated and equated to t to give the component y_r of the signal y, so that

$$y_r = [y_r(t_1, t_2, \ldots, t_r)]_{t_1=t_2=\cdots=t_r=t} = [\mathbf{L}^{-r} Y_r(s_1, s_2, \ldots, s_r)]_{t_1=t_2=\cdots=t_r=t} \quad \text{for } r = 1, 2, \ldots$$
$$(1.35)$$

In practice, this procedure is rarely used, because it is preferable to associate the time variables through the corresponding transform variables in the transform domain, rather than directly in the time domain [4, 15]. This can be accomplished by a sequence of relatively simple operations applied successively to reduce the order of the component transform $Y_r(s_1, s_2, \ldots, s_r)$. In the first operation, the transform variables s_{r-1} and s_r in the rth-order transform $Y_r(s_1, s_2, \ldots, s_r)$ are associated to obtain an $r-1$th-order transform $Y_r^{(r-1)}(s_1, s_2, \ldots, s_{r-1})$ that is valid for $t_{r-1} = t_r$. In the second, the variables s_{r-2} and s_{r-1} in $Y_r^{(r-1)}(s_1, s_2, \ldots, s_{r-1})$ are associated to obtain $Y_r^{(r-2)}(s_1, s_2, \ldots, s_{r-2})$ that is valid for $t_{r-2} = t_{r-1} = t_r$. This sequence of operations is continued until a first-order transform $Y_r^{(1)}(s_1)$ is obtained that is valid for $t_1 = t_2 = \cdots = t_r$. This is then inverse transformed to obtain the component $y_r(t)$ directly.

In the kth operation, the transform variables s_{r-k} and s_{r-k+1} in the $r-k+1$th transform $Y_r^{(r-k+1)}(s_1, s_2, \ldots, s_{r-k}, s_{r-k+1})$ are associated to obtain the $r-k$th transform $Y_r^{(r-k)}(s_1, s_2, \ldots, s_{r-k})$ through

$$Y_r^{(r-k)}(s_1, s_2, \ldots, s_{r-k}) = \sum_{\substack{\text{residues} \\ \text{of } s_{r-k+1} \text{ in } C_{r-k+1}}} Y_r^{(r-k+1)}(s_1, s_2, \ldots, s_{r-k-1}, s_{r-k} - s_{r-k+1}, s_{r-k+1})$$

$$(1.36)$$

where for all practical purposes the contour C_{r-k+1} includes the whole of the left half of the s_{r-k+1} plane.

In the symbolic computation, each operation is accomplished as an application of the residue theorem. The $r-1$ operations involved in reducing the rth-order transform $y_r(s_1, s_2, \ldots, s_r)$ to the first-order transform $Y_r^{(1)}(s_1)$ are implemented as the procedure

$$Yrs = \text{assocvar}[Yr] \qquad (1.37)$$

The component of the output signal y due to rth-order nonlinearities is then obtained from the procedure

$$yrt = \text{Inverse Laplace Transform}[Yrs, s, t] \tag{1.38}$$

1.6.2 Frequency-domain response

Harmonic responses are obtained directly from the multidimensional transfer functions [4, 16]. The steady-state response $y^{(ss)}$ of a nonlinear system to a single-harmonic input signal with amplitude A and frequency ω radians/second is given by the Fourier series

$$y^{(ss)} = \sum_{k=-\infty}^{\infty} c'_k e^{jk\omega t} \tag{1.39}$$

When $k \neq 0$ in eqn. 1.39, the sum of the components $c'_k e^{jk\omega t}$ and $c'_{-k} e^{-jk\omega t}$ gives the total kth-harmonic component with frequency $\omega > 0$, for which the amplitude and phase are the magnitude and phase of $c_k = 2c'_k$. This component is itself the sum of subcomponents contributed by all multidimensional transfer functions $W_{k+2r}(s_1, s_2, \ldots, s_{k+2r})$ for which $r \geq 0$. If the amplitude and phase of the contribution from $W_{k+2r}(s_1, s_2, \ldots, s_{k+2r})$ are the magnitude and phase of c_{kk+2r}, then

$$c_k = c_{kk} + c_{kk+2} + \cdots + c_{kk+2r} + \cdots \tag{1.40}$$

where

$$c_{kk+2r} = \frac{A^{k+2r}}{2^{k+2r-1}} \binom{k+2r}{k+r} W_{k+2r}(j\omega, \ldots, j\omega, -j\omega, \ldots, -j\omega) \quad \text{for } r = 0, 1, 2, \ldots \tag{1.41}$$

and in $W_{k+2r}(j\omega, \ldots, j\omega, -j\omega, \ldots, -j\omega)$ there are $k + r$ positive and r negative arguments.

When $k = 0$ in eqn. 1.39, there is only one constant component $c_0 = c'_0$. This component is also the sum of subcomponents, as defined in eqn. 1.40, but in this case

$$c_{00} = 0$$

$$c_{02r} = \frac{A^{2r}}{2^{2r}} W_{2r}(j\omega, \ldots, j\omega, -j\omega, \ldots, -j\omega) \quad \text{for } r = 1, 2, 3, \ldots \tag{1.42}$$

and in $W_{2r}(j\omega, \ldots, j\omega, -j\omega, \ldots, -j\omega)$ there are r positive and r negative arguments.

In the symbolic computation, eqns, 1.40 to 1.42 are implemented in the procedure

$$ck = \text{harmonic}[Wr, \text{harmonic Number}] \tag{1.43}$$

to give the amplitude and phase of the kth component.

1.7 Example

As an example, responses in both the time and frequency domains are obtained for the system in Section 1.5.

1.7.1 Time-domain response

If the system input u is a unit step signal with transform $1/s$, then from eqns. 1.27 and 1.31 the first two nonzero components of the step response transform Y are

$$Y_1(s_1) = \frac{9}{s_1(s_1 + 1)(s_1 + 9)} \tag{1.44}$$

and

$$Y_1(s_1, s_2, s_3) =$$

$$-\frac{3(s_1 + 10)(s_2 + 10)(s_3 + 10)}{(s_1 + s_2 + s_3 + 1)(s_1 + s_2 + s_3 + 9)(s_1 + 1)(s_1 + 9)(s_2 + 1)(s_2 + 9)(s_3 + 1)(s_3 + 9)} \tag{1.45}$$

The first-order component is obtained by inverse transforming $Y_1(s_1)$ in eqn. 1.44 to give

$$y_1(t) = 1 - \tfrac{9}{8} e^{-t} + \tfrac{1}{8} e^{-9t} \tag{1.46}$$

The third-order component is obtained by reducing the component transform $Y_3(s_1, s_2, s_3)$ in two stages by the association of variables and inverse-transforming the resulting first-order transform $Y_3^{(1)}(s_1)$. In the first stage, eqn. 1.36 is used to associate the variables s_3 and s_2 in $Y_3(s_1, s_2, s_3)$ to obtain $Y_3^{(2)}(s_1, s_2)$ as

$$Y_3^{(2)}(s_1, s_2) = \sum_{\substack{\text{residues} \\ \text{of } s_3 \text{ in } C_3}} Y_3(s_1, s_2 - s_3, s_3) = \sum_{\substack{\text{residues} \\ \text{of } s_3 \text{ in } C_3}}$$

$$-\frac{3(s_1 + 10)(s_2 - s_3 + 10)(s_3 + 10)}{(s_1 + s_2 + 1)(s_1 + s_2 + 9)(s_1 + 1)(s_1 + 9)(s_2 - s_3 + 1)(s_2 - s_3 + 9)(s_3 + 1)(s_3 + 9)}$$

$$= -\frac{6(s_1 + 10)(s_1 + 20)}{(s_1 + s_2 + 1)(s_1 + s_2 + 9)(s_1 + 1)(s_1 + 9)(s_2 + 2)(s_2 + 10)(s_2 + 18)} \tag{1.47}$$

In the second stage, eqn. 1.36 is used to associate the variables s_2 and s_1 in $Y_3^{(2)}(s_1, s_2)$ to obtain $Y_3^{(1)}(s_1)$ as

$$
\begin{aligned}
Y_3^{(1)}(s_1) &= \sum_{\substack{\text{residues} \\ \text{of } s_2 \text{ in } C_2}} Y_3^{(2)}(s_1 - s_2, s_2) \\
&= \sum_{\substack{\text{residues} \\ \text{of } s_2 \text{ in } C_2}} -\frac{6(s_1 - s_2 + 10)(s_1 - s_2 + 20)}{(s_1 + 1)(s_1 + 9)(s_1 - s_2 + 1)(s_1 - s_2 + 9)(s_2 + 2)(s_2 + 10)(s_2 + 18)} \\
&= -\frac{3(s_1^3 + 60s_1^2 + 1163s_1 + 7350)}{(s_1 + 1)(s_1 + 3)(s_1 + 9)(s_1 + 11)(s_1 + 19)(s_1 + 27)}
\end{aligned}
\tag{1.48}
$$

The third-order component is then obtained by inverse-transforming $Y_3^{(1)}(s_1)$ in eqn. 1.48 to give

$$
\begin{aligned}
y_3(t) &= -\frac{1041}{4160} e^{-t} + \frac{729}{2048} e^{-3t} - \frac{169}{960} e^{-9t} + \frac{729}{10240} e^{-11t} - \frac{9}{10240} e^{-19t} \\
&\quad + \frac{1}{79872} e^{-27t}
\end{aligned}
\tag{1.49}
$$

In the symbolic computation, the output transform components of every order up to the highest order specified are obtained from the multidimensional transfer functions and the input transform. For a fourth-order approximation, these are equivalent to eqns. 1.25, 1.44, 1.29 and 1.45. The association of variables procedure in eqn. 1.37 is then applied to the nonzero component transforms of order higher than one to reduce them to first-order transforms, equivalent to eqn. 1.48. The procedure in eqn. 1.38 is then applied to the first-order transforms to obtain the response components, equivalent to eqns. 1.46 and 1.49.

1.7.2 Frequency-domain response

The harmonic responses are obtained by applying eqns. 1.41 and 1.42 to the multidimensional transfer functions in eqn. 1.32. For a fourth-order approximation, there are contributions to the first harmonic from the first- and third-order transfer functions, and a contribution to the third harmonic from the third-order transfer function, so that

$$
\begin{aligned}
c_1 &= c_{11} + c_{13} \\
c_3 &= c_{33}
\end{aligned}
\tag{1.50}
$$

where

$$c_{11} = \frac{9A}{(j\omega + 1)(j\omega + 9)}$$

$$c_{13} = -\frac{9A^3(j\omega)(j\omega + 10)(j\omega)(j\omega + 10)(-j\omega)(-j\omega + 10)}{4(j\omega + 1)(j\omega + 9)(j\omega + 1)(j\omega + 9)(j\omega + 1)(j\omega + 9)(-j\omega + 1)(-j\omega + 9)}$$

$$= -\frac{9A^3(j\omega)(\omega^2)(j\omega + 10)(\omega^2 + 100)}{4(j\omega + 1)^2(\omega^2 + 1)(j\omega + 9)^2(\omega^2 + 81)}$$

$$c_{33} = -\frac{3A^3(j\omega)(j\omega + 10)(j\omega)(j\omega + 10)(j\omega)(j\omega + 10)}{4(3j\omega + 1)(3j\omega + 9)(j\omega + 1)(j\omega + 9)(j\omega + 1)(j\omega + 9)(j\omega + 1)(j\omega + 9)}$$

$$= -\frac{A^3(j\omega)^3(j\omega + 10)^3}{4(3j\omega + 1)(j\omega + 1)^3(j\omega + 3)(j\omega + 9)^3} \tag{1.51}$$

The amplitude and phase of the first two subcomponents of the first harmonic are therefore

$$|c_{11}| = \frac{9A}{\sqrt{(\omega^2 + 1)(\omega^2 + 81)}}$$

$$\angle c_{11} = -\tan^{-1}\omega - \tan^{-1}\frac{\omega}{9}$$

$$|c_{13}| = \frac{9A^3\omega^3\sqrt{(\omega^2 + 100)^3}}{4(\omega^2 + 1)^2(\omega^2 + 81)^2}$$

$$\angle c_{13} = -\frac{\pi}{2} + \tan^{-1}\frac{\omega}{10} - 2\tan^{-1}\omega - 2\tan^{-1}\frac{\omega}{9} \tag{1.52}$$

and the amplitude and phase of the first subcomponent of the third harmonic are given by

$$|c_{33}| = \frac{A^3\omega^3\sqrt{(\omega^2 + 100)^3}}{4\sqrt{(9\omega^2 + 1)(\omega^2 + 1)^3(\omega^2 + 9)(\omega^2 + 81)^3}}$$

$$\angle c_{33} = \frac{\pi}{2} + 3\tan^{-1}\frac{\omega}{10} - \tan^{-1}3\omega - 3\tan^{-1}\omega - \tan^{-1}\frac{\omega}{3} - 3\tan^{-1}\frac{\omega}{9} \tag{1.53}$$

In the symbolic computation, the multidimensional transfer functions up to the highest order specified are obtained. For a fourth-order approximation, these are equivalent to eqn. 1.32. The procedure in eqn. 1.43 is then applied to obtain the contributions from the multidimensional transfer functions to each specified harmonic, equivalent to eqns. 1.51 to 1.53.

1.8 Graphical user interface

The application of symbolic computation removes most of the known disadvantages of this method of nonlinear system modelling and analysis, and provides a basis for its extension and exploitation. However, its full potential cannot be achieved without the provision of a graphical user interface [17], which is required, particularly for large and complex systems, to construct the system block diagram, to pass the resulting system definition to the symbolic computation software, and to control the subsequent analysis. A cost-effective solution is to link an existing graphical user interface to the symbolic computation software. A natural choice in this case is to link the graphical user interface for the simulation software SIMULINK [3], in which virtually any dynamic system model can be represented, to the symbolic computation software *Mathematica* in which the modelling and analysis procedures already exist. The linkage is made and controlled by MATLAB [18] and *MathLink* [19] software, as shown in Figure 1.5.

The blocks in the models constructed through the SIMULINK user interface are restricted to those shown in Figures 1.1 and 1.2, supplemented by inport and outport blocks to represent links to external inputs and outputs. The SIMULINK model of a nonlinear system is constructed from these blocks in the usual way

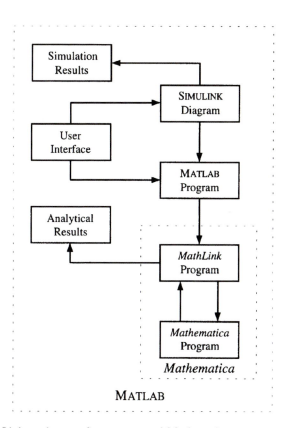

Figure 1.5   Linkages between SIMULINK and Mathematica

and saved to a file, which is then translated by a MATLAB program into the system list that defines the *Mathematica* model, as in eqn. 1.21. The additional information needed to run the *Mathematica* program is input through a MATLAB graphical user interface. *Mathlink* software provides the connections that allow *Mathematica* programs to be run and the results returned to MATLAB. Analytical and numerical results can be displayed, plotted and compared with the results of SIMULINK simulations.

The root window of the MATLAB graphical user interface is shown in Figure 1.6.

The window has three buttons and five pull-down menus. The buttons open the following dialogue boxes:

- **Input**–specifies the input signal(s) to be used for analysis and simulation, by means of a dialogue box
- **System**–specifies the SIMULINK model to be used for analysis and simulation, by either enabling a new model to be constructed or selecting an existing model
- **Output**–specifies the highest order of approximation required and the simulation parameters to be used.

The options provided by the pull-down menus are:

- **File**–provides facilities to load and save MATLAB files, to clear the workspace and to quit
- **Run**–executes the *Mathematica* program and the SIMULINK simulation
- **Display**–displays the results obtained from the *Mathematica* program and the SIMULINK simulation
- **Window**–provides for switches between windows
- **Help**–provides a comprehensive help facility.

Help is also provided through a help button in each pop-up window. Other information, such as warnings, is provided through pop-up windows.

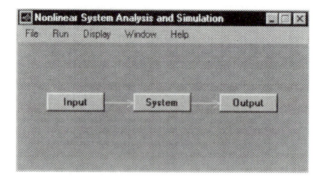

Figure 1.6   Root window of MATLAB graphical user interface

1.9 Example

To illustrate the results that may be obtained and displayed by means of the graphical user interface, the nonlinear system in Figure 1.3 was constructed in SIMULINK. The components of the unit step response were then obtained by symbolic computation from *Mathematica*, with the order of approximation specified as six, and the simulated response was obtained from SIMULINK. Figure 1.7 shows the displayed results, which include the response components y_1, y_3 and y_5, together with the sum of these components and the simulated response. The latter are virtually indistinguishable.

The subcomponents of the first and third harmonics were then obtained by symbolic computation from *Mathematica*. Figure 1.8 shows the displayed results for the first harmonic, which include the subcomponents c_{11}, c_{13} and c_{15}, together with the sum of these components that is the sixth-order approximation of the component c_1.

Figure 1.9 shows the displayed results for the third harmonic, which include the subcomponents c_{33} and c_{35}, together with the sum of these components that is the sixth-order approximation of the component c_3.

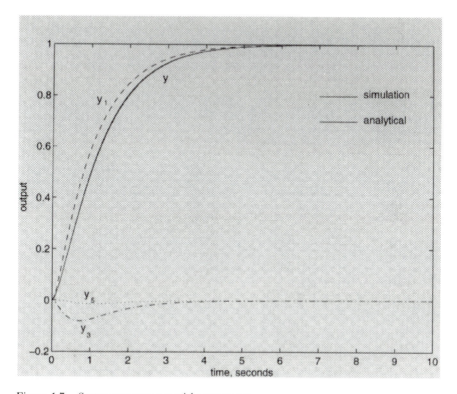

Figure 1.7 System step response with components

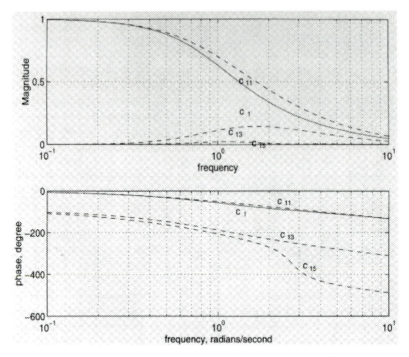

Figure 1.8   First-harmonic components

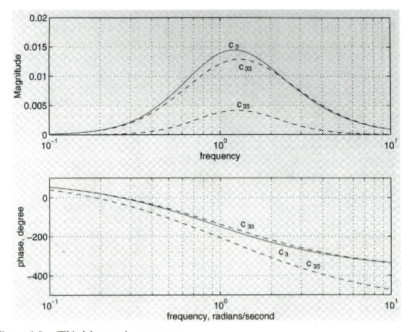

Figure 1.9   Third-harmonic components

1.10 References

1 VOLTERRA, V.: 'Theory of functionals and integro-differential equations' (Blackie, London, 1930)
2 BRILLIANT, M. B.: 'Theory of the analysis of nonlinear systems'. Research Laboratory of Electronics Report 345, Massachusetts Institute of Technology, USA, 1958
3 SIMULINK User's Guide (The MathWorks, Natick, USA, 1992)
4 GEORGE, D. A.: 'Continuous nonlinear systems'. Research Laboratory of Electronics Report 355, Massachusetts Institute of Technology, USA, 1959
5 BANSAL, V. S.: 'Volterra series analysis of a class of nonlinear time-varying systems using multi-linear parametric transfer functions', *Proc. IEE*, 1969, **116**, pp. 1957–1960
6 LUBBOCK, J. K., and BANSAL, V. S.: 'Multidimensional Laplace transforms for the solution of nonlinear equations', *Proc. IEE*, 1969, **116**, pp. 2075–2082
7 MULLINEUX, N., REED, J. R., and RICHARDSON, R. G.: 'Multidimensional Laplace transforms and nonlinear problems', *Int. J. Electr. Eng. Educ.*, 1973, **11**, pp. 5–17
8 BARKER, H. A., and KO, Y. W.: 'The application of a computer algebra system to the analysis of a class of nonlinear systems'. Proceedings of the IFAC symposium on *Nonlinear Control System Design*, Capri, Italy, 1989, pp. 88–93
9 PAVELLE, R., and WANG, P. S.: 'Macsyma from F to G', *Symbol. Comput.*, 1985, **1**, pp. 69–100
10 BARKER, H. A., and KO, Y. W.: 'Modelling nonlinear systems by multidimensional transforms and computer algebra'. Proceedings of the Fourteenth IASTED International Conference on *Modelling, Identification and Control*, Innsbruck, Austria, 1995, pp. 45–48
11 WOLFRAM, S.: 'Mathematica: a system for doing mathematics by computer' (Addison-Wesley, Reading, 1991)
12 BARKER, H. A., and ZHUANG, M.: 'Linking simulation and computer algebra software for dynamic system simulation and analysis'. Proceedings of the Fifteenth IASTED International Conference on *Modelling, Identification and Control*, Innsbruck, Austria, 1996, pp. 369–371
13 BARKER, H. A., and ZHUANG, M.: 'A graphical environment for nonlinear system modelling and analysis', *UKACC International Conference, Control '96*, IEE Conf. Pub. 427, 1996, pp. 1326–1331
14 BARKER, H. A., and ZHUANG, M.: 'Control system analysis using *Mathematica* and a graphical user interface', *Comput. Control Eng. J.*, 1997, **8**, pp. 64–69
15 REDDY, D. C., and JAGAN, N. C.: 'Multidimensional transforms: new technique for the association of variables', *Electron. Lett.*, 1971, **7**, pp. 278–279
16 TSANG, K. M.: 'Spectral analysis for nonlinear systems'. PhD thesis, University of Sheffield, 1988
17 BARKER, H. A., CHEN, M., GRANT, P. W., JOBLING, C. P., and TOWNSEND, P.: 'A man-machine interface for computer-aided design and simulation of control systems', *Automatica*, 1989, **25**, pp. 311–316
18 MATLAB User's Guide (The MathWorks, Natick, USA, 1992)
19 *Mathlink* Reference Guide (Wolfram Research, Champaign, USA, 1993)

Chapter 2

Symbolic computation for manipulation of hierarchical bond graphs

P. J. Gawthrop and D. J. Ballance

2.1 Introduction

Bond graphs [1−7] provide a graphical description of physical dynamic systems. As discussed by Gawthrop and Ballance [8], we believe that the advent of symbolic computing gives significant extra power to the automation of system modelling using bond graphs. In particular, bond graph based software tools based on symbolic computing (including MTT [9, 10] and Archer [11]) have proved useful in providing representations and tools for system *understanding* as opposed to mere modelling and simulation.

In the basic bond graph literature [1−6], there are two sorts of bond graph:

- a bond graph
- a word bond graph.

A bond graph provides a complete description of the system; a word bond graph provides a top-level summary of the system and is useful for the initial conception of the scope of the system model. Software tools such as MTT [9] and Archer [11]—at least until recently—provide for the analysis of a system described by a bond graph but *not* a system described by a word bond graph. This precludes a top-down approach to system modelling which makes it hard (both conceptually and practically) to model complex systems.

However, modelling and simulation software has, for some time, included mechanisms for hierarchical system descriptions—see, for example, the discussions in [12, 13]. Indeed, bond graph-based simulation tools such as CAMAS [14] have used a hierarchical approach, and these ideas have fed into the development of MTT. In particular, as discussed in [12, 13], the concept of a *port* provides the natural interface between components of such a hierarchical description.

This chapter details some recent developments in hierarchical bond graph modelling in the context of *symbolic* modelling which allow for a (multilevel) word bond graph system description whilst allowing a bond graph description at a lower level. The advantages of such an approach can be summarised as follows:

- Each bond graph (or word bond graph) fragment is small enough to be comprehended by the engineer.
- Common components can be reused (with different parameters if required) within a system model or, via a library, are available for use in building a range of systems.
- A word bond graph can be directly interpreted by software tools, thus retaining the conceptual advantages of the word bond graph in a software implementation.

The chapter is organised as follows. After a brief introduction to modelling with bond graphs, modelling with hierarchical bond graphs is discussed. The concepts of representations, languages, transformations and translations are discussed in Section 2.4, where a list of the different representations available in MTT is also given. Section 2.5 considers an application and discusses the way in which symbolic computing is used to transform from a core bond graph-based representation to a number of different representations. The essential feature of this is that, since the core representation is held in a *symbolic* form, information loss about the real system in the model is minimised and therefore transformations to other representations can be handled automatically by symbolic computation.

2.2 Modelling with bond graphs

The bond graph approach to modelling is an energy-based technique where energy flows are the basis for modelling. The bond graph itself is a graphical representation which shows how energy flows around a system. In addition it can show causality, and a separate file contains the constitutive relationships which define in detail the way in which components interact. Being energy based, bond graphs are able to represent all types of dynamic systems, including, but not limited to, electrical, mechanical, hydraulic, thermodynamic and chemical process systems. More importantly they can represent, in a single unified approach, systems which contain subsystems drawn from all of these domains.

A bond graph provides an unambiguous symbolic representation from which other representations can be derived, e.g., linearised state equations, nonlinear simulation code, system transfer functions, or differential equations. From the core bond graph representation it is also possible to derive other related representations automatically, e.g., sensitivity functions, observers, estimators or controllers.

Bond graphs have been chosen for modelling because they have a number of advantages over conventional block diagram or code-based modelling techniques:

- They provide a visual representation of the system.
- Being an energy-based methodology which works on the principle of conservation of energy it is difficult to accidentally introduce extra energy into a system and therefore it is hard, for example, to produce perpetual motion machines.
- The bonds are symbols which contain meaning. These, together with the constitutive relationships, accurately and unambiguously define the system symbolically. Symbolic computation is then required to manipulate the bond graphs to produce symbolic or numerical output in a number of different formats.
- Bond graphs separate the structure of a system from causality and can therefore use the same model whatever external causality is applied. For example, a motor can be driven as a motor (current or voltage fed) or as a generator. This would require different models in a block diagram or procedural code-based simulation language; however, the system, can be represented by a single bond graph.
- Each bond in a bond graph contains a bidirectional flow of information. This ensures that it is easy to model systems which produce a 'back force' on the input of the system depending on the output without having to introduce artificial feedback loops; for example, modelling of a motor connected to a gearbox does not produce the same problems with inertias as it does in block diagram notation.
- Bond graphs can be viewed as a way of setting up symbolic system equations in a structured, reliable and flexible way.
- Hierarchical bond graphs are a high-level language which can aid the conceptual modelling process. The process of splitting up systems into smaller components is not essential; however, in many spheres, e.g., software engineering, it is recognised that splitting up a large complex system into smaller components (or subroutines) is useful both to ensure reusability of code and to reduce errors.

Bond graphs consist of a number of components, which are described below.

2.2.1 Variables

Energy methods are the fundamental key to modelling with bond graphs. Associated with any energy flow or transfer are two variables—the flow and effort variables—the product of which is energy. The effort and flow (sometimes called across and through) variables for different domains are shown in Table 2.1.

Table 2.1   Effort and flow variables in different domains

Domain	Effort	Flow
Mechanical	Force	Velocity
Mechanical	Torque	Ang. velocity
Electrical	Voltage	Current
Hydraulic	Pressure	Flow rate
Thermal	Temperature	Entropy flow

2.2.2 Bonds

A simple bond is shown in Figure 2.1. The bond is the line with the half arrowhead. Each bond represents an energy flow in the system, and the conventional direction of energy flow (or sign convention for positive energy flow) is in the direction of this arrow. In addition each energy flow has associated with it a flow/effort pair (multiplied together they equate to power), and the flow variable is conventionally associated with the side of the bond with the half arrow.

2.2.3 Junctions

There are two types of junctions in bond graphs: the **0** junction has a common effort variable for all bonds associated with it and the flow variables sum to zero, while the **1** junction has common flow variables and the effort variables sum to zero. **0** and **1** junctions are shown in Figure 2.2. The equations for the **0** junction are

$$e_1 = e_2 = e_3 \qquad f_1 - f_2 - f_3 = 0$$

while those for the **1** junction are

$$f_1 = f_2 = f_3 \qquad e_1 - e_2 - e_3 = 0$$

Note that there can be any number of bonds per junction and that the sign is determined by the direction of the half arrow. Also note that the equations are written in a noncausal manner.

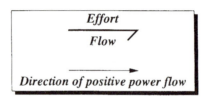

Figure 2.1   A simple bond

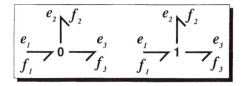

Figure 2.2   Bond graph junctions

2.2.4 Comaonents

Components, along with junctions and bonds, are the main building blocks of bond graphs. There are two main types of basic components—one-port components which connect to one bond, and two-port components which connect to two bonds. More recently, multiport components have been introduced, and the advent of hierarchical bond graph modelling tools has enabled subsystem models to be developed and considered as multiport components which can be included in higher-level bond graphs.

2.2.4.1 One-port components

Figure 2.3 shows four one-port components. These are

  R: resistor, damper, flow resistance, attenuator, ...
  C: capacitor, spring, compressibility, ...
  I: inductor, mass, fluid intertia, ...
  SS: current/voltage source, force/velocity source, pressure/flow source, ...

The relationship between the effort and flow variables, is determined by the constitutive relationship of the component (see Section 2.2.6). The SS (source/sensor) components provide the links from a bond graph to the external environment or boundary conditions of a system.

2.2.4.2 Two-port components

Two-port components are shown in Figure 2.4. The two main forms of two-port component are transformers, where the flows and efforts on either side are related to each other, and gyrators, where the effort on one side is linked to the

Figure 2.3   One-port components

Figure 2.4   Two-port components

flow on the other and vice versa. Transformers and gyrators can link different energy domains, e.g., a motor links the electrical and mechanical domains. Examples of transformers and gyrators are:

TF: transformer, lever, gearbox, piston, . . .
GY: DC motor, gyros, ...

2.2.5 *Parameters*

Numerical values of parameters are stored in the `numpar` parameter file. Numerical values are used when required but parameters remain in symbolic form at all other times.

2.2.6 *Constitutive relationships*

Constitutive relationships (CRs) define the relationship between the parameters of a component and the flow and effort of the bonds attached to the component. The constitutive relationship for a resistor is given by the equation

$$1 = iR/V$$

Note that the relationship is, where possible, given in a noncausal manner. The CRs are stored in the `cr` file and are read into `reduce` once causality has been established.

2.2.7 *Causality*

One of the main features of modelling in bond graphs is that the causality is not specified when the model is being developed. The causality is specified only for source/sensor (SS) components and can then automatically be propagated throughout a bond graph. Causality is indicated on a bond graph by a causal stroke which is a short line at right angles to the main bond at one end or the other. The position of this bond determines the causality. Figure 2.5 shows two examples of causal strokes on a bond attached to a resistance. A causal stroke at one end of a bond imposes the flow at that end on the flow at the other and makes the effort at the end with the causal stroke equal to the effort at the other end. Therefore in Figure 2.5a the effort at the bottom end of the bond, e_1, is determined by the effort at the top end, e_2 the effort at the top end is determined by the constitutive relationship of the R component, and the flow at the top end of the component, f_2, is determined by the flow at the bottom end, f_1. Figure 2.5a therefore represents a resistance with flow as input and effort as output, i.e., $e_1 := rf_1$. Likewise a similar argument can be given for Figure 2.5b which gives

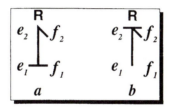

Figure 2.5   Causal strokes on bonds

that effort is the input and flow is the output, i.e., $f_1 := e_1/r$. Note that these equations are now written in a causal manner. An extended version of causality, using bicausal bonds, is discussed elsewhere [15, 16].

2.3 Hierarchical bond graphs

Hierarchical bond graphs extend the reusability of subsystem models and the clarity of bond graphs, reduce the number of bonds on any particular diagram, allow modelling to be carried out at a higher level, and aid with the construction of large scale models.

As stated in the Introduction, in the basic bond graph literature [1–6] there are two sorts of bond graph:

- a bond graph
- a word bond graph.

The latter concept, the word bond graph, provides a top-level system description—omitting the details. For example, Figure 2.6 gives the word bond graph of a conceptual motor/generator comprising an (electrical) power supply driving a motor, which in turn drives an electrical generator, which in turn drives a load. Each bond shows flow of power—electrical or mechanical as appropriate.

The traditional bond graph approach is to convert the word bond graph directly into a bond graph (in one step); this Section provides a *conceptual* framework for hierarchically creating an executable word bond graph by a process of stepwise refinement. Section 2.4 describes the different representations and languages used and the software developed to support this process.

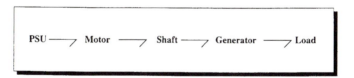

Figure 2.6   System MotorGenerator: word bond graph

A basic requirement for hierarchical bond graph modelling is to provide a mechanism for connecting a subsystem to its environment—the encapsulating system. As discussed in [12, 13] the concept of a *port* provides the most useful approach for dynamic system modelling. In the bond graph context, a port provides a (power bond) connection to the environment. With this in mind, it is logical to use the source-sensor SS component [17] to this end—this is the approach espoused in this Chapter.

In the context of this chapter, a hierarchical bond graph is defined as a bond graph which, in addition to standard bond graph components and bonds, may contain two additional types of component:

- components which are themselves described by a hierarchical bond graph
- ports (represented by SS components).

Thus arbitrary multiport components can themselves be defined as hierarchical bond graphs and connected to other bond graph components to form a hierarchical system description.

An example of such a component is the *Motor* component appearing in the word bond graph of Figure 2.6. This is described graphically as the hierarchical bond graph component of Figure 2.7, which contains:

- two ports ([in] and [out])
- a hierarchical bond graph component of type DC called *Motor*.

The DC component is described graphically as the hierarchical bond graph component of Figure 2.8, which contains:

- two ports ([in] and [out])
- two standard *I* components
- two standard *R* components
- two standard **1** junction components
- a standard gyrator GY component.

Other parts of this system are described in Section 2.5.

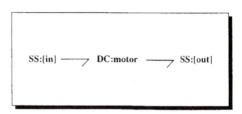

Figure 2.7   Subsystem motor: hierarchical bond graph

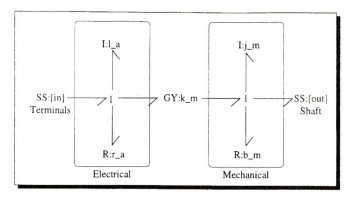

Figure 2.8   Subsystem DC: hierarchical bond graph

2.4 Representations, languages and software

Bond graphs are symbolic objects. In modelling with bond graphs, and in converting bond graph models to other representations (e.g. linearised state space equations), it is therefore clear that it is necessary to undertake symbolic computation. The MTT (model transformation tools) software has been written to enable transformations to take place between different representations of the same system. There is also a requirement for different languages to be used to describe the same representations. Symbolic computation is used in both the transformations between different representations and in the translation between different languages. Sometimes it is necessary to translate a representation to a different language before it can be transformed to a different representation. A conceptual diagram of this is shown in Figure 2.9. Different software tools are used for different translations and transformations. In general a transformation may involve loss of information about the system. In general a translation does not involve loss of any information.

Tables 2.2 and 2.3 give details of the different languages and representations used in MTT.

MTT is based on the UNIX make utility and each representation/language combination has a rule associated with it. Each rule has a list of dependent files (which are, in turn, dependent upon other files) and a *method* which details how to derive the representation/language combination from the dependent files. The command line syntax of MTT is such that each MTT command specifies the system, the representation required and the language of the output. Thus to view the Nyquist frequency response of a system MotorGenerator the command

```
mtt MotorGenerator nyfr view
```

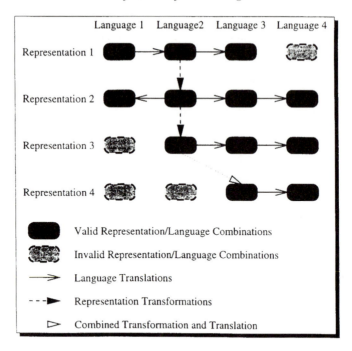

Figure 2.9   Transformations and translations in MTT

Table 2.2   System representations and languages

Language	Description
c	C code for simulation
dat	data file
fig	Xfig format diagram
gdat	data file in format for graphing
h	C header file
m	MATLAB mfile
ps	a PostScript view of the information
r	reduce code
sdat	data file for import into a spreadsheet
sh	executable shell code
tex	a description in $\LaTeX$ suitable for inclusion in reports
txt	a plain text file
view	an on-screen view of the object

is given. This will result in a series of transformations to generate the Nyquist frequency response, followed by translations to produce an output on the shell screen

```
MTT (Model Transformation Tools) version 2.7 ($Date:
1998/01/23$)
This is free software with ABSOLUTELY NO WARRANTY.
Type 'mtt warranty' for details
Creating MotorGenerator_nyfr.dat
Creating MotorGenerator_nyfr.gdat
Creating MotorGenerator_nyfr.ps
```

followed by a window with the Nyquist frequency response being displayed on the workstation.

Different tools are used to achieve different transformation and translations: for example, reduce is used to translate differential algebraic equations in reduce language to the L^{A}T$_E$X language; it is also used to transform constrained state equations to linear state-constrained matrices, both in the reduce language, while octave is used to transform a descriptor matrix representation to a step response. A list of required programs and utilities is given in Table 2.4.

2.5 Example: a motor generator

This Section contains an example of hierarchical bond graph modelling and automatic transformations and translations. The objective set has been to view a graph of the numerical solution of the ordinary differential equation representation of the model. A number of intermediate steps in the process are shown and the MTT commands used and details of the tools used to achieve the transformations and translations are also discussed.

2.5.1 Acausal bond graph

The word bond graph for the motor generator is shown in Figure 2.10. It represents the MotorGenerator system at the highest level. It is edited in xfig and defines the main components of the system. It is viewable using the MTT command:

```
mtt MotorGenerator abg view
```

The subsystems are listed in Section 2.5.1.1.

2.5.1.1 Subsystems

MTT identifies the following subsystems of the MotorGenerator bond graph:

- Generator: a simple DC generator (Section 2.5.1.2) consisting of:
  —DC: DC motor (or generator) (Section 2.5.1.3); no subsystems

Table 2.3   System representations and languages

Mnemonic	Description	Language								
		c	dat	m	ps	r	tex	txt	view	other
abg	acausal bond graph			•	•		•		•	fig
cbg	causal bond graph			•						fig
cr	constitutive relationship for each subsystem							•		
cse	constrained-state equations			•	•	•	•		•	
csm	constrained-state matrices			•	•	•	•		•	
dae	differential-algebraic equations			•	•	•	•		•	
daes	dae solution—state		•	•	•	•			•	
daeso	dae solution—output		•	•	•				•	
def	definitions—system orders etc.									
desc	verbal description of system	•								
dm	descriptor matrices									
ese	elementary system equations			•	•	•	•		•	
fr	frequency response		•	•	•	•	•		•	
input	numerical input declaration			•	•					
ir	impulse response—state		•	•	•			•	•	
iro	impulse response—output		•	•	•			•	•	
lbl	label file							•	•	
lmfr	loglog modulus frequency response		•	•	•				•	
lpfr	semilog phase frequency response		•	•	•				•	
nifr	Nichols style frequency response		•	•	•				•	
numpar	numerical parameter declaration	•								
nyfr	Nyquist style frequency response		•	•	•				•	
obs	observer equations for CGPC			•	•	•	•		•	

Abbreviation	Description	Extension
ode	ordinary differential equations	
odes	ode solution—state	
odes	ODE simulation header file	h
odeso	ode solution—output	
odeso	ode solution—output for spreadsheet	sdat
odess	ode numerical steady states—states	
odesso	ode numerical steady states—outputs	
params	symbolic parameter setting	
rbg	raw bond graph	
rep	report	
rfe	robot-form equations	fig
sabg	stripped acausal bond graph	
simp	simplification information	
sm	state matrices	
smc	controller form state matrices etc.—siso only	
smo	observer form state matrices etc.—siso only	
sms	ode (in state matrix form) solution—state and output	
smss	SM simulation header file	h
sr	step response—state	
sro	step response—output	
ss	steady-state equations	
ssk	state-space controller gain—siso only	
ssl	state-space observer gain—siso only	
sspar	steady-state definition	
struc	structure—list of inputs, outputs and states	
sub	executable subsystem list	sh
sub	LᴬTEX subsystem list	h
sympar	symbolic parameters	
tf	transfer function	

Table 2.4   Programs and utilities used by MTT

Program	Description
comm	Gnu compare sorted file by line
dvips	dvi to Postscript file converter
fig2dev	File conversion for fig files
gawk	Gnu awk
gcc	Gnu C compiler
ghostview	Postscript viewer
gmake	The GNU make utility. Used to determine dependencies and keep files up to date
gnuplot	Gnu graph plotting software
grep	Gnu grep
latex	The L^AT_EX document preparation system produces dvi files
latex2html	Converter from L^AT_EX to HTML format
octave	A matrix manipulation package similar to MATLAB
perl	The perl programming language (needed for latex2html)
reduce	A symbolic Algebra package
sed	Gnu sed
sh	The UNIX Bourne shell
xdvi	dvi file viewer
xfig	A drawing program that runs under X windows

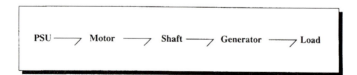

Figure 2.10   System MotorGenerator: acausal bond graph

- Load: an inertia (Section 2.5.1.4); no subsystems
- Motor: a simple DC motor (Section 2.5.1.5) consisting of:
  —DC: DC motor (or generator) (Section 2.5.1.3); no subsystems
- PSU: power supply (Section 2.5.1.6); no subsystems
- Shaft: a flexible shaft (Section 2.5.1.7); no subsystems.

2.5.1.2 System Generator

The acausal bond graph of system Generator is displayed in Figure 2.11. Its label file gives details of the names of the parameters to be used with each of the components in the bond graph. An extract from it is given below. The Generator subsystem itself has one subsystem:

- DC: DC motor (or generator) (Section 2.5.1.3)

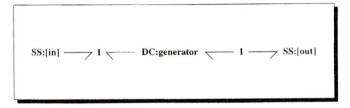

Figure 2.11 System Generator: acausal bond graph

Label file information:

```
%% Label file for system Generator (Generator_lbl.txt)

% Component type DC
generator lin k_g; l_g; r_g; j_g; b_g

% Component type SS
[in] external external
[out] external external
```

2.5.1.3 System DC

The acausal bond graph of system DC is displayed in Figure 2.12 and its label file is listed below. There are no further subsystems. DC is a two-port component representing a DC motor. It has the five parameters listed in Table 2.5.

Label file information:

```
% SUMMARY DC: DC motor (or generator)
%DESCRIPTION Port [in]: Electrical (in)
%DESCRIPTION Port [out]: Mechanical (out)
%DESCRIPTION Parameter 1: Motor gain (k_m)
%DESCRIPTION Parameter 2: Armature inductance (l_a)
%DESCRIPTION Parameter 3: Armature resistance (r_a)
%DESCRIPTION Parameter 4: Inertia (j_m)
%DESCRIPTION Parameter 5: Friction coefficient (b_m)

%% Label file for system DC (DC_lbl.txt)

%% Each line should be of one of the following forms:
% a comment (i.e. starting with %)
% Component-name CR_name arg1, arg2,..argn
% blank

%Motor gain
k_m lin flow, $1

% Electrical components
%Inductance
l_a lin effort, $2
```

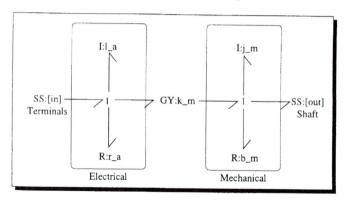

Figure 2.12   System DC: acausal bond graph

```
%Resistance
r_a lin flow, $3

% Mechanical components
%Inertia
j_m lin flow, $4
%Friction
b_m lin flow, $5
```

2.5.1.4 System Load

The acausal bond graph of system Load is displayed in Figure 2.13 and its label file information is given below. There are no further subsystems.

Label information:

```
%% Label file for system Load (Load_lbl.txt)
% Component type R
r_l lin flow, r_l

% Component type SS
v_2 external 0
```

Table 2.5   DC motor parameters

Index	Parameter
1	Motor gain (k_m)
2	Armature inductance (l_a)
3	Armature resistance (r_a)
4	Inertia (j_m)
5	Friction coefficient (b_m)

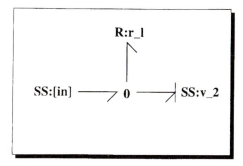

Figure 2.13 System Load: acausal bond graph

2.5.1.5 System Motor

The acausal bond graph of system Motor is displayed in Figure 2.14. An extract from the label file is given below. The Motor subsystem itself has one subsystem:

- DC: DC motor (or generator) (Section 2.5.1.3)

Label file information:

```
%% Label file for system Motor (Motor_lbl.txt)

% Component type DC
motor lin k_m; l_m; r_m; j_m; b_m

% Component type SS
[in] external external
[out} external external
```

2.5.1.6 System PSU

The acausal bond graph of system PSU is displayed in Figure 2.15 and its label file information is given below. Note that although the representation is acausal, because causality for the whole system has not been completed, this bond graph

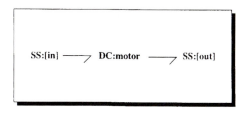

Figure 2.14 System Motor: acausal bond graph

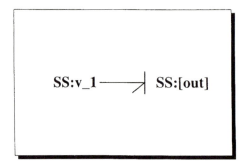

Figure 2.15   System PSU: acausal bond graph

does have a causal stroke since the PSU is defined as a voltage source. There are no further subsystems.

Label information:

```
%% Label file for system PSU (PSU_lbl.txt)

% Component type SS
v_1 external internal
```

2.5.1.7 System Shaft

The acausal bond graph of system Shaft is displayed in Figure 2.16 and its label file information is given below. There are no further subsystems.

Label information:

```
%% Label file for system Shaft (Shaft_lbl.txt)

% Component type C
k lin flow, k
```

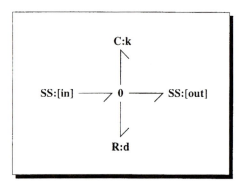

Figure 2.16   System Shaft: acausal bond graph

```
% Component type R
d lin flow, d
```

```
% Component type SS
[in] external external
[out] external external
```

2.5.2 System structure

A causal bond graph implies a description of the system structure. The structure (struc) representation describes the structure of these equations in terms of the inputs, outputs, states and nonstates of the system.

The structure representation is generated by the MTT command

```
mtt MotorGenerator struc view
```

The effect of this command is to produce the following MTT output:

```
Creating MotorGenerator_rbg.m
Creating MotorGenerator_cmp.m
Creating MotorGenerator_fig.fig
Creating MotorGenerator_sabg.fig
Creating MotorGenerator_sub.sh
Creating Generator_rbg.m
Creating Generator_cmp.m
Creating Generator_fig.fig
Creating Generator_sabg.fig
Creating Generator_sub.sh
Creating DC_rbg.m
Creating DC_cmp.m
Creating DC_fig.fig
Creating DC_sabg.fig
Creating DC_sub.sh
Creating DC_abg.m
Creating Generator_abg.m
Creating Shaft_abg.m
Creating MotorGenerator_abg.m
  ⋮
Creating MotorGenerator_cbg.m
Creating MotorGenerator_type.sh
INFORMATION: Final causality of MotorGenerator is 100% complete.
Creating MotorGenerator_ese.r
Creating MotorGenerator_def.r
Creating MotorGenerator_struc.txt
Creating MotorGenerator_struc.tex
Creating MotorGenerator_struc.doc
Creating MotorGenerator_struc.dvi
Creating view of MotorGenerator_struc
```

As can be seen from the output, MTT generates a number of intermediate files for each system in the hierarchical bond graph to produce a causal bond graph for the whole MotorGenerator system (cbg.m). It firstly takes the raw bond graph and converts it to the stripped bond graph (sabg) using a gawk script. This is then repeated for each subsystem until it is possible to produce the acausal bond graph in MATLAB mfile (m) language (abg.m) using a Bourne shell script. The causality is then completed using MATLAB or octave. The elementary system equations (ese) and definitions (def) are then created in reduce format using MATLAB or octave. The structure file is also created at this point and converted to L^AT_EX for output. This is shown in Tables 2.6−2.8.

Table 2.6   System inputs

	Component	System	Repetition
1	v_1	MotorGenerator_1	1

Table 2.7   System outputs

	Component	System	Repetition
1	v_2	MotorGenerator_5	1

Table 2.8   System states

	Component	System	Repetition
1	l_a	MotorGenerator_2_motor	1
2	j_m	MotorGenerator_2_motor	1
3	k	MotorGenerator_3	1
4	l_a	MotorGenerator_4_generator	1
5	j_m	MotorGenerator_4_generator	1

2.5.3 System ordinary differential equations

The ordinary differential equations of the system are created in tex language by the MTT command:

```
mtt -o MotorGenerator ode tex
```

This creates the following MTT output:

```
Creating MotorGenerator_cr.txt
Creating Generator_cr.txt
Creating DC_cr.txt
Creating Load_cr.txt
Creating Motor_cr.txt
Creating DC_cr.txt
Creating PSU_cr.txt
Creating Shaft_cr.txt
Creating MotorGenerator_cr.r
Creating MotorGenerator_sympar.txt
Creating Generator_sympar.txt
Creating DC_sympar.txt
Creating Load_sympar.txt
Creating Motor_sympar.txt
Creating DC_sympar.txt
Creating PSU_sympar.txt
Creating Shaft_sympar.txt
Creating MotorGenerator_sympar.r
Creating MotorGenerator_dae.r
Copying MotorGenerator_dae.r to MotorGenerator_ode.r
Creating MotorGenerator_ode.tex
```

In creating the system ordinary differential equations a number of steps (transformations and translations) have to be made. These are:

1. For each of the subsystems translate the label files and create text files detailing the constitutive relationships (using gawk).
2. Create a single file with all of the constitutive relationships for the whole system and convert to reduce format.
3. For each of the subsystems create a symbolic parameter file from the label file (using gawk).
4. Create a symbolic parameter file for the whole system and convert it to reduce format.
5. Convert elementary system equations to differential algebraic equations using reduce.
6. In this case the differential algebraic equations are the same as the ordinary differential equations, so the file is copied across.

The final line produces the output given below:

$$\dot{x}_1 = \frac{((l_m u_1 - x_1 r_m) j_m k_m - l_m x_2)}{(j_m k_m l_m)}$$

$$\dot{x}_2 = \frac{((kx_1 - k_m l_m x_3) j_m - b_m k k_m l_m x_2)}{(j_m k k_m l_m)}$$

$$\dot{x}_3 = \frac{((j_g x_2 - j_m x_5) dk - j_g j_m x_3)}{(d j_g j_m k)}$$

$$\dot{x}_4 = \frac{(-((r_g + r_l) j_g k_g x_4 + l_g x_5))}{(j_g k_g l_g)} \tag{2.1}$$

$$\dot{x}_5 = \frac{((kx_4 + k_g l_g x_3) j_g - b_g k k_g l_g x_5)}{(j_g k k_g l_g)}$$

$$y_1 = \frac{(x_4 r_l)}{l_g} \tag{2.2}$$

2.5.4 System state matrices

To create the system representation in state matrices in both L^A^TEX format and mfile (m) format, the MTT commands

```
mtt -o MotorGenerator sm tex
```

and

```
mtt -o MotorGenerator sm m
```

are used.

This produces the following MTT output:

```
Creating MotorGenerator_cse.r
Creating MotorGenerator_ss.r
Creating MotorGenerator_csm.r
Creating MotorGenerator_sm.r
Creating MotorGenerator_sm.m
```

To create the system state equations MTT must do a number of manipulations of the system equations in reduce. This first of all involves transformation of the differential algebraic equations to constrained state equations. These are then further processed to produce the constrained state matrices and the linearised state matrices using reduce. The final output here is in mfile format.

The matrices are given in eqns. 2.3—2.6, while the corresponding mfile is given below:

$$A = \begin{pmatrix} \dfrac{(-r_m)}{l_m} & \dfrac{(-1)}{(j_m k_m)} & 0 & 0 & 0 \\[2ex] \dfrac{1}{(k_m l_m)} & \dfrac{(-b_m)}{j_m} & \dfrac{(-1)}{k} & 0 & 0 \\[2ex] 0 & \dfrac{1}{j_m} & \dfrac{(-1)}{(dk)} & 0 & \dfrac{(-1)}{j_g} \\[2ex] 0 & 0 & 0 & \dfrac{(-(r_g + r_l))}{l_g} & \dfrac{(-1)}{(j_g k_g)} \\[2ex] 0 & 0 & \dfrac{1}{k} & \dfrac{1}{(k_g l_g)} & \dfrac{(-b_g)}{j_g} \end{pmatrix} \qquad (2.3)$$

$$B = \begin{pmatrix} 1 \\ 0 \\ 0 \\ 0 \\ 0 \end{pmatrix} \qquad (2.4)$$

$$C = \begin{pmatrix} 0 & 0 & 0 & \dfrac{r_l}{l_g} & 0 \end{pmatrix} \qquad (2.5)$$

$$D = (0) \qquad (2.6)$$

```
function [A,B,C,D] = MotorGenerator_sm
%function [A,B,C,D] = MotorGenerator_sm
%Linearised state matrices for system MotorGenerator;
%File MotorGenerator_sm.m;
%Generated by MTT;
% Set the parameters;
global ...
b_g ...
j_g ...
k_g ...
l_g ...
r_g ...
r_l ...
b_m ...
j_m ...
k_m ...
l_m ...
r_m ...
d ...
```

```
k . . .
 ;
A = zeros (5,5);
B = zeros (5,1);
C = zeros (1,5);
D = zeros (1,1);
A(1,1)=(-r_m)/l_m;
A(1,2)=(-1)/(j_m*k_m);
A(2,1)=1/(k_m*l_m);
A(2,2)=(-b_m)/j_m;
A(2,3)=(-1)/k;
A(3,2)=1/j_m;
A(3,3)=(-1)/(d*k);
A(3,5)=(-1)/j_g;
A(4,4)=(-(r_g+r_l))/l_g;
A(5,5)=(-1)/(j_g*k_g);
A(5,3)=1/k;
A(5,4)=1/(k_g*l_g);
A(5,5)=(-b_g)/j_g;
B(1,1)=1;
C(1,4)=r_l/l_g;
```

2.5.5 System frequency response

The MTT commands

```
mtt -o MotorGenerator lmfr ps
```

and

```
mtt -o MotorGenerator lpfr ps
```

are used to produce the modulus and phase diagrams for the frequency response of the system. These are shown in Figures 2.17 and 2.18, respectively. The MTT output is:

```
Creating MotorGenerator_def.m
Creating MotorGenerator_dm.r
Creating MotorGenerator_numpar.m
Creating MotorGenerator_dm.m
Creating MotorGenerator_fr.m
Creating MotorGenerator_lmfr.m
Creating MotorGenerator_lpfr.m
Creating MotorGenerator_nyfr.m
Creating MotorGenerator_nifr.m
Creating MotorGenerator_lmfr.dat
Creating MotorGenerator_lmfr.gdat
Creating MotorGenerator_lmfr.ps
```

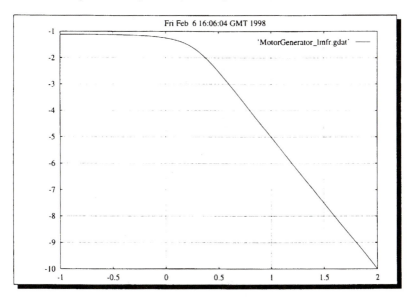

Figure 2.17   System MotorGenerator, representation lmfr

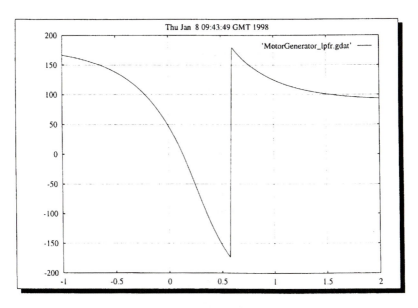

Figure 2.18   System MotorGenerator, representation lpfr

To create the graph of the magnitude of the frequency response, MATLAB or octave is used. First of all the system definitions are translated from reduce format to mfile format. The linear descriptor matrices are then created from the differential algebraic equations and then converted to mfile format. MATLAB or octave is used to calculate the frequency responses and the output is then translated to the format required for the gnuplot program.

The output produced by this MTT command (MotorGenerator_lmfr.ps) is shown in Figure 2.17. The corresponding phase is shown in Figure 2.18.

2.5.6 *Numerical parameters and initial states*

The numerical parameters used in the simulation are displayed in text format with the MTT command:

```
mtt -c -o MotorGenerator numpar txt
```

This produces the following output file:

```
# Numerical parameter file (MotorGenerator_numpar.txt)
# Generated by MTT at Wed Jan 7 12:07:36 GMT 1998

# %%%%%%%%%%%%%%%%%%%%%%%%%%%%%%%%%%%%%%%%%%%%%%%%%%%%%%%%%%%%%
# %% Version control history
# %%%%%%%%%%%%%%%%%%%%%%%%%%%%%%%%%%%%%%%%%%%%%%%%%%%%%%%%%%%%%
# %% $Id$
# %% $Log$
# %%%%%%%%%%%%%%%%%%%%%%%%%%%%%%%%%%%%%%%%%%%%%%%%%%%%%%%%%%%%%

# Parameters
b_g = 1.0; # Parameter b_g for Generator
j_g = 1.0; # Parameter j_g for Generator
k_g = 1.0; # Parameter k_g for Generator
l_g = 1.0; # Parameter l_g for Generator
r_g = 1.0; # Parameter r_g for Generator
r_l = 1.0; # Parameter r_l for Load
b_m = 1.0; # Parameter b_m for Motor
j_m = 1.0; # Parameter j_m for Motor
k_m = 1.0; # Parameter k_m for Motor
l_m = 1.0; # Parameter l_m for Motor
r_m = 1.0; # Parameter r_m for Motor
d = 1.0; # Parameter d for Shaft
k = 1.0; # Parameter k for Shaft

# Initial states
x(1) = 0.0; # Initial state for MotorGenerator_2_motor (1_a)
x(2) = 0.0; # Initial state for MotorGenerator_2_motor (j_m)
x(3) = 0.0; # Initial state for MotorGenerator_3 (k)
```

```
x(4) = 0.0; # Initial state for MotorGenerator_4_generator(l_a)
x(5) = 0.0; # Initial state for MotorGenerator_4_generator(j_m)
```

2.5.7 System response

The parameters for the solution of the ordinary differential equations are given in the file MotorGenerator_odes.h, while their solution is shown graphically in Figure 2.19.

The MTT commands

```
mtt -c -o MotorGenerator odes h
```

and

```
mtt -c -o MotorGenerator odeso ps
```

produce the following output:

```
Creating MotorGenerator_sympar.c
Creating MotorGenerator_ode.c
Creating MotorGenerator_odes.c
Creating MotorGenerator_numpar.c
Creating MotorGenerator_sympar.h
Creating MotorGenerator_input.c
Creating MotorGenerator_odes.m
Creating MotorGenerator_odeso.m
```

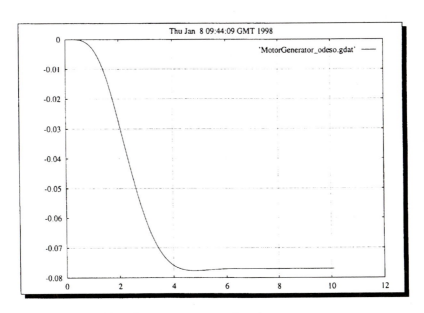

Figure 2.19   System MotorGenerator, representation odeso

In this case the system files are converted to C programs, which are then solved using standard numerical differential equation algorithms. The file MotorGenerator_sympar.h gives details of the parameters for the solution of the ordinary differential equations, and the output is then created in mfile format and converted to Postscript notation for viewing:

```
/*
%% Parameter file for system MotorGenerator
%% (MotorGenerator_odes.h)
%% This file provides the params for simulation:
*/
#define DT 0.1 /* Time step (for printing) */
#define LAST 10.0 /* Last time */
#define STEPFACTOR 1000 /* Integration steps per time step */
```

2.6 Conclusion

In this chapter we have shown how symbolic computation is more than simply the use of symbolic algebra computer packages. The essential feature of modelling with hierarchical bond graphs is that information is kept in a symbolic format which represents the structure of the system and is transformed and translated into the desired representations and languages when required. Many different programs are used to achieve the manipulation of hierarchical bond graphs but in most cases the calculations and manipulations are performed symbolically rather than numerically, thus ensuring that the loss of information at any stage is minimised.

The use of hierarchical bond graphs allows for the reuse of many components and the creation of libraries of bond graph models of subsystems. This aids the translation from initial conceptual modelling into a detailed simulation or design model and would not be possible without the use of significant symbolic computation.

2.7 References

1 KARNOPP, D. C., and ROSENBERG, R. C.: 'System dynamics: a unified approach' (Wiley, 1975)
2 THOMA, J.: 'Introduction to bond graphs and their applications' (Pergamon Press, 1975)
3 WELLSTEAD, P. E.: 'Introduction to physical system modelling' (Academic Press, 1979)
4 ROSENBERG, R. C., and KARNOPP, D. C.: 'Introduction to physical system dynamics' (McGraw-Hill, 1983)
5 KARNOPP, D. C., MARGOLIS, D. L., and ROSENBERG, R. C.: 'System dynamics: a unified approach' (Wiley, 1990)
6 GAWTHROP, P. J., and SMITH, L. P. S.: 'Metamodelling: bond graphs and dynamic systems' (Prentice Hall, 1996)

7 'The bond graph compendium'. Online via the WWW. URL: http://www.ece.arizona.edu/~cellier/bg.html/ also mirrored at: http://www.eng.gla.ac.uk/bg/.

8 GAWTHROP, P. J., and BALLANCE, D. J.: 'Symbolic algebra and physical-model-based control', *Comput. Control J.*, 1997, **8**,(2), pp. 70–76

9 GAWTHROP, P. J.: 'MTT: Model transformation tools'. Proceedings of international conference on *Bond Graph Modeling and Simulation* (ICBGM'95), Society for Computer Simulation, Las Vegas, January 1995, pp. 197–202

10 GAWTHROP, P. J.: 'MTT: Model transformation tools'. Online WWW home page, 1997. URL: http://www.eng.gla.ac.uk/~peterg/software/MTT/.

11 AZMANI, A., and DAUPHIN-TANGUY, G.: 'Archer: a program for computer aided modelling and analysis', in BEEDVELD, P. C. and DAUPHIN-TANGUY, G. (Eds.), 'Bond graphs for engineers' (IMACS, North Holland, 1992), pp. 263–277

12 CELLIER, F. E.: 'Continuous system modelling' (Springer-Verlag, 1991)

13 DE VRIES, T. J. A.: 'Conceptual design of controlled electro-mechanical systems'. PhD thesis, Universiteit Twente, 1994

14 BROENINK, J. F., BEKKINK, J., and BREEDVELD, P. C.: 'Multibond-graph version of the camas modelling and simulation environment', in BEEDVELD, P. C. and DAUPHIN-TANGUY, G. (Eds.), 'Bond graphs for engineers' (IMACS, North Holland, 1992), pp. 253–262

15 GAWTHROP, P. J.: 'Bicausal bond graphs'. Proceedings of international conference on *Bond Graph Modeling and Simulation* (ICBGM'95), Society for Computer Simulation, Las Vegas, January 1995, pp. 83–88

16 GAWTHROP, P. J.: 'Physical interpretation of inverse dynamics using bond graphs', *Bond Graph Dig.*, 1998, **2**,(1), 23pp

17 GAWTHROP, P. J., and SMITH, L.: 'Causal augmentation of bond graphs', *J. Franklin Inst.*, 1992, **329**,(2), pp. 291–303

Chapter 3

A survey of customised computer algebra programs for multibody dynamic modelling

I. C. Brown and P. J. Larcombe

3.1 Introduction

3.1.1 General comments

The application of symbolic mathematical computing to control problems has begun to be recognised as a significant benefit of modern technological achievements, the results from which now appear in high ranking journals and international conference proceedings with increasing regularity. They are testimony to the growing profile of research that embraces algebraic computation, and they affirm the acceptability of such work as a worthwhile addition to the control profession throughout industry and academia. The uptake of computer algebra across the broad spectrum of control, however, has been relatively slow compared with some scientific disciplines, and there is evidently much scope within control engineering to find new uses for this potent resource.

Control of a physical system is provided for by a dynamic representation of it. This chapter gives a concise account of the consistent spread of customised symbolic computer code over the last quarter of a century or so as a means to construct mathematically faithful dynamic models of multibody systems. After some introductory remarks on multibody systems dynamics and computer algebra, Section 3.2 details briefly some of the classic, and more recent, mathematical formulations and mechanisms for generating dynamic models of which the majority are amenable to symbolic computation. Sections 3.3 and 3.4 deal with the type of customised routines produced for this area of modelling, together with their application and historical development. A summary concludes the discussion, and includes a tabulation of known dynamics programs to help clarify the presentation.

Once the dynamic equations of a system are known analytically, they remain valid for all system motion and obviate the need for their recalculation in tandem with time discretisation. With this in mind, computer algebra has long been viewed as a tool for computer-based dynamic formulation and analysis, and has proved itself to be an indispensable aid as symbolic packages have become more robust (and affordable) and the hardware on which they run improved. Its exploitation reflects strongly the maturity of the field of multibody modelling, and provides an interesting topic for investigation that is deserving of treatment. For our purposes, the term 'multibody system' refers in the main to one of mechanical type, although multibody systems occur in different forms throughout a variety of disciplines.

3.1.2 Multibody dynamics and computer algebra

Dynamic modelling is a vast subject possessing numerous facets, and has been actively researched for a number of years (there have been many textbooks and articles devoted to it, and such is the present day breadth of related work that a dedicated journal, *Multibody System Dynamics*, was released by Kluwer academic publishers in 1997). Accordingly, the role of symbolic mathematical software has become established as one of great influence, having implications for commercial packages that purport to offer user-friendly representations of rigid and elastic body systems. Access to an accurate mathematical description of a system's dynamics remains an integral part of the associated control design and simulation processes. In many instances, however, the sheer complexity of component interaction renders manual derivation of governing equations of motion impossible. The desire to improve by automation the analytical procedures involved has sustained research addressing this specific issue, and today's reliable computer algebra packages lend themselves readily to this task. The full symbolic retention of all contributing variables and their error-free (infinite precision) manipulation can also yield insights which are not forthcoming from classic numeric approaches to systems modelling and control design employing sometimes heuristic, and inherently approximate, techniques. The following exposition, in discussing symbolic computational modelling in relation to multibody dynamics, captures the essentials of this area of study and provides a reference listing which the reader may find useful if pursuing it to a greater depth. An effort to tabulate and classify modelling codes has also been made.

The fundamentals of modelling—mathematical abstraction of a system, followed by analysis and/or simulation which is then interpreted with regard to its original form—are longstanding and of universal acceptance, having brought about inestimable scientific and technological advances. In practice, dynamic modelling has been used in predictive engineering to minimise costs in product development, whilst academically it has served as fruitful ground for the application of the theories of rigid and nonrigid mechanics, numerical analysis and control. The review by Sharp [1] makes the point that, given the speed of the latest computers, the model building problem is not one of principle but one of detail. Numeric *solution* of complex equations presents far less difficulty nowadays than it once did, but in general the *formulation* of complete and correct dynamic models is not easily achieved for it often demands a certain level of intimate physical knowledge about the system from the modeller. It is for this reason that software tools with a modular approach to the construction of dynamic models are sought keenly by engineers (who are neither expert dynamicists nor accomplished mathematicians) in order that their own job requirements and creative aspirations are satisfied.

The fact that some of the many mathematical terms in the dynamics of a system play only a small part in affecting its behaviour led early workers in the modelling field to consider their neglect so as to ensure efficiency in formulation (for simulation and control) whilst simultaneously keeping sufficient model accuracy. Of necessity, the first work on the formation of models was centred on relevant theory and numerical computation. As then, the mathematical structure of dynamic equations, and the manner in which they are both realised and handled, remain important concerns. The theme of this chapter is the shift in emphasis towards the incorporation of computer algebra throughout multibody dynamic modelling, and the way it has taken place.

In its most basic form, a multibody system can be said to consist of a set of mass components (rigid or flexible), the connections between which can, without loss of generality, be described by two types of single-degree-of-freedom joint — revolute (rotational) and prismatic (sliding). In the event that a joint of more than one degree-of-freedom connects two bodies, it can usually be modelled as a combination of them. Multibody systems take many forms (ground and underwater vehicles, robots, aircraft, spacecraft, industrial machinery, etc.), and require analysis via modelling techniques. As opposed to kinematics — which is the study of movement without consideration of the forces that cause it — dynamics accommodates externally applied forces and, from the resulting equations, time-dependent motion can be determined and a control methodology implemented as appropriate.

The kinematics of a system refers to all geometric and time-based properties of motion. It suffices to say that this is itself a vigorous area of research in which computer algebra has been utilised, and the reader is referred to the seminal paper by Denavit and Hartenberg [2] which laid the mathematical foundation for much of the subsequent work on kinematics. With reference to its kinematic parameters, specification of the mass distribution throughout a system allows dynamic formulation which, in closed form, is a set of differential or differential-algebraic equations. In generating the equations of motion of a general three-dimensional system by symbolic computation (inclusive of all constituent bodies, contact forces, springs, dampers, and so on), there are certain considerations of major importance: (i) the choice of algorithm to achieve this in terms of its suit-ability, and (ii) the proficiency of the computer with regard to its speed and symbolic storage capability. Algebraic computation is very memory-intensive and can place a severe burden on a computer. Because of this, as well as the number of bodies that a system contains, its topological features—such as whether or not it is open-chain (as in a manipulator arm) or closed-chain (as in the classic slider-crank mechanism)—affect the automated operations, since any inherent kinematic simplicity reduces processing time whilst complex internal structures will obviously tend to increase it. Often the working conditions of a real system give rise to constraints which add a level of complexity to any mathematical dynamic model. Naturally, during the assimilation by dynamicists of programs that cater for the many types of subtleties of systems, manually prepared models have set some of the standards against which computer-generated code can be measured.

3.2 Dynamic formulations and modelling approaches

Over the years—beginning with work focused on open- and closed-chain assemblages of elements connected with robotics, biomechanics and the space industry during the 1960s and 1970s—many different methods have been used to derive the dynamic equations of rigid multibody systems. Some algorithms have been developed by the consistent improvement of existing ones or by combining the attributes of others, and a vast amount of literature has been produced accordingly. The main formulations used are the following:

- in the Newton-Euler approach [3, 4] a multibody system is broken down into its disparate elements. The orientation of the system is defined using relative co-ordinates, and each element is allocated its own 'moving' reference frame. Each of these frames is related, in turn, to the previous one in a chain running back to the base element which receives a 'world' reference frame. Kinematics information can thereby be propagated by forward recursion from base to end element. A similar backward recursion—working from the end-effector to the base—determines the forces and moments exerted on each body, from which the torques applied to the joints are calculated. The resulting set of compact recursive equations constitutes the Newton-Euler equations, and a final closed form is available.
- the Lagrange formulation [5, 6] is a holistic approach whereby (once the energies of individual components have been found) the system is dealt with as a single entity. Generalised co-ordinates are defined, in terms of which all energies are expressed. The total kinetic energy is calculated (that is, the sum of the kinetic energies for each mass) and similarly the total potential energy, with the difference between the two functions known as the Lagrangian. Lagrange's equations—one for each co-ordinate—require the Lagrangian to be differentiated with respect to co-ordinates, velocities and time, and are a system of second-order (usually highly nonlinear and coupled) differential equations describing motion.

- D'Alembert's principle [7, 8] also decomposes the system into isolated bodies. It is based on the assertion that each mass is acted on by an 'inertia force' described by a vector whose line of action passes through the mass and is equal and opposite to the mass multiplied by the inertial acceleration. This assumption implies that the system is a null system (that is, the vector summation of all forces and of all moments about any point is zero). With judicious choice of moment centres to minimise calculations, an explicit set of closed-form differential equations is obtained.
- Kane's method [9, 10] is an algorithmic approach based on a development of Lagrange's form of d'Alembert's principle (virtual work). As with the Newton-Euler method, relative co-ordinates are utilised. The introduction of partial linear and angular velocities, and generalised speeds, leads to the automatic elimination of nonworking constraint forces and torques. The formulation also makes use of intermediate variables to stand for recurring combinations of algebraic terms (known as the 'Z values'). The application of the principle of virtual work produces a set of equations in terms of the linear and angular velocities, with the equations being uncoupled with respect to joint force and moment components.

We should also mention here the use of look-up tables in providing a reference mechanism for the dynamics of a system. The production of a finite (multidimensional) table of values of precomputed system information allows dynamic data to be available during simulation and control; interpolation between known node points suffices to cover the whole 'workspace'. An early discussion paper on this is [11], although the idea first arose in modelling the human cerebellum in 1975.

As stated, the above list represents the most commonly arising types of dynamic formulation. Others are available which have either been largely ignored, employed in a limited fashion or have fallen out of use (as in the case of look-up tables). Briefly, they include the following:

- graph theory (the vector network approach) [12]: a visual approach that first consists of the identification of the system nodes (points of interconnection), followed by the representation of all elements by displacement vectors. It is postulated that the sum of forces at every point, and the sum of displacements for every closed loop, are zero, and this leads to development of equations of motion using relationships between forces and displacements.
- Gibbs-Appell equations [13]: obtained from a method bearing similarities to that of Lagrange, where the acceleration energy is computed instead of the kinetic energy. Dynamic equations result from differentiation of the energy function (use of potential energy is optional).
- Hamilton's canonical equations: these are produced by differentiation of the Hamiltonian, which is a function of the generalised co-ordinates and generalised momenta.

- Boltzmann-Hamel equations: arising from an approach that follows the same procedure as Lagrange's method, but involves the introduction of dependent variables instead of (independent) generalised co-ordinates.

For a comparison of the last two methods with the main ones see Kane and Levinson [14].

All of these formulations are necessarily equivalent, but they differ in the manner in which they are put into practice and the ways in which dynamic equations are created. The choice of any method of derivation is determined by the context of the problem for which a model is required. The above are outlined in connection with their application to rigid body systems, largely in keeping with their origins. Flexible systems are more complicated to model and have led to the development of particular techniques of formulation. Some of the rigid body dynamic analysis methodologies, however, have been built upon to treat systems that contain deformable elements and as such have provided a foundation for the whole field of multibody modelling.

3.2.1 Bond graphs

We end this section with reference to an interesting modelling technique that has existed for some time and has, since the advent of algebraic computation, seen the latter adapted to it. Resembling something of a typical engineering schematic, the bond graph is especially well suited to deal with aspects of dynamics which span, for instance, the mechanical-hydraulic-electrical-magnetic spheres. (Coupling the properties from these and others is at the heart of the field of mechatronics, where to establish a system's full dynamics is notoriously difficult.) A bond graph is, very briefly, a portrayal of the exchange of energy through a system, and defines a flow sequence around it which provides a succinct and unambiguous dynamic representation. When encoded on a computer it lies at the human interface level of a hierarchical modelling environment. Driven by 'effort' and 'flow' variables—two fundamental parameters with analogies in different physical domains—to which are assigned causality, constituent system elements and their energy interplay are delineated in bond graph language using entities such as 'ports', 'bonds' and 'junctions', and preserve a system's physical integrity entirely.

We note that bond graphs, as a modelling device and unifying notation, date back to the first work of Paynter [15], from which many developments have been made during the subsequent two decades and beyond (see, for example, texts [16–24]). The marriage of bond graphs and computer algebra is, though, quite new, and it seems generally agreed that Nolan [25,26] was probably the first to bring it about, emphasising that its conceptual background is very much founded on work in graph/network theory and rule/knowledge-based programming using artificial intelligence languages such as LISP, C and PROLOG which are able to process data symbolically. We do not expand on this here, other than to remark that modern symbolic mathematical languages, through their structure and functionality, are an ideal tool to translate the bond graph description of a system into a mathematical one. As an example, a fairly recent application may be found in what has been termed physical-model-based control [27], in which a system-specific controller—using computer algebra embedded in a sophisticated modelling arena and employed to perform model transformation—is designed from a generic bond graph.

An excellent outlet for dissemination of bond graph research has been the *Journal of the Franklin Institute*, in which a copious compilation of related literature citations appeared for the first time in 1991 [28]. This will be of interest to anyone wishing to read up on what is a somewhat under-represented, but very powerful and elegant, method of modelling whose protagonists have started to take advantage of mainstream algebraic packages and their versatility. Because of the suitability now of hardware and software, the creation of a bond graph and readable symbolic dynamic code therefrom is no longer the problem it once was, and it is perhaps the combination of the initially awkward language of the bond graph and the physical understanding of systems needed to put bond graph theory into practice that prevents more people from considering this technique as a means to model multibody systems.

3.3 Customised dynamics programs: type and application

In referring to customised programs we essentially mean software that is directed at particular classes of multibody system, although within this the range covered is diverse. Whilst our interest lies primarily with symbolic programs, their evolution has to be set in the context of the wider picture of the field and is affected by developments in numeric-based codes. Before this, we note the variability in the level of modelling a package may offer. Rigidly structured code may be written to perform analysis on a single type of application, taking into account only characteristics specific to it. By nature, such a program will be of limited scope, with the likelihood that many aspects of the system must satisfy preset criteria. Such a restriction, however, may result in efficient running of the program. Although harder to construct, code that permits the examination of more general systems has wider appeal. The increased complexity of routines that give the operator a wider modelling remit may suffer a lower operational efficiency, but this is offset by their greater flexibility. This factor may be critical to the utility and success of any modelling package released commercially.

Classification can be made in different ways, but customised dynamics programs fall into two natural categories—numeric and symbolic; the pros and cons of each are well established and are outlined briefly below. In this section it can be seen that the trend over time has been away from fully numeric methods of formulation towards those with a strong or total symbolic flavour to them. This is due overwhelmingly to rapid improvements in computer hardware technology in general, and to those in computer algebra software in particular. A summary of the literature survey undertaken for this chapter is shown in Tables

3.1 and 3.2 of Section 3.5, which give respective listings of 'named' and 'unnamed' dynamics codes. They indicate the type of each program (symbolic or numeric), and both the method(s) of dynamic formulation and the computer language(s) used.

3.3.1 Numeric methods

The numeric formulation of the equations of motion of a system requires that all of the basic parameters be predetermined quantitatively. In other words, each single application of the software pertains to one particular case of the system being modelled, with no provision for any additional generality. If a system is controlled on-line, it is necessary to reconstruct the dynamic equations rapidly as its configuration changes. This requires not only an efficient formulation but also considerable processing power, and execution times can be lengthy. Most numeric-based programs contain, or have access to, routines to deal with the integration of the equations, and can produce output in graphical form depicting variable time histories from which animations can be formed.

During the embryonic years of symbolic processing, numeric formulations were standard. As those multibody systems considered by dynamicists grew in complexity, the calculations incurred became more involved and provided an impetus for the improved algorithms available today. In spite of this, the requirement to repeatedly reassemble dynamic equations numerically to allow for any changes in defining model parameters is inescapable, and developments in the field of computer algebra have given modellers an alternative medium for the construction of mathematical dynamic models which do not have this constraint.

3.3.2 Symbolic methods

These are based on the symbolic upholding of all those system parameters desired, giving rise to the formulation of a single, all-encompassing dynamic description which can be referred to when needed (that is to say, substitution of actual numeric values for variables can be delayed as appropriate). A distinguishing feature of symbolic-based methods is the mathematically comprehensible output they generate (especially true when the software incorporates advanced algebraic simplification procedures), so that the significance of individual terms, or groups of terms, may be identified. This brings with it the opportunity to gain insights into the model that would otherwise not be available. In their infancy, symbolic methods were computationally cumbersome and slow to implement, but they are now able to compete seriously with their numeric counterparts and in many aspects surpass them; numeric simulation and data processing can be effected after symbolic dynamic formulation with little difficulty.

There are two ways in which a symbolic approach to dynamic formulation can be employed. Programs can be written exclusively in one of several mainstream computer algebra languages (such as REDUCE and MAPLE), of which there are currently over half a dozen on the market. Alternatively, a general programming language (such as Fortran and PASCAL) can be adapted, into which symbolic operations can be built. Writing in a dedicated algebraic language, whilst convenient, involves certain problems that cannot be avoided. Many functions are not under the control of the programmer (only calls for them are), and computational overload can arise from 'intermediate expression swell'—a phenomenon with a reputation in computer algebra circles and one to which dynamics calculations are vulnerable. Also, some simplification rules (such as trigonometric and hyperbolic identities) sometimes have to be added in, and this can be problematic if the language is sensitive to their ordering and form. The approach does carry certain advantages, however, for the symbolic capability of the program is limited only by the language used, and the subsequent code is relatively compact and simple in its structure. Programs written in general (non-specialist) languages derive their algebraic facilities from rudimentary routines defined by the coder. Only those processes required for the problem in hand need to be included, and they are available for modification. Difficulties with the simplification of expressions and symbolic 'blow up' exist here too, but the main disadvantage of this approach resides in the level of programming skills needed to produce a program of this type. Increasingly, modern general-purpose symbolic mathematical languages contain ever more powerful and rigorously tested capabilities, and the effort being put into them means that personalised symbolic routines in this sense have become virtually outmoded.

3.3.3 Applications

Numeric and symbolic multibody dynamics has been accelerated by research in several domains, developments having taken place in a roughly parallel fashion amongst them. This is mainly due to the fact that many different types of multibody system possess similar characteristics at a very fundamental level, and so new ideas in one area can quite often be cross-fertilised and applied to another. With respect to the progression of multibody dynamics, the following list highlights the most important systems on which research has been conducted:

- mechanical systems, such as production machinery, vehicle suspensions and gear systems. Dynamics programs that deal with these have been in existence for the longest time. DRAM [29] is an example of one, written for the computer-aided design of systems undergoing large displacements which are typified by machine suspensions.
- aerospace systems, such as satellites and interplanetary spacecraft. The development of programs for this class of system arose from the Cold War 'race for space' between the USA and the USSR. The complex nature of spacecraft led to the study of high-degree-of-freedom systems, and satellite interactions produced tree topology models containing multiple branches. An example program is MBDY [30], which was produced to analyse open-chain space systems such as the Viking Probe.

- robotic systems, such as manipulators used in hazardous operating conditions and industrial robots carrying out continuous processes. The most common robotic construction consists of a set of links arranged in open series. These can move relative to each other, and determining the behaviour of the endmost link (usually a gripper or tool) is normally the most important concern. Robotics formulation programs are often used for system control, an example being ROBMAT [31] which derives equations of motion for control of rigid open-chain manipulators.

- ground vehicle systems, such as trains and motor cars. Dynamics software deals with these most usually for simulation purposes and design testing. Whilst in many cases general multibody programs are used, an example of a dedicated vehicle dynamics package is AUTOSIM [32]. Requiring symbolic system parameters as input, the output is a self-contained program for simulation in any of several target computer languages.

Having given some space over to a short introduction to multibody dynamics and an explanation of the profitable use of algebraic processing as a computer-based modelling tool, the following section addresses the real theme of the chapter as given by the title. In recognition of the progress made in customised packages, we document the main areas of application and aspects of routines produced for dynamic modelling; citations are tabled as appropriate, to condense the literature and to help the reader gain an overview of the field.

3.4 Multibody dynamics: historical development

3.4.1 Early dynamics programs

The first customised dynamics program seen was DYANA [33], developed by General Motors Corporation in 1960 and applicable to unconstrained, multi-degree-of-freedom, one-dimensional arrangements (specifically that of a spring-mass-damper, but analogous systems such as electrical networks were within its scope). Despite the simplicity of the models coped with by the software, it achieved pioneering status bearing in mind the relatively elementary standards of computers at that time. The next major innovation was SADII [34], which was capable of simulating dual-spin spacecraft (so named because one spinning section that stabilised the attitude of the system was connected by a rotary joint to a nonmoving platform). The ability to include the effects of a control system was a key part of the software which was the forerunner of the numeric space-craft simulators DISCOS [35] and N-BOD [36].

Driven by a surge in the development of computer technology, the early 1970s produced the first wave of customised multibody dynamics programs in the form of numeric simulation packages. These included DAMN [37] (later to become DRAM [29] for obvious aesthetic reasons), VECNET [12] (which seeded the two spin-off programs DYNIS [38] and PLANET [39]), IMP [40] (later IMP-UM), MEDUSA [41], MBDY [30], ADAMS [42] and DYMAC

[43]. As technology was still comparatively primitive during this period, models were conceptually simple and output of a basic form in terms of its readability and sophistication. The majority of programs addressed planar (two-dimensional) systems, the exceptions being MBDY, IMP and VECNET/DYNIS. MBDY was written for complex spacecraft dynamics and so operated as a three-dimensional modelling package. It extended previous work carried out on exclusively rigid elements to allow for the possibility that the terminal body (that at the end of a chain) could be flexible. IMP, developed at Ford (and later imported to the University of Michigan, hence IMP-UM), was the first program of its kind which was able to perform spatial (three-dimensional) simulation. It could be applied to a range of mechanisms through the application of graph theory. The machinery simulators VECNET and DYNIS adopted a similar approach, with much of their processing time spent on producing graphical output.

Graph theory was not particularly well suited to the numerically-oriented computer processes of the time, and interest waned in favour of other, more appropriate, mechanisms for dynamic modelling (better, and more readily available, hardware and software has since tended to invigorate work on dynamics using bond graphs). The most widely applied mathematical formulation at this time was the method of Lagrange, having become the standard option following work carried out in the 1960s by Uicker [5] and Kahn [6]. It was to be superseded by the recursive Newton-Euler formulation (of which the iterative nature made it ideal for numeric computation), and later also by Kane's equations.

3.4.2 *Pioneering symbolic dynamics programs*

The first symbolic programs to emerge in multibody dynamics were TOAD [44] and OSSAM [45] in 1973. TOAD was written for manipulator arm modelling, giving both kinematic and dynamic information as output. It was executed as five separate modules—two to formulate and display nonsimplified equations, two more to simplify expressions, and an edit routine to format the output. The language used was FORMAC (the PL1 symbolic interpreter), the computational and memory overheads resulting from FORMAC itself making TOAD very slow. The program scope was one limited to open-chain systems, with the further restriction that arms with multiple branches were not allowed. For a (standard) 6 degree-of-freedom robot model it took over 7 hours to generate the equations of motion, with 90% of that time being spent on algebraic simplification. In contrast, OSSAM was a general-purpose multibody dynamics program. Its author made several general points relating to the field of symbolic modelling:

- calculations involved in the study of dynamics are lengthy, tedious and so error-prone; their automation was desirable.
- there was a lack (at that time) of efficient trigonometric simplification procedures.

- there was the potential for direct generation of simulation code from programs (OSSAM could fulfil this, but only after significant editing of the dynamic equations).
- existing symbolic languages were slow, and there were possible inefficiencies in their use due to the inclusion of facilities not required for specific problems.

The above led to the development of a purpose-built symbolic processing program written in LISP. Although limited in the number of problems it could handle, the undertaking was an enormous one and thus an impressive application of computing power then at the disposal of the scientific community. This was a noteworthy contribution to symbolic multibody modelling 25 years ago, and it is easy to forget the innovations made in the application of computers during the time when they were still relatively unrefined.

3.4.3 Second-generation symbolic programs

Gradually, other symbolic software began to appear as researchers drew on the algebraic implementation of different formalisms. EGAM [46] (previously known as EDYLMA and catering only for revolute joints, but now able to analyse prismatic connections) was created in 1976 to construct Lagrange's equations for systems with tree topologies. This was followed in 1977 by a program developed by Levinson [47] that mimicked Kane's method of dynamic formulation. Both of these programs were written in FORMAC, and their computational demands led to slow run times. Also in 1977, Schiehlen and Kreuzer [48] produced the symbolic multibody code written in Fortran that was later to become NEWEUL based on the Newton-Euler approach.

During the early to mid 1980s, the number of symbolic codes being generated began to rival that associated with numeric programs. Dynamicists acknowledged the advantages of automated algebraic manipulation, and foresaw the processing power of the future. At this time there was a rapid increase in modelling software directed at manipulators as robotics became the cutting edge of dynamics research. The progression of multibody dynamics is, in character, readily seen by charting activity in this area; we give further details.

DYMIR [49] (written in REDUCE) could form a fully symbolic three-link robot dynamic model (including elastic links when the program was later improved). A Fortran program MIRE [50] went further in being able to simulate robots, in addition to which it possessed a reference library of commonly found mechanisms. AMIR [51] extended robotic modelling capabilities by addressing closed chains. Simplification of formulation was achieved by Luh and Lin [52], who produced a program allowing the user to identify and remove negligible dynamic terms. This was echoed in the operation of EMDEG [53], which contained algorithms for pattern matching and subsequent simplification of some terms. Luh and Lin proposed a Newton-Euler-based symbolic program to generate closed-form models, and Vukobratovic and Kircanski [54], amongst others, later implemented this via the software EXPERT PROGRAM.

It excluded, however, centripetal and Coriolis effects (which become significant for high-speed robot control). The algorithm was later augmented and gave rise to SYM [55]. ARM [56] was written in 1984 with two sections to it—a C 'composer' to produce a robot model, and a LISP 'performer' to execute the symbolic manipulation and generate closed-form equations. The simulation tool VAST [57] was later created to interface with ARM.

As obstacles encountered in the modelling of closed-chain and high-degree-of-freedom open-chain systems were overcome, attention focused on improving the efficiency of the automated processes through enhanced formulations and faster algorithms. Symbolic/numeric robotics packages such as DYNAMAN [58], ROBPACK [59], ROBSIM [60] and SYMORO [61]—as well as those symbolic routines from Neto *et al.* [62], Leu and Hemati [63] and Izaguirre and Paul [64]—explored these aspects of computation. Of note are the two parallel REDUCE programs of Koplik and Leu [65] to compare the Newton-Euler and Lagrange methods; they showed that the former uses less CPU time at the expense of increased storage requirements. Concurrent symbolic programs dealing with other applications included the general-purpose PASCAL package MESA VERDE [66], and the spacecraft simulation codes SD/EXACT [67] (written in Fortran and subsequently named SD/FAST) and SYMBOD [68] (coded in MACSYMA). NEWEUL was shown to be extendible from its original purpose (modelling terrestrial bodies) to accommodate space vehicles [69].

It is no surprise to find that some of the second-generation programs were able to make their mark as a direct result of advances in software technology spanning the decades of the 1970s and 1980s. The number of instances where general computer algebra languages (as opposed to personalised symbolic routines) were utilised increased, and their role in multibody dynamics was firmly consolidated.

3.4.4 Continuing use of numeric programs

Naturally, although symbolic-based dynamic modelling gained in popularity, the development of numeric packages was still ongoing. ADAMS [70] (much enhanced since its inception in 1973) and DADS [71] (further updated to cater for three-dimensional systems in [72]) were two commercially available general-purpose packages used substantially in industry (particularly with regard to motor vehicles). Generalised numeric-based codes such as these and TREETOPS [73] carried out the full dynamic modelling procedure from model description to simulation. Other numeric programs for narrower fields of application included MEDYNA [74] (dealing with road, rail and magnetically levitated vehicles) and FRESH [75]. The latter was a specialised code devoted to 4-wheel ground vehicles, where computational speed and efficiency could be aided by the manual simplification and reintroduction of the equations of motion prior to simulation. The use of human expertise to complement computing power is discussed in the following sub-section.

3.4.5 Programs reliant on user knowledge

Many of the ground-breaking symbolic dynamics packages were applicable only to a restricted type of multibody system. They were also quite slow to run and generated large intermediate expressions *en route*. Modification of some of these packages and the introduction of new ideas led to better formalisms and more acceptable running times. An interesting approach was made in 1986 by Faessler [10]. Using a purpose-built Fortran symbolic processor, Faessler suggested shifting the kinematic portion of dynamic analysis back to the operator, with particular expressions having to be supplied by the analyst (this was an extension of the interactive aspects of FRESH). He provided examples of programs for the user to implement from initial input, and proposed possible future work in this area claiming that for a one-off dynamic formulation more effort was needed to write code than to engage in a hand derivation; since the human did some of the work, the level of manipulation required by the computer remained fairly low and problems with software bugs (still then a feature of some languages) lessened. The notion was repeated to some extent in 1988 by Schaechter and Levinson [76] with the fully operational program AUTOLEV which was able to produce a dynamic model and complete code for simulation. It took some of the automatic processes of other packages and, through an interactive interface, returned them to the dynamicist similarly. Both AUTOLEV and Faessler's work made use of Kane's formulation, being ideal for this type of program because of the structure to the algorithm. As a result, storage requirement was considerably reduced during processing, although the reliance on human input meant that a highly proficient operator was needed to utilise the software (carrying its own risk of error).

3.4.6 Modern programs

The late 1980s saw reports of symbolic software such as DYNAM [77] (which was said to be computationally faster than any contemporary numeric program) and ARDEG [78] (which contained several advanced simplification algorithms, each pertaining to a specific type of dynamic term). OOPSS [79] demonstrated an alternative representation of systems through object-oriented programming. With some of the obvious similarities between modelling processes and object-oriented design procedures (and with the prevalence of such languages as Visual Basic and C^{++} today), it is likely that the two will be combined in modelling practices of the future.

Turning our attention to the current situation, we begin to see the true power that latter day computer technology affords to multibody dynamics modellers. The modelling and simulation package AUTOSIM [32] is written specifically for ground vehicles. Based on Kane's method, the output (as mentioned earlier) is a self-contained numeric simulation program in one of several possible languages. The package also possesses the ability to describe behaviour due to unconventional forces and to automatically generate constraint equations. As an indication of the performance level of modern symbolic processors, AUTOSIM compares well with previous automobile industry packages (DADS, ADAMS and FRESH) and has been shown to be much faster to run (up to 50 times quicker than, for example, DADS). Its computation rate has in fact proven to be three times that of real-time working speeds. Continuing, SIMPACK [80] was created for the simulation of multibody systems with flexible elements and closed loops, allowing for the intervention of the user at many stages. The system elements are permitted large movements relative to each other, making the software well suited to handling, for instance, sizeable ground vehicles. Parallel processing has since been embraced by it [81], and has been shown to deliver large reductions in computing time. The more robust and tested codes now in common usage are continually updated as a matter of course and also absorb improvements to hardware. Examples of this can be seen in [81], where enhancements to well established programs such as ADAMS and DADS are detailed.

To illustrate the requirements of a modern dynamics program we look at two recent innovations. JAMES [82] was developed at Aerospatiale (Cannes La Bocca, France) for use in flexible multibody systems modelling. From given system parameters it produces symbolic equations of motion for simulation. If needed, it can linearise and convert the equations to state space form to facilitate the design of control laws. MOTIONMC [83] is a package with similar attributes which is also capable of formulating syntactically correct Fortran simulation code. Written in MAPLE (in both cases), these are comprehensive dynamics packages that perform extremely complex calculations at high speed.

3.5 Summary

Until the 1960s classical mechanics, rigid body theory and their application were characterised by strong restrictions on model complexity. The nonlinearities present in the dynamic description of many systems proved to be insurmountable problems due to the relatively poor numerical methods available for dealing with equations of motion. It was mainly the demand for better models of satellites and spacecraft, and the simultaneous developments in computer power, which at this time led to a new branch of mathematics called multibody system dynamics. The longstanding results of mechanics were reviewed and extended as the basis of practical algorithms—multibody formalisms—and multibody dynamics first took shape as a discipline.

With regard to customised dynamics packages it is hoped that an informative account of the field has been provided. As for most areas of research that are strongly reliant on computers, the future of this one is assured but the long-term form into which it will evolve is uncertain; interestingly, it would appear that from the late 1980s onwards work on ground vehicle dynamics modelling has increased relative to other specialist areas. In completing the discussion, we summarise developments and offer one or two concluding remarks. Using the given key, Tables 3.1 and 3.2 (with additional references [84–97]) reveal something of the maturity of multibody dynamics. The programs are listed chronologically with respect to the order of the documents in which they are reported (each paper is the earliest we found for that package). The tables do not, therefore, necessarily list the programs in the order that they were conceived. Several trends emerge from the literature.

Key for Tables 3.1 and 3.2

T	Type	N	Numeric
		S	Symbolic

F	Formulation	L	Lagrange
		N	Newton-Euler
		K	Kane
		D	D'Alembert
		G	Graph/Network theory
		B	Bond graphs

Table 3.1   Chronological table of named customised dynamics programs

Acronym	Name in full	Language	Ref.	T	F
DYANA	DYnamics ANAlyzer program	Fortran	[33]	N	D
SADII	—	Fortran	[34]	N	L
DYMAC	DYnamics of MAChinery	Fortran	[43]	N	D
DAMN	Dynamic Analysis of Mechanical Networks	Fortran	[37]	N	L
IMP	Integrated Mechanisms Program	Fortran	[40]	N	G
MEDUSA	MachinE Dynamics Universal System Analyzer	Fortran	[41]	N	B
VECNET	VECtor-NETwork simulation program	Fortran	[12]	N	G
ADAMS	Automatic Dynamic Analysis of Mechanical Systems	Fortran	[42]	N	L
DRAM	Dynamic Response of Articulated Machinery	Fortran	[29]	N	L
DYNIS	DYNamic Interactive Simulation	Fortran	[38]	N	G
OSSAM	Ohio State Symbolic Algebraic Manipulator	LISP	[45]	S	L
TOAD	Tele Operators Arm Design	FORMAC	[44]	S	L
MBDY	MultiBoDY	Fortran	[30]	N	L
PLANET	PLAne-motion simulation using NETwork methods	Fortran	[39]	N	G
EGAM	Equation Generator of Articulated Mechanisms	FORMAC	[46]	S	L

Table 3.1 (Continued)

Acronym	Name in full	Language	Ref.	T	F
NEWEUL	NEWton-EULer formulation	Fortran	[48]	S	N
DISCOS	Dynamic Interaction Simulation of COntrols and Structures	Fortran	[35]	N	L
N-BOD	—	Fortran	[36]	N	N
DYMIR	DYnamic Models of Industrial Robots	REDUCE	[49]	S	L
MIRE	—	Fortran	[50]	S	L
KADAM	Kinematic And Dynamic Analysis of Mechanisms	Fortran	[84]	N	D
AMIR	Automatic Modelling of Industrial Robots	PASCAL	[51]	S	L
DADS	Dynamic Analysis and Design System	Fortran	[71]	N	L
SYMBOD	SYMbolic multiBODy	MACSYMA	[68]	S	N
ARM	Algebraic Robot Modeller	C/LISP	[56]	S	L/N
DYSPAM	DYnamics of SPAtial Mechanisms	Fortran	[85]	N	D
EXPERT PROGRAM	—	Fortran	[54]	S	N
FRESH	—	Fortran	[75]	N	L
VAST	Versatile robot Arm Simulation Tool	C	[57]	N	L
MESA VERDE	MEchanism, SAtellite, VEhicle and Robot Dynamic Equations	PASCAL	[66]	S	D
RESTRI	RESeau TRIdimensionnel (3-dimensional network)	Fortran	[86]	N	G
TREETOPS	TREE TOPologieS	Not known	[73]	N	K
COMPAMM	COMPuter Analysis of Mechanisms and Machines	Fortran	[87]	N	K
DYNAMAN	DYNamics and Analysis of MANipulators	MACSYMA	[58]	S	N
EMDEG	Efficient Manipulator Dynamic Equation Generator	LISP	[53]	S	K/L
MEDYNA	MEhrkorper-DYNAmik (multibody dynamics)	Fortran	[74]	N	L
ROBPACK	ROBot PACKage	SIMULA	[59]	N	N
ROBSIM	ROBot SIMulation system	Fortran	[60]	N	K/L
SD/EXACT	Symbolic Dynamics/EXACT	Fortran	[67]	S	K
SYMORO	SYmbolic MOdelling of RObots	Fortran	[61]	S	N
DYNAM	—	PASCAL	[77]	S	L
DYNOCOMBS	DYNamics Of COnstrained Multi Body Systems	Fortran	[88]	N	K
AUTOLEV	—	BASIC	[76]	S	K
PIODYM	—	Not known	[89]	S	N
ROBOTRAN	ROBOt TRANslator	PASCAL	[90]	S	D

Table 3.1 (Continued)

Acronym	Name in full	Language	Ref.	T	F
SYM	SYMbolic	C/Fortran	[55]	S	N
ARDEG	Automatic Robot Dynamic Equations Generator	BASIC	[78]	S	L
BAMMS	Bondgraph based Algorithms for Modelling Multibody Systems	Fortran	[91]	N	B
MECANO	—	Fortran	[92]	N	L
OOPSS	Object Oriented Planar System Simulator	MACSYMA	[79]	S	L
AUTODYN	—	Fortran	[80]	N	D
AUTOSIM	—	LISP	[32]	S	K
NUBEMM	NUmerische BErechnung Mechanischer Mehrkorpersysteme (numerical calculation of mechanical multibody systems)	Fortran	[80]	S	N
PLEXUS	—	Fortran	[80]	N	N
SIMPACK	SIMulation PACKage	Fortran	[80]	N	D
SPACAR	—	Fortran	[80]	N	K
MOTIONMC	—	MAPLE	[83]	S	D
A'GEM	Automatic Generation of Equations of Motion	Not known	[93]	N	L
FASIM	FAhrdynamik-SIMulation (vehicle dynamics simulation)	Fortran	[93]	N	D
MEFLAMO	MEchanism and Feedback LAw MOdelling	C	[94]	S	L
REVS	Reduced Equation Vehicle Simulation	Not known	[93]	N	K
ACIDYM	Analyse CInematique et DYnamique des Mecanismes (kinematic and dynamic analysis of mechanisms)	PASCAL	[81]	N	L/N/D
ALASKA	Advanced LAgrangian Solver in Kinetic Analysis	Fortran	[95]	N	L
CMSP	—	Fortran	[95]	N	L
DYMSTIP	DYnamics of Multibody Systems with Time-varying Inertial Properties	Fortran	[81]	N	N
DYNAMITE	DYNAmics of MultIbody sysTEms	Not known	[81]	N	L
LMS	—	C/Fortran	[95]	N	N
MEXX	MEXanical systems eXtrapolation integrator	Fortran	[81]	N	L
NUSTAR	—	Fortran	[95]	N	N
JAMES	—	MAPLE	[82]	S	L
ROBOTICA	—	Mathematica	[96]	S	L

Table 3.1 *(Continued)*

Acronym	Name in full	Language	Ref.	T	F
ROBMAT	ROBotics with MAThematica	Mathematica	[31]	S	L/N
MASS	Multibody Analysis Simulation System	LISP	[97]	S	L

Table 3.2   Chronological table of unnamed customised dynamics programs

Author(s)	Details	Language	Ref.	T	F
Luh & Lin	Simplified algorithm (removal of negligible terms)	Fortran	[52]	S	N
Faessler	Examples of symbolic code and proposals for future work	Fortran	[10]	S	K
Izaguirre & Paul	Simplified algorithm (no Coriolis or centripetal terms)	LISP	[64]	S	N
Koplik & Leu	2 programs comparing approaches	REDUCE	[65]	S	N/L
Leu & Hemati	Symbolic dynamic equations generator	MACSYMA	[63]	S	L
Neto *et al.*	Symbolic dynamic equations generator	PROLOG	[62]	S	L
Levinson	Symbolic dynamic equations generator	FORMAC	[47]	S	K

Our initial observation concerns the preference for methods of dynamic formulation. Early programs generally made use of Lagrange's method, along with graph theory and d'Alembert's Principle to a lesser degree. Dominance of the Lagrange formulation decreased gradually as greater interest was shown in the Newton-Euler method during the late 1970s and early 1980s. This was due to the development of the associated recursive algorithms which reduced the order of computations. The latest (relative) shift in implementation has been to Kane's equations, which arose from work conducted in the 1960s but which were not introduced explicitly until 1980 (and then not becoming established until the middle to late 1980s). Whilst more computationally demanding than that of Newton-Euler, Kane's method yields closed-form equations and is reasonably efficient for many systems. Also noticeable from the tables is the swing from numeric to symbolic formulations. Lack of algebraic software, and more so the original *raison d'être* for the computer, led early multibody dynamicists to pursue numeric paths in modelling. It was inevitable that, as the field of mathematical symbolic processing moved forward, they would take on board computer algebra with full justification and demonstrate its worth. We remark that it has been our intention here to show the salient features of the gradual progression in modelling practices, rather than to attempt to provide a definitive historical commentary on what is now a huge discipline.

Finally, we note the clear change in the algebraic basis of dynamics software. As reported by Fritzson and Fritzson [98], early use of symbolic programming was confined mainly to code written in common languages such as LISP, C and Fortran for personalised algebraic manipulation routines. Despite their relative accessibility, dedicated symbolic languages were then regarded as nonspecialist commodities and insufficiently advanced to be of any real benefit to the serious modeller. However, even during the formative years of symbolic dynamic modelling, many dynamicists understood their potential and persisted with computer algebra whilst simultaneously calling for new capabilities to be included in general-purpose packages. These demands, in conjunction with those from users involved in other types of work, led to the improvements that have enabled modern day researchers in dynamic modelling to rely on algebraic processing as a vital part of their day to day occupation. More generally, other computer related technological developments have empowered workers in the field to the point where fully integrated dynamic modelling, analysis, simulation and visualisation of systems is achievable.

As an example of one initiative brought about by interest in systems modelling, in 1987 the German Research Council (DFG) funded a five year project into multibody dynamics with the dual goals of (i) investigating the choice and range of available software, and (ii) developing an integrated collection of general-purpose multibody system programs. The results of this work are presented in [81], the main thrust of which has given rise to a modular suite of routines written in Fortran and utilising the object-oriented RSYST database (which stores data such as system definitions and simulation results, and serves as a standardised interface for data exchange between modules); the program NEWEUL formed the dynamic basis for the venture. Currently, a large project, being run under the management of a European consortium, aims to develop a new language for multibody modelling called Modelica. The work will unify modelling representations over many application domains, and will involve a combination of numeric and algebraic processing techniques hinged around the concepts of object-oriented programming and noncausal modelling. The ideas behind Modelica stem from a lack of interoperability which exists among the different modelling and simulation mediums of today and becomes more severe with the increased complexity and heterogeneity of those systems required to be analysed. The realisation of a next generation modelling capability is an ambitious objective of this multinational design effort.

3.5.1 Concluding remarks

An extensive review by Schiehlen [99] contains perspectives on past, current and future issues pertaining to multibody systems in general. From the article one senses that the subject is both diverse and lively, with fundamental research needed on a number of topics. These include data models for computer-aided design, parameter identification, optimal system specification, real-time simulation and animation, integration routines, contact and impact problems, multibody system and fluid interaction, strength analysis, mechatronics and control, and nonholonomic systems. In addition to application in dynamic modelling, it is safe to assume that computer algebra will be exploited in such work, especially since some of the new and exciting challenges arising provide interesting applied mathematical problems with an implicit high algebraic content.

Over the course of our discussion occasional reference has been made to flexible systems. Modelling of these began in the early 1970s and was motivated by the need to simulate industrial and technological systems where the effects of flexibility were significant. Traditionally, rigid body analysis had been applied throughout much of their modelling because of the bulky solids used in the majority of systems design, but the introduction of new lightweight materials into mechanical systems and the demand for higher speeds of operation made assumptions of rigidity in many dynamic formulations false. The deformation properties of a flexible body mean that the dynamic description of a flexible multibody system is more complex than for a rigid one and requires different treatment. Within this, it is apparent from the survey paper of Shabana that the role of computer-based symbolic manipulation is an area of current research interest [100]. It should be understood that the number of co-ordinates needed to model a system which is inclusive of flexible components is greater than in the completely rigid version (as is the degree of nonlinearity in the mathematical form of its dynamics), making the need for computers even more clearly defined. Note that the numerical routines used for solving the governing equations of motion of flexible systems remain similar to those successful for rigid systems.

As can be seen, computer algebra has been utilised to produce automated dynamic models of multibody systems as a prerequisite to simulation, control synthesis and dynamic analysis. The field of computer algebra has its own aspects and lines of research, and is responsible for the formidable computational systems that are currently available. The existence of meaningful applications has had a massive impact on the production of symbolic packages in terms of their performance and portability, and the proliferation of customised programming code in the context of dynamic modelling is evidence of but one of the many uses to which they may be put with great effect. It is hoped that the above exposition, which makes no claim to be exhaustive, has demonstrated this adequately.

We end the chapter here, on the following note given in the spirit of the wider remit of this book. The opportunities afforded by computer algebra within control have emerged over the last few years as ones with clear promise, although research that is underpinned by symbolic languages is still at a fairly low level. It is almost inevitable, however, that symbolic computation will in time come to be viewed as a standard part of some control engineering practices—relevant to both the academic and the industrial employee—maintaining a level of mathematical manipulation in model formulation and analysis which in the past was unimaginable. The integration of modelling methodologies and control theory within a computational environment permitting dual symbolic-numeric investigation of a problem through directed toolboxes is a goal towards which important steps are being taken. The future thus looks bright for this type of desktop workspace, especially if independent code and subroutines can be easily attached as an aid to the operator. As symbolic systems impose themselves and receive due appreciation, scientists and researchers everywhere are becoming awakened to the ways in which the technology of computer algebra can be brought to bear on otherwise intractable problems, leading to new results and understanding in the process. In the light of this, and expected advances in both hardware and software, control engineers are well advised to embrace these packages fully, and with enthusiasm. Some of those involved in multibody modelling have already done so.

3.6 References

1 SHARP, R. S.: 'The application of multibody computer codes to road vehicle dynamics modelling problems', *Proc. Inst. Mech. Eng. D, J. Automob. Eng.*, 1994, **208**, pp. 55–61

2 DENAVIT, J., and HARTENBERG, R. S.: 'A kinematic notation for lower-pair mechanisms based on matrices', *J. Appl. Mech.*, 1955, **22**, pp. 215–221

3 WALKER, M. W., and ORIN, D. E.: 'Efficient dynamic computer simulation of robotic mechanisms', *J. Dyn. Syst. Meas. Control*, 1982, **104**, pp. 205–211

4 LUH, J. Y. S., WALKER, M. W., and PAUL, R. P. C.: 'On-line computational scheme for mechanical manipulators', *J. Dyn. Syst. Meas. Control*, 1980, **102**, pp. 69–76

5 UICKER, J. J.: 'On the dynamic analysis of spatial linkages using 4×4 matrices'. PhD Dissertation, Northwestern University, Illinois, USA, 1965

6 KAHN, M. E.: 'The near minimum time control of open loop articulated kinematic chains'. Memo AIM 106, Artificial Intelligence Laboratory, Stanford University, USA, 1969

7 LEE, C. S. G., LEE, B. H., and NIGAM, R.: 'Development of the generalized d'Alembert equations of motion for mechanical manipulators'. Proceedings of the IEEE Conference on *Decision and Control*, San Antonio, USA, 1983, pp. 1205–1210

8 SONG, S. M., and LIN, Y. J.: 'An alternative method for manipulator kinetic analysis: the d'Alembert method'. Proceedings of the IEEE International Conference on *Robotics and Automation*, Philadelphia, USA, 1988, pp. 1361–1366

9 KANE, T. R., and LEVINSON, D. A.: 'The use of Kane's dynamical equations in robotics', *Int. J. Robot. Res.*, 1983, **2**, (3), pp. 3–21

10 FAESSLER, H.: 'Computer-assisted generation of dynamical equations for multibody systems', *Int. J. Robot. Res.*, 1986, **5**, (3), pp. 129–141

11 RAIBERT, M. H.: 'Analytical equations vs. table look-up for manipulation: a unifying concept'. Proceedings of the IEEE International Conference on *Decision and Control*, New Orleans, USA, 1977, pp. 576–579

12 SAVAGE, G. J., and ANDREWS, G. C.: 'Visual simulation of dynamic three dimensional systems using the vector network', *Simulation*, 1972, **19**, pp. 187–191

13 RICHARD, M. J., and LEVESQUE, B.: 'An efficient dynamic formulation for a manipulator based on Appell's equations'. Proceedings of the ASME *Design Technical Conference*, Chicago, USA, 1990, pp. 43–49

14 KANE, T. R., and LEVINSON, D. A.: 'Formulation of equations of motion for complex spacecraft', *J. Guid. Control*, 1980, **3**, (2), pp. 99–112

15 PAYNTER, H. M.: 'Analysis and design of engineering systems' (MIT Press, Cambridge, USA, 1961)

16 KARNOPP, D. C., and ROSENBERG, R. C.: 'Analysis and simulation of multiport systems: the bond graph approach to physical system dynamics' (MIT Press, Cambridge, USA, 1968)

17 KARNOPP, D. C., and ROSENBERG, R. C.: 'System dynamics: a unified approach' (Wiley, New York, USA, 1975)

18 THOMA, J. U.: 'Introduction to bond graphs and their applications' (Pergamon Press, Oxford, UK, 1975)

19 WELLSTEAD, P. E.: 'Introduction to physical system modelling' (Academic Press, London, UK, 1979)

20 BLUNDELL, A. J.: 'Bond graphs for modelling engineering systems' (Ellis Horwood, Chichester, UK, 1982)

21 ROSENBERG, R. C., and KARNOPP, D. C.: 'Introduction to physical system dynamics' (McGraw-Hill, New York, USA, 1983)

22 KARNOPP, D. C., MARGOLIS, D. L., and ROSENBERG, R. C.: 'System dynamics: a unified approach' (Wiley, New York, USA, 1990, 2nd edn.)

23 THOMA, J. U.: 'Simulation by bondgraphs: introduction to a graphical method' (Springer-Verlag, New York, USA, 1990)

24 GAWTHROP, P. J., and SMITH, L. P. S.: 'Metamodelling: bond graphs and dynamic systems' (Prentice Hall, Hemel Hempstead, UK, 1996)

25 NOLAN, P. J.: 'Symbolic analysis of bond graphs'. Proceedings of the European *Simulation* Conference, Edinburgh, UK, 1989, pp. 73–79

26 NOLAN, P. J.: 'Symbolic and algebraic analysis of bond graphs', *J. Franklin Inst.*, 1991, **328**, (5/6), pp. 1027–1046

27 GAWTHROP, P. J., and BALLANCE, D. J.: 'Symbolic algebra and physical-model-based control', *Comput. Control Eng. J.*, 1997, **8**, (2), pp. 70–76

28 BREEDVELD, P. C., ROSENBERG, R. C., and ZHOU, T. (Eds.): 'Bibliography of bond graph theory and application', *J. Franklin Inst.*, 1991, **328**, (5/6), pp. 1067–1109

29 CHACE, M. A., and SHETH, P. N.: 'Adaptation of computer techniques to the design of mechanical dynamic machinery'. ASME Paper 73-DET-58, 1973

30 HOOKER, W. W.: 'Equations of motion for interconnected rigid and elastic bodies: a derivation independent of angular momentum', *Celest. Mech.*, 1975, **11**, pp. 337–359

31 HORN, G., and LINGE, S.: 'Analytical generation of the dynamical equations for mechanical manipulators', *Model. Identif. Control*, 1995, **16**, (3), pp. 155–167

32 SAYERS, M. W., and MOUSSEAU, C. W.: 'Real-time vehicle simulation obtained with a symbolic multibody program'. Proceedings of the ASME Symposium on *Transportation Systems*, Dallas, USA, 1990, pp. 51–58

33 HARGREAVES, B.: 'GMR DYANA: the computing system and its applications', *Gen. Motors Eng. J.*, 1961, **8**, pp. 7–13

34 VELMAN, J. R.: 'Simulation results for a dual-spin spacecraft'. Proceedings of the Symposium on *Attitude Stabilization and Control of Dual-Spin Spacecraft*, El Segundo, USA, 1967, pp. 11–23

35 BODLEY, C. S., DEVERS, A. D., PARK, A. C., and FRISCH, H. P.: 'A digital computer program for the dynamic interaction simulation of controls and structure'. Technical Paper 1219, National Aeronautics and Space Administration, Washington, USA, 1978

36 FRISCH, H. P.: 'The N-BOD2 user's and programmer's manual'. Technical Paper 1145, National Aeronautics and Space Administration, Washington, USA, 1978

37 CHACE, M. A., and SMITH, D. A.: 'DAMN—digital computer program for the dynamic analysis of generalised mechanical systems', *Trans. Soc. Autom. Eng.*, 1971, **80**, pp. 969–991

38 SAVAGE, G. J., and ANDREWS, G. C.: 'DYNIS: a dynamic interactive simulation program for three-dimensional mechanical systems'. Proceedings of the *Man-Computer Communications* Seminar, Ottawa, Canada, 1973, pp. 25.1–25.10

39 ROGERS, R. J., and ANDREWS, G. C.: 'Simulating planar systems using a simplified vector-network method', *Mech. Mach. Theory*, 1975, **10**, pp. 509–517

40 SHETH, P. N., and UICKER, J. J.: 'IMP (Integrated Mechanisms Program): a computer-aided design analysis system for mechanisms and linkage', *J. Eng. Ind.*, 1972, **94**, pp. 454–464

41 DIX, R. C., and LEHMAN, T. J.: 'Simulation of the dynamics of machinery', *J. Eng. Ind.*, 1972, **94**, pp. 433–438

42 ORLANDEA, N.: 'Node-analogous, sparsity-oriented methods for simulation of mechanical dynamic systems'. PhD Dissertation, University of Michigan, USA, 1973

43 PAUL, B., and KRAJCINOVIC, D.: 'Computer analysis of machines with planar motion: part 2—dynamics', *J. Appl. Mech.*, 1970, **37**, pp. 703–712

44 STURGES, R.: 'Teleoperator arm design (TOAD)'. Technical Report E-2746, Charles Stark Draper Laboratory, Massachusetts Institute of Technology, USA, 1973

45 DILLON, R.: 'Computer assisted generation in linkage dynamics'. PhD Dissertation, Ohio State University, USA, 1973

46 LIEGEOIS, A., KHALIL, W., DUMAS, J. M., and RENAUD, M.: 'Mathematical and computer models of interconnected mechanical systems'. Proceedings of the International Symposium on the *Theory and Practice of Robots and Manipulators*, Warsaw, Poland, 1976, pp. 5–17

47 LEVINSON, D. A.: 'Equations of motion for multiple-rigid-body systems via symbolic manipulation', *J. Spacecraft Rockets*, 1977, **14**, pp. 479–487

48 SCHIEHLEN, W. O., and KREUZER, E. J.: 'Symbolic computerised derivation of equations of motion'. Proceedings of the IUTAM Symposium on the *Dynamics of Multibody Systems*, Munich, Germany, 1977, pp. 290–305

49 VECCHIO, L., NICOSIA, S., and NICOLO, F.: 'Automatic generation of dynamical models of manipulators'. Proceedings of the International Symposium on *Industrial Robots*, Milan, Italy, 1980, pp. 293–301

50 LIEGEOIS, A., FOURNIER, A., ALDON, M. J., and BORREL, P.: 'A system for computer aided design of robots and manipulators'. Proceedings of the International Symposium on *Industrial Robots*, Milan, Italy, 1980, pp. 441–452

51 DRAGANOIU, G., DAVIDOVICIU, A., MOANGA, A., and TUFIS, I.: 'Computer method for setting dynamical model of an industrial robot with closed kinematic chains'. Proceedings of the International Symposium on *Industrial Robots*, Paris, France, 1982, pp. 371–379

52 LUH, J. Y. S., and LIN, C. S.: 'Automatic generation of dynamic equations for mechanical manipulators'. Proceedings of the Joint *Automatic Control* Conference, Charlottesville, USA, 1981, pp. TA2D1–TA2D5

53 BURDICK, J. W.: 'An algorithm for generation of efficient manipulator dynamic equations'. Proceedings of the IEEE International Conference on *Robotics and Automation*, San Francisco, USA, 1986, pp. 212–218

54 VUKOBRATOVIC, M., and KIRCANSKI, N.: 'A method for computer aided construction of analytical models of robot manipulators'. Proceedings of the IEEE International Conference on *Robotics and Automation*, Atlanta, USA, 1984, pp. 519–528

55 KIRCANSKI, M., VUKOBRATOVIC, M., KIRCANSKI, N., and TIMCENKO, A.: 'A new program package for the generation of efficient manipulator kinematic and dynamic equations in symbolic form', *Robotica*, 1988, **6**, pp. 311–318

56 MURRAY, J. J., and NEUMANN, C. P.: 'ARM: an algebraic robot dynamic modelling program'. Proceedings of the IEEE International Conference on *Robotics*, Atlanta, USA, 1984, pp. 103–120

57 PFEIFER, M. S., and NEUMANN, C. P.: 'An adaptable simulator for robot arm dynamics', *Comput. Mech. Eng.*, 1984, **3**, (3), pp. 57–64

58 SREENATH, N., and KRISHNAPRASAD, P.: 'DYNAMAN: a tool for manipulator design and analysis'. Proceedings of the IEEE International Conference on *Robotics and Automation*, San Francisco, USA, 1986, pp. 836–841

59 OLSEN, H. B., and BEKEY, G. A.: 'Identification of robot dynamics'. Proceedings of the IEEE International Conference on *Robotics and Automation*, San Francisco, USA, 1986, pp. 1004–1010

60 WLOKA, D. W.: 'ROBSIM: a robot simulation system'. Proceedings of the IEEE International Conference on *Robotics and Automation*, San Francisco, USA, 1986, pp. 1859–1864

61 KLEINFINGER, J. F., and KHALIL, W.: 'Dynamic modelling of closed-loop robots'. Proceedings of the International Symposium on *Industrial Robots*, Brussels, Belgium, 1986, pp. 401–412

62 NETO, J. L. S., PEREIRA, A. E. C., and ALVES, J. B. M.: 'Symbolic computation applied to robot dynamic modelling'. Proceedings of the International Symposium on *Industrial Robots*, Brussels, Belgium, 1986, pp. 389–400

63 LEU, M. C., and HEMATI, N.: 'Automated symbolic derivation of dynamic equations of motion for robotic manipulators', *J. Dyn. Syst. Meas. Control*, 1986, **108**, pp. 172–179

64 IZAGUIRRE, A., and PAUL, R.: 'Automatic generation of the dynamic equations of robot manipulators using a LISP program'. Proceedings of the IEEE International Conference on *Robotics and Automation*, San Francisco, USA, 1986, pp. 220–226

65 KOPLIK, J., and LEU, M. C.: 'Computer generation of robot dynamic equations and the related issues', *J. Robot. Syst.*, 1986, **3**, (3), pp. 301–319

66 WITTENBURG, J., and WOLZ, U.: 'MESA VERDE: a symbolic program for nonlinear articulated-rigid-body dynamics'. ASME Paper 85-DET-151, 1985

67 ROSENTHAL, D. E., and SHERMAN, M. A.: 'High performance multibody simulations via symbolic equation manipulation and Kane's method', *J. Astronaut. Sci.*, 1986, **34**, (3), pp. 223–239

68 MACALA, G. A.: 'SYMBOD: a computer program for the automatic generation of symbolic equations of motion for systems of hinge connected rigid bodies'. Proceedings of the AIAA *Aerospace Science* Meeting, Reno, USA, 1983, Paper AIAA-83-0013

69 KREUZER, E. J., and SCHIEHLEN, W. O.: 'Generation of symbolic equations of motion for complex spacecraft using formalism NEWEUL', *Adv. Astronaut. Sci.*, 1983, **54**, pp. 21–36

70 ORLANDEA, N., and BERENYI, T.: 'Dynamic continuous path synthesis of industrial robots using ADAMS computer program', *J. Mech. Des.*, 1981, **103**, pp. 602–607

71 WEHAGE, R. A., and HAUG, E. J.: 'Generalized coordinate partitioning for dimension reduction in analysis of constrained dynamic systems', *J. Mech. Des.*, 1982, **104**, pp. 247–255

72 NIKRAVESH, P. E., and SRINIVASAN, M.: 'Generalized coordinate partitioning in static equilibrium analysis of large-scale mechanical systems', *Int. J. Numer. Methods Eng.*, 1985, **21**, pp. 451–464

73 SINGH, R. P., VANDERVOORT, R. J., and LIKINS, P. W.: 'Dynamics of flexible bodies in tree topology: a computer oriented approach', *J. Guid. Control Dyn.*, 1985, **8**, (5), pp. 584–590

74 JASCHINSKI, A., KORTUM, W., and WALLRAPP, O.: 'Simulation of ground vehicles with the multibody program MEDYNA'. Proceedings of the ASME Symposium on the *Simulation and Control of Ground Vehicles and Transportation Systems*, Anaheim, USA, 1986, pp. 315–341

75 WILSON, D. L., and BACHRACH, B. I.: 'The FRESH handling simulation: development and application'. Proceedings of the ASME International Conference on *Computers in Engineering*, Las Vegas, USA, 1984, pp. 586–590

76 SCHAECHTER, D. B., and LEVINSON, D. A.: 'Interactive computerised symbolic dynamics for the dynamicist'. Proceedings of the American *Control* Conference, Atlanta, USA, 1988, pp. 177–188

77 TOOGOOD, R. W.: 'Symbolic generation of robot dynamics equations'. Technical Report 87-05, Alberta Centre for Machine Intelligence and Robotics, University of Alberta, Canada, 1987

78 YIN, S., and YUH, J.: 'An efficient algorithm for automatic generation of manipulator dynamic equations'. Proceedings of the IEEE International Conference on *Robotics and Automation*, Scottsdale, USA, 1989, pp. 1812–1817

79 SREENATH, N., and KRISHNAPRASAD, P. S.: 'Multibody simulation in an object oriented programming environment', *in* GROSSMAN, R. (Ed.): 'Symbolic computation: applications to scientific computing' (Society for Industrial and Applied Mathematics, Philadelphia, USA, 1989), pp. 153–180

80 SCHIEHLEN, W. O. (Ed.): 'Multibody systems handbook' (Springer-Verlag, Heidelberg, Germany, 1990)

81 SCHIEHLEN, W. O. (Ed.): 'Advanced multibody system dynamics: simulation and software tools' (Kluwer, Dordrecht, The Netherlands, 1993)

82 RIDEAU, P.: 'Computer algebra and mechanics. The JAMES software', *in* COHEN, A. M. (Ed.): 'Computer algebra in industry' (Wiley, Chichester, UK, 1993), pp. 143–158

83 LIEH, J. H., and HAQUE, I.: 'Symbolic closed-form modelling and linearisation of multibody systems subject to control', *J. Mech. Des.*, 1991, **113**, pp. 124–132

84 WILLIAMS, R. J., and RUPPRECHT, S.: 'Dynamic force analysis of planar mechanisms', *Mech. Mach. Theory*, 1981, **16**, (4), pp. 425–440

85 SCHAFFA, R. B.: 'Dynamic analysis of spatial mechanisms'. PhD Dissertation, University of Pennsylvania, USA, 1984

86 RICHARD, M. J.: 'Dynamic simulation of constrained three dimensional multibody systems using vector network techniques'. PhD Dissertation, Queen's University, Kingston, Canada, 1985

87 GARCIA DE JALON, G., UNDA, J., AVELLO, A., and JIMENEZ, J. M.: 'Dynamic analysis of three-dimensional mechanisms in natural coordinates'. ASME Paper 86-DET-137, 1986

88 KAMMAN, J. W., and HUSTON, R. L.: 'User's manual for UCIN DYNOCOMBS-II'. Technical Report PB87-216594/AO5, National Technical Information Services, Springfield, USA, 1987

89 CHENG, P. Y., WENG, C. I., and CHEN, C. K.: 'Symbolic derivation of dynamic equations of motion for robot manipulators using Piogram symbolic method', *J. Robot. Autom.*, 1988, **4**, (6), pp. 599–609

90 BONIVERT, L., MAES, P., and SAMIN, J. C.: 'Simulation of the lateral dynamics of the GLT vehicle by means of ROBOTRAN: a model generator for robots'. Proceedings of the European *Simulation* Multiconference, Nice, France, 1988, pp. 123–127

91 VERHEUL, C. H., PACEJKA, H. B., and LOOS, H.: 'Research into the torque steer phenomenon using dynamic simulation', *Veh. Syst. Dyn.*, 1989, **18**, (5), pp. 588–602

92 CARDONA, A., and GERADIN, M.: 'Time integration of the equations of motion in mechanism analysis', *Comput. Struct.*, 1989, **33**, (3), pp. 801–820

93 SAUVAGE, G. (Ed.): 'The dynamics of vehicles on roads and on tracks' (Swets & Zeitlinger, Amsterdam, The Netherlands, 1992)

94 CHARBONNEAU, G., VINARNICK, S., NEEL, P., EVARISTE, C., and VIBET, C.: 'Symbolic modelling of controlled mechanisms', *Computer Methods App. Mech. Eng.*, 1992, **98**, (1), pp. 23–40

95 KORTUM, W., and SHARP, R. S. (Eds.): 'Multibody computer codes in vehicle system dynamics' (Swets & Zeitlinger, Amsterdam, The Netherlands, 1993)

96 NETHERY, J. F., and SPONG, M. W.: 'Robotica: a Mathematica package for robot analysis', *IEEE Robot. Autom. Mag.*, 1994, **1**, (1), pp. 13–20

97 STEJSKAL, V., and VALASEK, M.: 'Kinematics and dynamics of machinery' (Marcel Dekker, New York, USA, 1996)

98 FRITZSON, P., and FRITZSON, D.: 'The need for high-level programming support in scientific computing applied to mechanical analysis', *Comput. Struct.*, 1992, **45**, (2), pp. 387–395

99 SCHIEHLEN, W. O.: 'Multibody system dynamics: roots and perspectives', *Multibody Syst. Dyn.*, 1997, **1**, pp. 149–188

100 SHABANA, A. A.: 'Flexible multibody dynamics: review of past and recent developments', *Multibody Syst. Dyn.*, 1997, **1**, pp. 189–222

Part II

System analysis

## Chapter 4

Robust control under parametric uncertainty. Part I: analysis

L. H. Keel and S. P. Bhattacharyya

4.1 Introduction

In this chapter we deal with the problem of analysing and designing control systems containing uncertain parameters. We present a collection of recent results related to this problem and illustrate their use in the design of control systems. Since our main focus is on their use in control system design, the proofs of the results are omitted, and the reader is referred to the literature. These results were developed since the mid 1980s following the publication of a break-through in 1978 on the stability of an interval polynomial family. The results specifically discussed here are the calculation of the parametric stability margin, Kharitonov's theorem, the edge theorem, the generalized Kharitonov theorem and the determination of frequency-domain templates for systems with uncertain parameters. These results are structural in nature and complement the H_2 and H_∞ approaches to control design based on optimality. An important characteristic of many of these results is that the theory picks out a small subset of points or lines in the parameter space where the 'weakest' set of systems lies. With this set in hand one can evaluate robustness of stability, worst case stability margins and performances of the control system. These calculations therefore form an important part of the control engineer's toolkit. This chapter describes the theoretical results and in Chapter 8 we illustrate the application of these ideas to control system analysis and design through examples.

In the 1980s a new approach to control systems began to emerge. This was based on several sharp results on the stability and performance of control systems subject to multiple uncertain real parameters. This theory complements the standard control theory based on the H_∞, H_2, l^1 and μ optimal control methods, which, by and large did not deal with real parameter uncertainty. The results obtained in this area, which we may call real parametric robust control theory (RPRCT), reveal some new and interesting extremal properties of control systems that give insight into, and aid in, the control design process. These results have mainly been reported in [1−4]. The purpose of this chapter is to present the central results of this theory, without proof, and to point out how the mathematics can be used to obtain useful information from the point of view of control systems.

We first show how the stability radius in the space of uncertain parameters may be calculated. The calculation is exact when these parameters appear linearly or affinely in the closed-loop characteristic polynomial coefficients. When these coefficients appear multilinearly in the characteristic polynomial coefficients, the stability radius may be calculated to any degree of accuracy. Several applications of this calculation are demonstrated through examples.

Next we present some extremal results that are very useful for analysis and design. These are, respectively, Kharitonov's theorem, the edge theorem and the generalised Kharitonov theorem. These results allow us to evaluate the robust stability and performance of various systems subject to real parameter uncertainty in a computationally efficient manner. We then show how these results can aid in the construction of frequency-domain templates. Several control system applications of these results are given in Chapter 8, where we show how classical design techniques can be robustified. This combines the advantages of classical and modern control. We also show, in Chapter 8, how design problems can be formulated as linear programming problems using these basic results.

The results presented here are the beginnings of a more complete theory of robust control under parametric uncertainty and which we hope will stimulate further research into these topics. Some future research directions are given in the final Section.

4.2 Notation and preliminaries

In this section we introduce some basic notation and terminology that will be used throughout.

The stability of linear time-invariant control systems is characterised by the root locations of the characteristic polynomial. Consider the standard feedback control system shown in Figure 4.1 consisting of a plant and controller connected in a feedback loop. The plant and controller are assumed to be linear, time-invariant dynamic systems with respective real rational transfer function matrices $G(s)$ and $C(s)$. Let $\mathbf{p}$ denote a vector of physical plant parameters and $\mathbf{x}$ a vector of adjustable controller or design parameters. Write

$$C(s) = N_c(s, \mathbf{x})D_c^{-1}(s, \mathbf{x}) \qquad \text{and} \qquad G(s) = D_p^{-1}(s, \mathbf{p})N_p(s, \mathbf{p})$$

Figure. 4.1   Standard feedback system

where N_c, D_c, N_p and D_p are polynomial matrices in the complex variable s. Now the **characteristic polynomial** of the closed-loop system is written as

$$\delta(s, \mathbf{x}, \mathbf{p}) = \det[D_c(s, \mathbf{x})D_p(s, \mathbf{p}) + N_c(s, \mathbf{x})N_p(s, \mathbf{p})]$$

The stability of the control system is equivalent to the condition that the roots of the characteristic polynomial all lie in a certain prescribed region S of the complex plane. For continuous-time systems the stability region S is the open left half, $\mathbb{C}^-$, of the complex plane and for discrete-time systems it is the open unit disc, that is, a circle of radius unity, denoted $\mathbb{D}^1$, centred at the origin. In the control literature stability of continuous time systems or left half-plane stability is referred to as *Hurwitz stability*, and stability of discrete time systems or unit circle stability is referred to as *Schur stability*.

To clarify the above notation, a PID controller, for example, has transfer function

$$C(s, \mathbf{x}) = K_P + \frac{K_I}{s} + K_D s$$

where the controller parameter vector is

$$\mathbf{x} = [K_P \quad K_I \quad K_D]$$

Suppose the plant has transfer function $G(s)$ parametrised as

$$G(s, \mathbf{p}_1) = \frac{\mu(s - \alpha)}{(s - \beta)(s - \gamma)} = \frac{a_1 s + a_0}{b_2 s^2 + b_1 s + b_0} = G(s, \mathbf{p}_2) \qquad (4.1)$$

with

$$\mathbf{p}_1 = [\mu \quad \alpha \quad \beta \quad \gamma]$$

and

$$\mathbf{p}_2 = [a_0 \quad a_1 \quad b_0 \quad b_1 \quad b_2]$$

The characteristic polynomial is representable as

$$\delta(s, \mathbf{p}_1, \mathbf{x}) = s(s - \beta)(s - \gamma) + \mu(s - \alpha)(K_P s + K_I + K_D s^2)$$

or

$$\delta(s, \mathbf{p}_2, \mathbf{x}) = s(b_2 s^2 + b_1 s + b_0) + (a_1 s + a_0)(K_P s + K_I + K_D s^2)$$

4.2.1 Parametric uncertainty

In general, mathematical models represent approximations to the real world and therefore it is appropriate to assume that the parameters appearing in such models actually lie in a range or interval of numerical values representing the uncertainty associated with that parameter. As linear models are supposed to account for the nonlinear behaviour of the systems these intervals may be large.

In the example treated above the uncertainty in the plant model may be expressed in terms of uncertainty in the gain μ and the pole and zero locations α, β, γ (see eqn. 4.1). Alternatively it may be expressed in terms of the transfer function coefficient a_0, a_1, b_0, b_1, b_2. Each of these sets of plant parameters is subject to variation and may be assumed to lie in intervals.

In many control systems the plant parameters may vary over a wide range about a nominal value $\mathbf{p}^0$. *Robust parametric stability* refers to the ability of a control system to maintain stability despite such large variations. During the design phase, the parameters $\mathbf{x}$ of a controller are regarded as adjustable variables, and robust stability with respect to these parameters is also desirable to allow for adjustments to nominal design to accommodate other design constraints.

If the controller is given, the **maximal** range of variation of the parameter $\mathbf{p}$, measured in a suitable norm, for which closed-loop stability is preserved, is the **parametric stability margin**. In other words

$$\rho_x := \sup\{\alpha : \delta(s, \mathbf{x}, \mathbf{p}) \text{ stable}, \ \|\mathbf{p} - \mathbf{p}^0\| < \alpha\}$$

is the parametric stability margin of the system with the controller $\mathbf{x}$. This is a quantitative measure of the performance of the controller $\mathbf{x}$. Since ρ represents the *maximal perturbation*, it is indeed a legitimate quantitative measure by which one can compare the robustness of two proposed controller designs $\mathbf{x}_1$ and $\mathbf{x}_2$. This calculation is an important aid in analysis and design, much as gain and phase margins or the value of a cost function or performance index is in optimal control. In the next sub-section we describe an important computational tool that can be used to evaluate ρ.

4.2.2 Boundary crossing and zero exclusion

The fundamental notions of *boundary crossing* and *zero exclusion* play an important role in robust control. They depend on continuity of the roots of the polynomial on a parameter. For example, in the space of coefficients of a polynomial of degree n consider a path connecting a stable polynomial to an unstable one. Assuming the degree remains invariant on this path, that is, the number of roots is preserved, the first unstable polynomial encountered on this path must have some roots on the boundary of the stability region and the rest of the roots in the interior of the stability region. This result, called the *boundary crossing theorem*, was rigorously proved in [2]. The computational version of this theorem is known as the *zero exclusion condition* and is described below.

Consider the family of polynomials $\delta(s, \mathbf{p})$ of degree n, where the real parameter $\mathbf{p}$ ranges over a connected set Ω. Let the stability region in the complex plane be denoted as $\mathcal{S}$ with boundary $\partial\mathcal{S}$. Suppose it is known that one member of the family is stable. Then a useful technique of verifying robust stability of the family is to ascertain that no member of the family has a root on the stability boundary $\partial\mathcal{S}$. This can be done by checking that

$$\delta(s^*, \mathbf{p}) \neq 0, \quad \text{for all } \mathbf{p} \in \Omega, s^* \in \partial\mathcal{S}$$

This can also be written as the **zero exclusion condition**

$$0 \notin \delta(s^*, \Omega), \quad \text{for all } s^* \in \partial\mathcal{S}$$

The parametric stability margin may be computed by finding the smallest perturbation of $\mathbf{p}^0$ which results in a root just crossing the boundary, equivalently when the zero exclusion just begins to fail. The above condition can be easily verified when the uncertainty set Ω is a box and the parameter $\mathbf{p}$ appears linearly or multilinearly in the characteristic polynomial coefficients. In the first case the image set $\delta(s^*, \Omega)$ is itself a convex polygon and in the latter case it lies in the convex hull of the image of Ω. In these cases the zero exclusion condition can be verified easily and so can stability margins. Motivated by such examples, the majority of robust parametric stability results are directed towards the *linear* and *multilinear* dependency cases, where fortunately fit many practical applications.

The robust controller synthesis problem, which is the problem of determining $\mathbf{x}$ to achieve stability and a prescribed level of parametric stability margin ρ, is unfortunately as yet unsolved. In an engineering sense, however, many effective techniques exist for robust parametric controller design. In particular the exact calculation of ρ_x can itself be used in an iterative loop to adjust $\mathbf{x}$ to robustify the system. In the following Section we derive the procedure to compute ρ in some detail.

4.3 Real parameter stability margin

In this Section we show how the parametric stability margin can be computed in the case in which the characteristic polynomial coefficients depend affinely on the uncertain parameters. In such cases we may write

$$\delta(s, \mathbf{p}) = a_1(s)p_1 + \cdots + a_l(s)p_l + b(s) \tag{4.2}$$

where $a_i(s)$ and $b(s)$ are real polynomials and the parameters p_i are real. Write $\mathbf{p}$ for the vector of uncertain parameters, $\mathbf{p}^0$ the nominal parameter vector and $\Delta\mathbf{p}$ the perturbation vector. In other words

$$\mathbf{p} = [p_1, p_2, \ldots, p_l] \qquad \mathbf{p}^0 = [p_1^0, p_2^0, \ldots, p_l^0]$$
$$\Delta\mathbf{p} = [p_1 - p_1^0, p_2 - p_2^0, \ldots, p_l - p_l^0]$$
$$= [\Delta p_1, \Delta p_2, \ldots, \Delta p_l]$$

The characteristic polynomial can be written as

$$\delta(s, \mathbf{p}^0 + \Delta\mathbf{p}) = \underbrace{\delta(s, \mathbf{p}^0)}_{\delta^0(s)} + \underbrace{a_1(s)\Delta p_1 + \cdots + a_l(s)\Delta p_l}_{\Delta\delta(s, \Delta\mathbf{p})} \tag{4.3}$$

Let $s*$ denote a point on the stability boundary $\partial\mathcal{S}$. For $s* \in \partial\mathcal{S}$ to be a root of $\delta(s, \mathbf{p}^0 + \Delta\mathbf{p})$ we must have

$$\delta(s*, \mathbf{p}^0) + a_1(s*)\Delta p_1 + \cdots + a_l(s*)\Delta p_l = 0 \tag{4.4}$$

In many instances it is important to consider weighted perturbations, to account for, say, scaling factors, units or normalisation. Letting $w_i > 0$, $i = 1, \ldots, l$, denote a set of weights, rewrite the above equation as follows:

$$\delta(s*, \mathbf{p}^0) + \frac{a_1(s*)}{w_1}w_1\Delta p_1 + \cdots + \frac{a_l(s*)}{w_l}w_l\Delta p_l = 0 \tag{4.5}$$

The minimum norm solution of this equation gives us $\rho(s*)$:

$$\rho(s*) = \inf\left\{\|\Delta\mathbf{p}\|^w : \delta(s*, \mathbf{p}^0) + \frac{a_1(s*)}{w_1}w_1\Delta p_1 + \cdots + \frac{a_l(s*)}{w_l}w_l\Delta p_l = 0\right\}$$

The equation corresponding to loss of degree is

$$\delta_n(\mathbf{p}^0 + \Delta\mathbf{p}) = 0 \tag{4.6}$$

If a_{in} denotes the coefficient of the nth-degree term in the polynomial $a_i(s)$, $i = 1, 2, \ldots, l$, the above equation becomes

$$\underbrace{a_{1n}p_1^0 + a_{2n}p_2^0 + \cdots + a_{ln}p_l^0}_{\delta_n(\mathbf{p}^0)} + a_{1n}\Delta p_1 + a_{2n}\Delta p_2 + \cdots + a_{ln}\Delta p_l = 0 \tag{4.7}$$

or, after introducing the weight $w_i > 0$,

$$\underbrace{a_{1n}p_1^0 + a_{2n}p_2^0 + \cdots + a_{ln}p_l^0}_{\delta_n(\mathbf{p}^0)} + \frac{a_{1n}}{w_1}w_1\Delta p_1 + \frac{a_{2n}}{w_2}w_2\Delta p_2 + \cdots + \frac{a_{ln}}{w_l}w_l\Delta p_l = 0 \tag{4.8}$$

The minimum norm $\|\Delta\mathbf{p}\|^w$ solution of this equation gives us ρ_d.

Eqn. 4.8 is real and can be rewritten in the form

$$\underbrace{\left[\frac{a_{1n}}{w_1} \cdots \frac{a_{ln}}{w_l}\right]}_{A_n} \underbrace{\begin{bmatrix} w_1\Delta p_1 \\ \vdots \\ w_l\Delta p_l \end{bmatrix}}_{t_n} = \underbrace{-\delta_n^0}_{b_n} \tag{4.9}$$

In eqn. 4.5, two cases may occur depending on whether s^* is real or complex. If $s^* = s_r$ where s_r is real, we have the single equation

$$\underbrace{\left[\frac{a_1(s_r)}{w_1} \dots \frac{a_l(s_r)}{w_l}\right]}_{A(s_r)} \underbrace{\left[\begin{array}{c} w_1 \Delta p_1 \\ \vdots \\ w_l \Delta p_l \end{array}\right]}_{t(s_r)} = \underbrace{-\delta^0(s_r)}_{b(s_r)} \tag{4.10}$$

We let x_r and x_i denote the real and imaginary parts of a complex number x, i.e.

$$x = x_r + jx_i \qquad \text{with } x_r, x_i \text{ real}$$

so that

$$a_k(s^*) = a_{kr}(s^*) + ja_{ki}(s^*)$$

and

$$\delta^0(s^*) = \delta^0_r(s^*) + j\delta^0_i(s^*)$$

If $s^* = s_c$, where s_c is complex, eqn. 4.5 is equivalent to two equations which can be written as follows:

$$\underbrace{\left[\begin{array}{ccc} \dfrac{a_{1r}(s_c)}{w_1} & \dots & \dfrac{a_{lr}(s_c)}{w_l} \\ \dfrac{a_{1i}(s_c)}{w_1} & \dots & \dfrac{a_{li}(s_c)}{w_l} \end{array}\right]}_{A(s_c)} \underbrace{\left[\begin{array}{c} w_1 \Delta p_1 \\ \vdots \\ w_l \Delta p_l \end{array}\right]}_{t(s_c)} = \underbrace{\left[\begin{array}{c} -\delta^0_r(s_c) \\ -\delta^0_i(s_c) \end{array}\right]}_{b(s_c)} \tag{4.11}$$

These equations completely determine the parametric stability margin in any norm. Let $t^*(s_c)$, $t^*(s_r)$ and t^*_n denote the minimum norm solutions of eqns. 4.11, 4.10 and 4.9, respectively. Thus,

$$\|t^*(s_c)\| = \rho(s_c) \tag{4.12}$$

$$\|t^*(s_r)\| = \rho(s_r) \tag{4.13}$$

$$\|t_n^*\| = \rho_d \tag{4.14}$$

If any of the above equations (eqns. 4.9–4.11) does not have a solution, the corresponding value of $\rho(\cdot)$ is set equal to infinity.

Let $\partial \mathcal{S}_r$ and $\partial \mathcal{S}_c$ denote the real and complex subsets of $\partial \mathcal{S}$:

$$\partial \mathcal{S} = \partial \mathcal{S}_r \cup \partial \mathcal{S}_c$$

$$\rho_r := \inf_{s_r \in \partial \mathcal{S}_r} \rho(s_r) \quad \text{and} \quad \rho_c := \inf_{s_c \in \partial \mathcal{S}_c} \rho(s_c)$$

Finally, the real parametric stability margin is

$$\rho = \inf\{\rho_r, \rho_c, \rho_d\} \tag{4.15}$$

4.3.1 ℓ_2 real parametric stability margin

If the length of the perturbation vector is measured by a weighted ℓ_2 norm the minimum ℓ_2 norm solution of eqns. 4.9, 4.10 and 4.11 is desired. Consider first eqn. 4.11 and assume that $A(s_c)$ has full row rank $= 2$. The minimum norm solution vector $t^*(s_c)$ can be calculated as follows:

$$t^*(s_c) = A^T(s_c)[A(s_c)A^T(s_c)]^{-1}b(s_c) \tag{4.16}$$

Similarly, if eqns. 4.10 and 4.9 are consistent (i.e. $A(s_r)$ and A_n are nonzero vectors), we can calculate the solution as

$$t^*(s_r) = A^T(s_r)[A(s_r)A^T(s_r)]^{-1}b(s_r) \tag{4.17}$$

$$t_n^* = A_n^T[A_nA_n^T]^{-1}b_n \tag{4.18}$$

If $A(s_c)$ has less than full rank two cases can occur. If the equation is consistent we can simply replace the two equations with a single equation and proceed as before. In case eqns. 4.11 are inconsistent we simply set

$$\rho(s_c) = \infty$$

These calculations are illustrated with the examples below.

Example 1 (ℓ_2 Schur stability margin)

Consider the discrete-time control system with the controller and plant specified respectively by their transfer functions:

$$C(z) = \frac{z+1}{z^2}, \qquad G(z, \mathbf{p}) = \frac{(-0.5 - 2p_0)z + (0.1 + p_0)}{z^2 - (1 + 0.4p_2)z + (0.6 + 10p_1 + 2p_0)}$$

The characteristic polynomial of the closed-loop system is

$$\delta(z, \mathbf{p}) = z^4 - (1 + 0.4p_2)z^3 + (0.1 + 10p_1)z^2 - (0.4 + p_0)z + (0.1 + p_0)$$

The nominal value of $\mathbf{p}^0 = [p_0^0, \; p_1^0, \; p_2^0] = [0, 0.1, 1]$ and the polynomial is Schur stable for the nominal parameter $\mathbf{p}^0$. The perturbation is denoted as usual by the vector

$$\Delta\mathbf{p} = [\Delta p_0 \quad \Delta p_1 \quad \Delta p_2]$$

Rewrite

$$\delta(z, \mathbf{p}^0 + \Delta\mathbf{p}) = (-z + 1)\Delta p_0 + 10z^2\Delta p_1 - 0.4z^3\Delta p_2$$
$$+ (z^4 - 1.4z^3 + 1.1z^2 - 0.4z + 0.1)$$

and note that the degree remains invariant ($= 4$) for all perturbations so that $\rho_d = \infty$. The stability region is the unit circle. For $z = 1$ to be a root of $\delta(z, \mathbf{p}^0 + \Delta\mathbf{p})$ (see eqn. 4.10), we have

$$\underbrace{[0 \quad 10 \quad -0.4]}_{A(1)} \underbrace{\begin{bmatrix} \Delta p_0 \\ \Delta p_1 \\ \Delta p_2 \end{bmatrix}}_{t(1)} = \underbrace{-0.4}_{b(1)}$$

Thus,

$$\rho(1) = \|t*(1)\|_2 = \left\|A^T(1)[A(1)A^T(1)]^{-1}b(1)\right\|_2 = 0.04$$

Similarly, for the case of $z = -1$ (see eqn. 4.10), we have

$$\underbrace{[2 \quad 10 \quad 0.4]}_{A(-1)} \underbrace{\begin{bmatrix} \Delta p_0 \\ \Delta p_1 \\ \Delta p_2 \end{bmatrix}}_{t(-1)} = \underbrace{-4}_{b(-1)}$$

and $\rho(-1) = \|t*(-1)\|_2 = 0.3919$. Thus $\rho_r = 0.04$.

For the case in which $\delta(z, \mathbf{p}^0 + \Delta\mathbf{p})$ has a root at $z = e^{i\theta}$, $\theta \neq \pi$, $\theta \neq 0$, using eqn. 4.11, we have

$$\underbrace{\begin{bmatrix} -\cos\theta + 1 & 10\cos 2\theta & -0.4\cos 3\theta \\ -\sin\theta & 10\sin 2\theta & -0.4\sin 3\theta \end{bmatrix}}_{A(\theta)} \underbrace{\begin{bmatrix} \Delta p_0 \\ \Delta p_1 \\ \Delta p_2 \end{bmatrix}}_{t(\theta)}$$

$$= -\underbrace{\begin{bmatrix} \cos 4\theta - 1.4\cos 3\theta + 1.1\cos 2\theta - 0.4\cos\theta + 0.1 \\ \sin 4\theta - 1.4\sin 3\theta + 1.1\sin 2\theta - 0.4\sin\theta \end{bmatrix}}_{b(\theta)}$$

Thus,

$$\rho(e^{i\theta}) = \|t*(\theta)\|_2 = \left\|A^T(\theta)[A(\theta)A^T\theta)]^{-1}b(\theta)\right\|$$

Figure 4.2 shows the plot of $\rho(e^{i\theta})$.

Therefore, the ℓ_2 parametric stability margin is

$$\rho_c = 0.032 = \rho_b = \rho*$$

Example 2

Consider the continuous-time control system with the plant,

$$G(s, \mathbf{p}) = \frac{2s + 3 - \frac{1}{3}p_1 - \frac{5}{3}p_2}{s^3 + (4 - p_2)s^2 + (-2 - 2p_1)s + (-9 + \frac{5}{3}p_1 + \frac{16}{3}p_2)}$$

and the proportional integral (PI) controller

$$C(s) = 5 + \frac{3}{s}$$

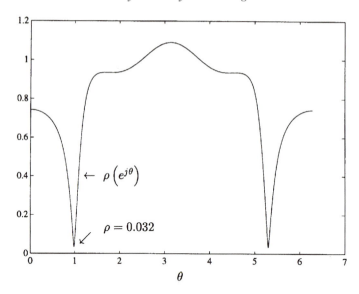

Figure 4.2. $\rho(\theta)$ (*Example 1*)

The characteristic polynomial of the closed-loop system is

$$\delta(s, \mathbf{p}) = s^4 + (4 - p_2)s^3 + (8 - 2p_1)s^2 + (12 - 3p_2)s + (9 - p_1 - 5p_2)$$

We see that the degree remains invariant under the given set of parameter variations and therefore $\rho_d = \infty$. The nominal values of the parameters are

$$\mathbf{p}^0 = [p_1^0, p_2^0] = [0, 0]$$

Then

$$\Delta\mathbf{p} = [\Delta p_1 \quad \Delta p_2] = [p_1 \quad p_2]$$

The polynomial is stable for the nominal parameter $\mathbf{p}^0$. Now we want to compute the ℓ_2 stability margin for this polynomial with weights $w_1 = w_2 = 1$. We first evaluate $\delta(s, \mathbf{p})$ at $s = j\omega$:

$$\delta(j\omega, \mathbf{p}^0 + \Delta\mathbf{p}) = (2\omega^2 - 1)\Delta p_1 + (j\omega^3 - 3j\omega - 5)\Delta p_2 + \omega^4 - 4j\omega^3 - 8\omega^2$$
$$+ 12j\omega + 9$$

For the case of a root at $s = 0$ (see eqn. 4.10), we have

$$\underbrace{[-1 \quad -5]}_{A(0)} \underbrace{\begin{bmatrix} \Delta p_1 \\ \Delta p_2 \end{bmatrix}}_{t(0)} = \underbrace{-9}_{b(0)}$$

Thus,

$$\rho(0) = \|t^*(0)\|_2 = \left\|A^T(0)[A(0)A^T(0)]^{-1}b(0)\right\|_2 = \frac{9\sqrt{26}}{26}$$

For a root at $s = j\omega, \omega > 0$, using the formula given in 4.11, we have with $w_1 = w_2 = 1$

$$\underbrace{\begin{bmatrix} (2\omega^2 - 1) & -5 \\ 0 & (\omega^2 - 3) \end{bmatrix}}_{A(j\omega)} \underbrace{\begin{bmatrix} \Delta p_1 \\ \Delta p_2 \end{bmatrix}}_{t(j\omega)} = \underbrace{\begin{bmatrix} -\omega^4 + 8\omega^2 - 9 \\ 4\omega^2 - 12 \end{bmatrix}}_{b(j\omega)} \qquad (4.19)$$

Here, we need to determine if there exists any ω for which the rank of the matrix $A(j\omega)$ drops (it is easy to see from the matrix $A(j\omega)$ that the rank drops when $\omega = (1/\sqrt{2})$ and $\omega = \sqrt{3}$):

(i)   rank $A(j\omega) = 1$. For $\omega = (1/\sqrt{2})$, we have rank $A(j\omega) = 1$ and rank $[A(j\omega), b(j\omega)] = 2$, and so there is no solution to eqn. 4.19. Thus

$$\rho\left(j\frac{1}{\sqrt{2}}\right) = \infty$$

For $\omega = \sqrt{3}$, rank $A(j\omega) = $ rank $[A(j\omega), b(j\omega)]$, and we do have a solution to eqn. 4.19. Therefore

$$\underbrace{\begin{bmatrix} 5 & -5 \end{bmatrix}}_{A(j\sqrt{3})} \underbrace{\begin{bmatrix} \Delta p_1 \\ \Delta p_2 \end{bmatrix}}_{t(j\sqrt{3})} = \underbrace{6}_{b(j\sqrt{3})}$$

Consequently

$$\rho(j\sqrt{3}) = \|t^*(j\sqrt{3})\|_2 = \left\|A^T(j\sqrt{3})[A(j\sqrt{3})A^T(j\sqrt{3})]^{-1}b(j\sqrt{3})\right\|_2 = \frac{3\sqrt{2}}{5}$$

(ii)   rank $A(j\omega) = 2$. In this case eqn. 4.19 has a unique solution and

$$\rho(j\omega) = \|t^*(j\omega)\|_2 = \left\|A^T(j\omega)[A(j\omega)A^T(j\omega)]^{-1}b(j\omega)\right\|_2$$

$$= \sqrt{\left(\frac{\omega^4 - 8\omega^2 - 11}{2\omega^2 - 1}\right)^2 + 16}$$

Figure 4.3 shows the plot of $\rho(j\omega)$ for $\omega > 0$. The values of $\rho(0)$ and $\rho(j\sqrt{3})$ are also shown in Figure 4.3. Therefore,

$$\rho(j\sqrt{3}) = \rho_b = \frac{3\sqrt{2}}{5} = \rho^*$$

is the stability margin.

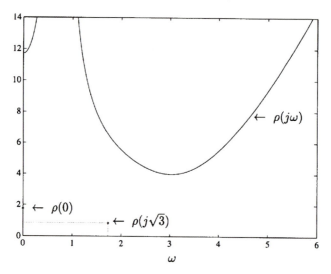

Figure 4.3 $\rho(j\omega)$ *(Example 2)*

4.3.2 ℓ_2 stability margin for time-delay systems

The results given above for determining the largest stability ellipsoid in parameter space for polynomials can be extended to quasipolynomials. This extension is useful when parameter uncertainty is present in systems containing time delays. As before, we deal with the case where the uncertain parameters appear linearly in the coefficients of the quasipolynomial.

Let us consider real quasipolynomials

$$\delta(s, \mathbf{p}) = p_1 Q_1(s) + p_2 Q_2(s) + \cdots + p_l Q_l(s) + Q_0(s) \tag{4.20}$$

where

$$Q_i(s) = s^{n_i} + \sum_{k=1}^{n_i} \sum_{j=1}^{m} a_{kj}^i s^{n_i - k} e^{-\tau_j^i s}, \qquad i = 0, 1, \ldots, l \tag{4.21}$$

and we assume that $n_0 > n_i$, $i = 1, 2, \ldots, l$, and that all parameters occurring in eqns. 4.20 and 4.21 are real. Control systems containing time delays often have characteristic equations of this form (see Example 3).

The uncertain parameter vector is denoted $\mathbf{p} = [p_1, p_2, \ldots, p_l]$. The nominal value of the parameter vector is $\mathbf{p} = \mathbf{p}^0$, the nominal quasipolynomial $\delta(s, \mathbf{p}^0) = \delta^0(s)$ and $\mathbf{p} - \mathbf{p}^0 = \Delta\mathbf{p}$ denotes the deviation or perturbation from the nominal. The parameter vector is assumed to lie in the ball of radius ρ centred at $\mathbf{p}^0$:

$$\mathcal{B}(\rho, \mathbf{p}^0) = \{\mathbf{p}: \|\mathbf{p} - \mathbf{p}^0\|_2 < \rho\} \tag{4.22}$$

The corresponding set of quasipolynomials is

$$\mathbf{\Delta}_\rho(s) := \{\delta(s, \mathbf{p}^0 + \Delta\mathbf{p}) \colon \|\mathbf{p}\|_2 < \rho\} \tag{4.23}$$

As it turns out, the Boundary Crossing Theorem can be applied to this class of quasipolynomials. From the fact that in the family eqn. 4.20 the e^{-st} terms are associated with the lower degree terms it follows that it is legitimate to say that each quasipolynomial defined above is Hurwitz stable if all its roots lie inside the left half of the complex plane. In other words stability can be lost only by a root crossing the $j\omega$ axis. Accordingly, for every $-\infty < \omega < \infty$ we can introduce a set in the parameter space

$$\mathbf{\Pi}(\omega) = \{\mathbf{p} \colon \delta(j\omega, \mathbf{p}) = 0\}$$

This set corresponds to quasipolynomials that have $j\omega$ as a root. Of course, for some particular ω this set may be empty. If $\mathbf{\Pi}(\omega)$ is nonempty we can define the distance between $\mathbf{\Pi}(\omega)$ and the nominal point $\mathbf{p}^0$:

$$\rho(\omega) = \inf_{\mathbf{p} \,\in\, \mathbf{\Pi}(\omega)} \{\|\mathbf{p} - \mathbf{p}^0\|\}$$

If $\mathbf{\Pi}(\omega)$ is empty for some ω we set the corresponding $\rho(\omega) := \infty$.

Theorem 1   *The family of quasipolynomials $\mathbf{\Delta}_\rho(s)$ is Hurwitz stable if and only if the quasipolynomial $\delta^0(s)$ is stable and*

$$\rho < \rho^* = \inf_{0 \,\leq\, \omega < \infty} \rho(\omega)$$

The reader is referred to [2] for the proof of this theorem. We illustrate its use with an example.

Example 3   (ℓ_2 stability margin for time-delay system)

The model of a satellite attitude control system, Figure 4.4, containing a time delay in the loop, is shown.

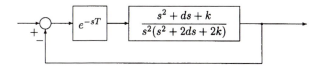

Figure 4.4   A satellite attitude control system with time delay (Example 3)

The characteristic equation of the system is the quasipolynomial

$$\delta(s, \mathbf{p}) = s^4 + 2ds^3 + (e^{-sT} + 2k)s^2 + e^{-sT}ds + e^{-sT}k$$

The nominal parameters are

$$\mathbf{p}^0 = [k^0 \quad d^0] = [0.245 \quad 0.0218973] \quad \text{and} \quad T = 0.1$$

The Hurwitz stability of the nominal system may be easily verified by applying the interlacing property. To compute the ℓ_2 real parametric stability margin around the nomial values of parameters, let

$$\Delta\mathbf{p} = [\Delta k \quad \Delta d]$$

We have

$$\delta(j\omega, \mathbf{p}^0 + \Delta\mathbf{p}) = \omega^4 - \omega^2 \cos\omega T - 2k^0\omega^2 - 2\Delta k\omega^2 + d^0\omega\sin\omega T$$
$$+ \Delta d\omega\sin\omega T + k^0\cos\omega T + \Delta k\cos\omega T$$
$$+ j(-2d^0\omega^3 - 2\Delta d\omega^3 + \omega^2\sin\omega T + d^0\omega\cos\omega T$$
$$+ \Delta d\omega\cos\omega T - k^0\sin\omega T - \Delta k\sin\omega T)$$

or

$$\underbrace{\begin{bmatrix} -2\omega^2 + \cos\omega T & \omega\sin\omega T \\ -\sin\omega T & -2\omega^3 + \omega\cos\omega T \end{bmatrix}}_{A(j\omega)} \underbrace{\begin{bmatrix} \Delta k \\ \Delta d \end{bmatrix}}_{t(j\omega)}$$

$$= \underbrace{\begin{bmatrix} -\omega^4 + \omega^2\cos\omega T + 2k^0\omega^2 - d^0\omega\sin\omega T - k^0\cos\omega T \\ 2d^0\omega^3 - \omega^2\sin\omega T - d^0\omega\cos\omega T + k^0\sin\omega T \end{bmatrix}}_{b(j\omega)}$$

Therefore

$$\rho(j\omega) = \|t^*(j\omega)\|_2$$
$$= \left\| A^T(j\omega)[A(j\omega)A^T(j\omega)]^{-1}b(j\omega) \right\|_2$$

From Figure 4.5, the minimum value of $\rho_b = 0.0146$.

The additional condition from the constant coefficient, corresponding to a root at $s = 0$, is

$$|\Delta k| < k^0 = 0.245$$

Thus, the ℓ_2 stability margin is 0.0146.

4.4 Extremal results in RPRCT

In this section we describe a group of results which may be considered to be the central results in this field. They are characterised by the important feature that they facilitate robust stability calculations by identifying *a priori* a small subset of parameters where stability or performance will be lost. We begin with the most spectacular of these results, namely Kharitonov's theorem, which gives a surprisingly simple necessary and sufficient condition for the robust stability of an interval family of polynomials.

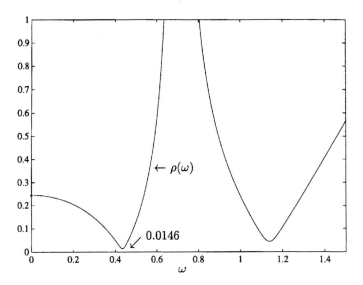

Figure 4.5 ℓ_2 *stability margin for a quasipolynomial (Example* 3)

4.4.1 Kharitonov's theorem

Consider the set $\mathscr{I}(s)$ of real polynomials of degree n of the form

$$\delta(s) = \delta_0 + \delta_1 s + \delta_2 s^2 + \delta_3 s^3 + \delta_4 s^4 + \cdots + \delta_n s^n$$

where the coefficients lie within given ranges,

$$\delta_0 \in [x_0, y_0], \qquad \delta_1 \in [x_1, y_1], \ldots, \delta_n \in [x_n, y_n]$$

Write

$$\underline{\delta} := [\delta_0, \delta_1, \ldots, \delta_n]$$

and identify a polynomial $\delta(s)$ with its coefficient vector $\underline{\delta}$. Introduce the box of coefficients

$$\mathbf{\Delta} := \{\underline{\delta} : \underline{\delta} \in \mathbb{R}^{n+1}, x_i \leq \delta_i \leq y_i, i = 0, 1, \ldots, n\} \qquad (4.24)$$

We assume that the degree remains invariant over the family, so that $0 \notin [x_n, y_n]$. Such a set of polynomials is called a real *interval* family and we loosely refer to $\mathscr{I}(s)$ or $\mathbf{\Delta}$ as an interval polynomial. We give Kharitonov's theorem below, without proof. The proof may be found in [2].

Theorem 2 (Kharitonov's theorem)   *Every polynomial in the family $\mathcal{I}(s)$ is Hurwitz if and only if the following four extreme polynomials are Hurwitz:*

$$K^1(s) = x_0 + x_1 s + y_2 s^2 + y_3 s^3 + x_4 s^4 + x_5 s^5 + y_6 s^6 + \cdots$$

$$K^2(s) = x_0 + y_1 s + y_2 s^2 + x_3 s^3 + x_4 s^4 + y_5 s^5 + y_6 s^6 + \cdots$$

$$K^3(s) = y_0 + x_1 s + x_2 s^2 + y_3 s^3 + y_4 s^4 + x_5 s^5 + x_6 s^6 + \cdots \qquad (4.25)$$

$$K^4(s) = y_0 + y_1 s + x_2 s^2 + x_3 s^3 + y_4 s^4 + y_5 s^5 + x_6 s^6 + \cdots$$

In the example below we show this result is useful in control systems.

Example 4

Consider the problem of checking the robust stability of the feedback system shown in Figure 4.6.

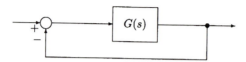

Figure 4.6   Feedback system (Example 4)

The plant transfer function is

$$G(s) = \frac{\delta_1 s + \delta_0}{s^2 (\delta_4 s^2 + \delta_3 s^3 + \delta_2)}$$

with coefficients being bounded as

$$\delta_4 \in [x_4, y_4], \qquad \delta_3 \in [x_3, y_3], \qquad \delta_2 \in [x_2, y2], \qquad \delta_1 \in [x_1, y_1], \qquad \delta_0 \in [x_0, y_0]$$

The characteristic polynomial of the family is written as

$$\delta(s) = \delta_4 s^4 + \delta_3 s^3 + \delta_2 s^2 + \delta_1 s + \delta_0$$

The Kharitonov polynomials are:

$$K^1(s) = x_0 + x_1 s + y_2 s^2 + y_3 s^3 + s_4 s^4,$$

$$K^2(s) = x_0 + y_1 s + y_2 s^2 + x_3 s^3 + x_4 s^4$$

$$K^3(s) = y_0 + x_1 s + x_2 s^2 + y_3 s^3 + y_4 s^4,$$

$$K^4(s) = y_0 + y_1 s + x_2 s^2 + x_3 s^3 + y_4 s^4$$

The problem of checking the Hurwitz stability of the family is therefore reduced to that of checking the Hurwitz stability of these four polynomials. In this particular case this in turn reduces to checking that the coefficients have the same sign (positive, say; otherwise multiply $\delta(s)$ by -1) and that the following inequalities hold over the vertices of $\boldsymbol{\Delta}$:

$$K^1(s) \quad \text{Hurwitz}: y_2 y_3 > x_1 x_4, \qquad x_1 y_2 y_3 > x_1^2 x_4 + y_3^2 x_0$$

$$K^2(s) \quad \text{Hurwitz}: y_2 x_3 > y_1 x_4, \qquad y_1 y_2 x_3 > y_1^2 x_4 + x_3^2 x_0$$

$$K^3(s) \quad \text{Hurwitz}: x_2 y_3 > x_1 y_4, \qquad x_1 x_2 y_3 > x_1^2 y_4 + y_3^2 y_0$$

$$K^4(s) \quad \text{Hurwitz}: x_2 x_3 > y_1 y_4, \qquad y_1 x_2 x_3 > y_1^2 y_4 + x_3^2 y_0$$

As another control application we give the example below.

Example 5

Consider the control system shown in Figure 4.7.

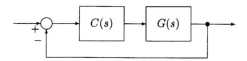

Figure 4.7   Feedback system with controller (Example 5)

The plant is described by the rational transfer function $G(s)$ with numerator and denominator coefficients varying independently in prescribed intervals. We refer to such a family of transfer functions $\mathbf{G}(s)$ as an *interval plant*. In the present example we take

$$\mathbf{G}(s) := \left\{ G(s) = \frac{n_2 s^2 + n_1 s + n_0}{s^3 + d_2 s^2 + d_1 s + d_0} : n_0 \in [1, 2.5], \ n_1 \in [1, 6], \ n_2 \in [1, 7], \right.$$

$$\left. d_2 \in [-1, 1], d_1 \in [-0.5, 1.5], d_0 \in [1, 1.5] \right\}$$

The controller is a constant gain, $C(s) = k$, that is to be adjusted, if possible, to robustly stabilise the closed-loop system. More precisely we are interested in determining the range of values of the gain $k \in [-\infty, +\infty]$ for which the closed-loop system is robustly stable, i.e. stable for all $G(s) \in \mathbf{G}(s)$.

The characteristic polynomial of the closed-loop system is

$$\delta(k, s) = s^3 + \underbrace{(d_2 + kn_2)}_{\delta_2(k)} s^2 + \underbrace{(d_1 + kn_1)}_{\delta_1(k)} s + \underbrace{(d_0 + kn_0)}_{\delta_0(k)}$$

Since the parameter $d_i, n_j, i = 0, 1, 2, j = 0, 1, 2,$ vary independently it follows that for each fixed k, $\delta(k, s)$ is an interval polynomial. Using the bounds given to describe the family $\mathbf{G}(s)$ we get the following coefficient bounds for positive k:

$$\delta_2(k) \in [-1 + k, 1 + 7k]$$

$$\delta_1(k) \in [-0.5 + k, 1.5 + 6k]$$

$$\delta_0(k) \in [-5 + k, 1.5 + 2.5k]$$

By applying Kharitonov's theorem we get the necessary and sufficient conditions for robust stability to be:

$$-1 + k > 0, \quad -0.5 + k > 0, \quad 1 + k > 0, \quad (-0.5 + k)(-1 + k) > 1.5 + 2.5k$$

From this it follows that the closed-loop system is robustly stable if and only if

$$k \in (2 + \sqrt{5}, +\infty)$$

4.4.2 *The edge theorem*

The interval family dealt with in Kharitonov's theorem is a very special type of polytopic family. Moreover, Kharitonov's theorem does not indicate where the roots of the polynomial family lie. The edge theorem deals with a general convex polytopic family of polynomials and gives a complete, exact and constructive characterisation of the root set of the family. Such a characterisation is obviously of value in the robustness and performance analysis of control systems. In the following we give the edge theorem without proof and refer the reader to the proof in [2].

Consider a family of nth degree real polynomials whose typical element is given by

$$\delta(s) = \delta_0 + \delta_1 s + \cdots + \delta_{n-1} s^{n-1} + \delta_n s^n \tag{4.26}$$

We will identify the polynomial in eqn. 4.26 with the vector

$$\underline{\delta} := [\delta_n, \delta_{n-1}, \dots, \delta_1, \delta_0]^T \tag{4.27}$$

Let $\Omega \subset \mathbb{R}^{n+1}$ be an m-dimensional *polytope*, that is, the convex hull of a finite number of points. We make the assumption that all polynomials in Ω have the same degree and the sign of δ_n is constant over Ω, say, positive.

A *supporting hyperplane* H is an affine set of dimension n such that $\Omega \cap H \neq \emptyset$, and such that every point of Ω lies on just one side of H. The *exposed sets* of Ω are those (convex) sets $\Omega \cap H$ where H is a supporting hyperplane. The one-dimensional exposed sets are called *exposed edges*.

For any $W \subset \Omega$, $R(W)$ is said to be the root space of W if

$$R(W) = \{s : \delta(s) = 0, \text{ for some } \underline{\delta} \in W\} \tag{4.28}$$

Finally, recall that the boundary of an arbitrary set S of the complex plane is designated by ∂S. We can now enunciate the edge theorem.

Theorem 3 (Edge theorem)   *Let $\Omega \subset \mathbb{R}^{n+1}$ be a polytope of polynomials as above. Then the boundary of $R(\Omega)$ is contained in the root space of the exposed edges of Ω.*

We illustrate the importance of this result in control systems with an example.

Example 6

Consider the unity feedback discrete-time control system with forward transfer function

$$G(z) = \frac{\delta_1 z + \delta_0}{z^2(z + \delta_2)}$$

The characteristic polynomial is

$$\delta(z) = z^3 + \delta_2 z^2 + \delta_1 z + \delta_0$$

Suppose that the coefficients vary in the intervals

$$\delta_2 \in [0.042, 0.158] \qquad \delta_1 \in [-0.058, 0.058] \qquad \delta_0 \in [-0.06, 0.056]$$

The boundary of the root space of the family can be generated by drawing the root loci along the 12 exposed edges of the box in coefficient space. The root space is inside the unit disc as shown in Figure 4.8. Hence the entire family is Schur stable.

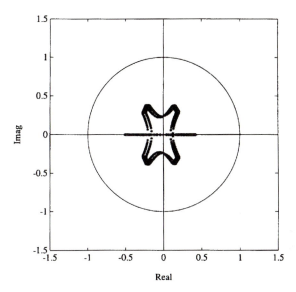

Figure 4.8   Root space of $\delta(z)$ (Example 6)

4.4.3 The generalised Kharitonov theorem

Kharitonov's theorem applies to polynomial families where the coefficients vary *independently*. In a typical control system problem, the closed-loop characteristic polynomial coefficients vary interdependently. For example, the closed-loop characteristic polynomial coefficients may vary only through the perturbation of the plant parameters while the controller parameters remain fixed. The generalised Kharitonov theorem described below deals with this situation and develops results that retain the flavour of Kharitonov's theorem, namely that the number of items to be checked for stability does not increase with number of parameters. In this respect it is important to remember that the edge theorem has the drawback that it requires checking all the exposed edges and this number increases exponentially with the number of parameters.

Consider real polynomials of the form

$$\delta(s) = F_1(s)P_1(s) + F_2(s)P_2(s) + \cdots + F_m(s)P_m(s) \tag{4.29}$$

Write

$$\underline{F}(s) := (F_1(s), F_2(s), \ldots, F_m(s)) \tag{4.30}$$

$$\underline{P}(s) := (P_1(S), P_2(S), \ldots, P_m(s)) \tag{4.31}$$

and introduce the notation

$$\langle \underline{F}(s), \underline{P}(s) \rangle := F_1(s)P_1(s) + F_2(s)P_2(s) + \cdots + F_m(s)P_m(s) \tag{4.32}$$

We will say that $\underline{F}(s)$ stabilises $\underline{P}(s)$ if $\delta(s) = \langle \underline{F}(s), \underline{P}(s) \rangle$ is *Hurwitz stable*. In this sub-section, stable means Hurwitz stable.

The polynomials $F_i(s)$ are fixed real polynomials, whereas $P_i(s)$ are real polynomials with coefficients varying independently in prescribed intervals.

Let $d^0(P_i)$ be the degree of $P_i(s)$:

$$P_i(s) := p_{i,0} + p_{i,1}s + \cdots + p_{i,d^0(P_i)}s^{d^0(P_i)} \tag{4.33}$$

and

$$\mathbf{p}_i := [p_{i,0}, p_{i,1}, \ldots, p_{i,d^0(P_i)}] \tag{4.34}$$

Let $\underline{n} = [1, 2, \ldots, n]$. Each $P_i(s)$ belongs to an interval family $\mathbf{P}_i(s)$ specified by the intervals

$$p_{i,j} \in [\alpha_{i,j}, \beta_{i,j}], \quad i \in \underline{m}; \; j = 0, \ldots, d^0(P_i) \tag{4.35}$$

The corresponding parameter box is

$$\mathbf{\Pi}_i := \{\mathbf{p}_i : \alpha_{i,j} \leq p_{i,j} \leq \beta_{i,j}, \qquad j = 0, 1, \ldots, d^0(P_i)\} \tag{4.36}$$

Let us denote the corresponding family of polynomial m-tuples

$$\mathbf{P}(s) := \mathbf{P}_1(s) \times \mathbf{P}_2(s) \times \cdots \times \mathbf{P}_m(s) \tag{4.37}$$

Let

$$\mathbf{p} := [\mathbf{p}_1, \mathbf{p}_2, \ldots, \mathbf{p}_m] \tag{4.38}$$

denote the global parameter vector and let

$$\mathbf{\Pi} := \mathbf{\Pi}_1 \times \mathbf{\Pi}_2 \times \cdots \times \mathbf{\Pi}_m \tag{4.39}$$

denote the global parameter uncertainty set. Now let us consider the polynomial eqn. 4.29 and rewrite it as $\delta(s, \mathbf{p})$ or $\delta(s, \underline{P}(s))$ to emphasise its dependence on the parameter vector $\mathbf{p}$ or the m-tuple $\underline{P}(s)$. We are interested in determining the Hurwitz stability of the set of polynomials

$$\boldsymbol{\Delta}(s) : = \{\delta(s, \mathbf{p}) : \mathbf{p} \in \boldsymbol{\Pi}\}$$
$$= \{\langle \underline{F}(s), \underline{P}(s)\rangle : \underline{P}(s) \in \mathbf{P}(s)\} \tag{4.40}$$

We call this a *linear interval polynomial* and adopt the convention

$$\boldsymbol{\Delta}(s) = F_1(s)\mathbf{P}_1(s) + F_2(s)\mathbf{P}_2(s) + \cdots + F_m(s)\mathbf{P}_m(s) \tag{4.41}$$

We make the standing assumptions that the elements of $\mathbf{p}$ perturb independently of each other, and that every polynomial in $\boldsymbol{\Delta}(s)$ is of the same degree. Henceforth we say that $\boldsymbol{\Delta}(s)$ *is stable if every polynomial in* $\boldsymbol{\Delta}(s)$ *is Hurwitz stable.* An equivalent statement is that $\underline{F}(s)$ *stabilises every* $\underline{P}(s) \in \mathbf{P}(s)$.

The solution developed below constructs an extremal set of line segments $\boldsymbol{\Delta}_\mathrm{E}(s) \subset \boldsymbol{\Delta}(s)$ with the property that the stability of $\boldsymbol{\Delta}_\mathrm{E}(s)$ implies stability of $\boldsymbol{\Delta}(s)$. This solution is constructive because the stability of $\boldsymbol{\Delta}_\mathrm{E}(s)$ can be checked, for instance by a set of root locus problems. The solution will be efficient since the number of elements of $\boldsymbol{\Delta}_\mathrm{E}(s)$ will be generated by first constructing an extremal subset $\mathbf{P}_\mathrm{E}(s)$ of the m-tuple family $\mathbf{P}(s)$. The extremal subset $\mathbf{P}_\mathrm{E}(s)$ is constructed from the Kharitonov polynomials of $\mathbf{P}_i(s)$. We describe the construction next.

4.4.3.1 Construction of the extremal subset

The Kharitonov polynomials corresponding to each $\mathbf{P}_i(s)$ are

$$K_i^1(s) = \alpha_{i,0} + \alpha_{i,1}s + \beta_{i,2}s^2 + \beta_{i,3}s^3 + \cdots$$
$$K_i^2(s) = \alpha_{i,0} + \beta_{i,1}s + \beta_{i,2}s^2 + \alpha_{i,3}s^3 + \cdots$$
$$K_i^3(s) = \beta_{i,0} + \alpha_{i,1}s + \alpha_{i,2}s^2 + \beta_{i,3}s^3 + \cdots$$
$$K_i^4(s) = \beta_{i,0} + \beta_{i,1}s + \alpha_{i,2}s^2 + \alpha_{i,3}s^3 + \cdots$$

and we denote them as

$$\mathcal{K}_i(s) := \{K_i^1(s), K_i^2(s), K_i^3(s), K_i^4(s)\} \tag{4.42}$$

For each $\mathbf{P}_i(s)$ we introduce four line segments, or convex combinations, joining pairs of Kharitonov polynomials as defined below:

$$\mathcal{S}_i(s) := \{[K_i^1(s), K_i^2(s)], [K_i^1(s), K_i^3(s)], [K_i^2(s), K_i^4(s)], [K_i^3(s), K_i^4(s)]\} \tag{4.43}$$

These four segments are called *Kharitonov segments*. They are illustrated in Figure 4.9 for the case of a polynomial of degree 2.

For each $l \in \{1, \ldots, m\}$ let us define

$$\mathbf{P}_\mathrm{E}^l(s) := \mathcal{K}_1(s) \times \cdots \times \mathcal{K}_{l-1}(s) \times \mathcal{S}_l(s) \times \mathcal{K}_{l+1}(s) \times \cdots \times \mathcal{K}_m(s) \tag{4.44}$$

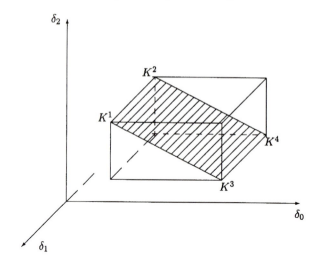

Figure 4.9   The four Kharitonov segments

A typical element of $\mathbf{P}_E^l(s)$ is

$$(K_1^{j_1}(s), K_2^{j_2}(s), \ldots, K_{l-1}^{j_{l-1}}(s), (1 - \lambda)K_l^1(s) + \lambda K_l^2(s), K_{l+1}^{j_{l+1}}(s), \ldots, K_m^{j_m}(s)) \quad (4.45)$$

with $\lambda \in [0, 1]$. This can be rewritten as

$$(1 - \lambda)(K_1^{j_1}(s), K_2^{j_2}(s), \ldots, K_{l-1}^{j_{l-1}}(s), K_l^1(s), K_{l+1}^{j_{l+1}}(s), \ldots, K_m^{j_m}(s))$$
$$+\lambda(K_1^{j_1}(s), K_2^{j_2}(s), \ldots, K_{l-1}^{j_{l-1}}(s), K_l^2(s), K_{l+1}^{j_{l+1}}(s), \ldots, K_m^{j_m}(s)) \quad (4.46)$$

Corresponding to the m-tuple $\mathbf{P}_E^l(s)$, introduce the polynomial family

$$\mathbf{\Delta}_E^l(s) := \{\langle \underline{F}(s), \underline{P}(s)\rangle : \underline{P}(s) \in \mathbf{P}_E^l(s)\} \quad (4.47)$$

The set $\mathbf{\Delta}_E^l(s)$ is also described as

$$\mathbf{\Delta}_E^l(s) = F_1(s)\mathcal{K}_1(s) + \cdots + F_{l-1}(s)\mathcal{K}_{l-1}(s) + F_l(s)\mathcal{S}_l(s) + F_{l+1}(s)\mathcal{K}_{l+1}(s)$$
$$+ \cdots + F_m(s)\mathcal{K}_m(s) \quad (4.48)$$

A typical element of $\mathbf{\Delta}_E^l(s)$ is the line segment of polynomials

$$F_1(s)K_1^{j_1}(s) + F_2(s)K_2^{j_2}(s) + \cdots + F_{l-1}(s)K_{l-1}^{j_{l-1}}(s) + F_l(s)[(1 - \lambda)K_l^1(s) + \lambda K_l^2(s)]$$
$$+ F_{l+1}(s)K_{l+1}^{j_{l+1}}(s) + \cdots + F_m(s)K_m^{j_m}(s) \quad (4.49)$$

with $\lambda \in [0, 1]$.

The *extremal subset* $\mathbf{P}(s)$ is defined by

$$\mathbf{P}_E(s) := \bigcup_{l=1}^{m} \mathbf{P}_E^l(s) \quad (4.50)$$

The corresponding *generalised Kharitonov segment* polynomials are

$$\Delta_E(s) := \bigcup_{l=1}^{m} \Delta_E^l(s)$$

$$= \{\langle \underline{F}(s), \underline{P}(s)\rangle : \underline{P}(s) \in \mathbf{P}_E(s)\} \tag{4.51}$$

The set of *m*-tuples of Kharitonov polynomials are denoted $\mathbf{P}_K(s)$ and referred to as the *Kharitonov vertices* of $\mathbf{P}(s)$:

$$\mathbf{P}_K(s) := \mathcal{K}_1(s) \times \mathcal{K}_2(s) \times \cdots \times \mathcal{K}_m(s) \subset \mathbf{P}_E(s) \tag{4.52}$$

The corresponding set of *Kharitonov vertex* polynomials is

$$\Delta_K(s) := \{\langle \underline{F}(s), \underline{P}(s)\rangle : \underline{P}(s) \in \mathbf{P}_K(s)\} \tag{4.53}$$

A typical element of $\Delta_K(s)$ is

$$F_1(s)K_1^{j_1}(s) + F_2(s)K_2^{j_2}(s) + \cdots + F_m(s)K_m^{j_m}(s) \tag{4.54}$$

The set $\mathbf{P}_E(s)$ is made up of one-parameter families of polynomial vectors. It is easy to see that there are $m4^m$ such segments in the most general case where there are four distinct Kharitonov polynomials for each $\mathbf{P}_i(s)$. The parameter space subsets corresponding to $\mathbf{P}_E^l(s)$ and $\mathbf{P}_E(s)$ are denoted by $\mathbf{\Pi}_l$ and

$$\mathbf{\Pi}_E := \bigcup_{l=1}^{m} \mathbf{\Pi}_l \tag{4.55}$$

respectively. Similarly, let $\mathbf{\Pi}_K$ denote the vertices of $\mathbf{\Pi}$ corresponding to the Kharitonov polynomials. Then, we also have

$$\Delta_E(s) := \{\delta(s, \mathbf{p}) : \mathbf{p} \in \mathbf{\Pi}_E\} \tag{4.56}$$

$$\Delta_K(s) := \{\delta(s, \mathbf{p}) : \mathbf{p} \in \mathbf{\Pi}_K\} \tag{4.57}$$

The set $\mathbf{P}_K(s)$ in general has 4^m distinct elements when each $\mathbf{P}_i(s)$ has four distinct Kharitonov polynomials. Thus $\Delta_K(s)$ is a discrete set of polynomials, $\Delta_E(s)$ is a set of line segments of polynomials, $\Delta(s)$ is a polytope of polynomials, and

$$\Delta_K(s) \subset \Delta_E(s) \subset \Delta(s) \tag{4.58}$$

With these preliminaries, we are ready to state the generalised Kharitonov theorem (GKT).

Theorem 4 (Generalised Kharitonov theorem (GKT)   *For a given m-tuple* $\underline{F}(s) = (F_1(s), \ldots, F_m(s))$ *of real polynomials:*

(I) $\underline{F}(s)$ *stabilises the entire family* $\mathbf{P}(s)$ *of m-tuples if and only if* $\underline{F}$ *stabilises every m-tuple segment in* $\mathbf{P}_E(s)$. *Equivalently,* $\Delta(s)$ *is stable if and only if* $\Delta_E(s)$ *is stable.*

(II)　*Moreover, if the polynimals $F_i(s)$ are of the form*

$$F_i(s) = s^{t_i}(a_i s + b_i) U_i(s) Q_j(s)$$

where $t_i \geq 0$ is an arbitrary integer, a_i and b_i are arbitrary real numbers, $U_i(s)$ is an anti-Hurwitz polynomial, and $Q_j(s)$ is an even or odd polynomial, then it is enough that $\underline{F}(s)$ stabilises the finite set of m-tuples $\mathbf{P}_K(s)$, or, equivalently, that the set of Kharitonov vertex polynomials $\mathbf{\Delta}_K(s)$ are stable.

(III)　*Finally, stablising the finite set $\mathbf{P}_K(s)$ is not sufficient to stabilise $\mathbf{P}(s)$ when the polynomials $F_i(s)$ do not satisfy the conditions in (II). Equivalently, the stability of $\mathbf{\Delta}_K(s)$ does not imply the stability of $\mathbf{\Delta}(s)$ when $F_i(s)$ do not satisfy the conditions in (II).*

We illustrate the control application of this theorem below.

Example 7

Consider the plant

$$G(s) = \frac{P_1(s)}{P_2(s)} = \frac{s^3 + \alpha s^2 - 2s + \beta}{s^4 + 2s^3 - s^2 + \gamma s + 1}$$

where

$$\alpha \in [-1, -2], \qquad \beta \in [0.5, 1], \qquad \gamma \in [0, 1]$$

Let

$$C(s) = \frac{F_1(s)}{F_2(s)}$$

denote the compensator. To determine if $C(s)$ robustly stabilises the set of plants given we must verify the Hurwitz stability of the family of characteristic polynomials $\mathbf{\Delta}(s)$ defined as

$$F_1(s)(s^3 + \alpha s^2 - 2s + \beta) + F_2(s)(s^4 + 2s^3 - s^2 + \gamma s + 1)$$

with $\alpha \in [-1, -2]$, $\beta \in [0.5, 1]$, $\gamma \in [0, 1]$. To construct the generalised Kharitonov segments, we start with the Kharitonov polynomials. There are two Kharitonov polynomials associated with $P_1(s)$:

$$K_1^1(s) = K_1^2(s) = 0.5 - 2s - s^2 + s^3$$

$$K_1^3(s) = K_1^4(s) = 1 - 2s - 2s^2 + s^3$$

and also two Kharitonov polynomials associated with $P_2(s)$:

$$K_2^1(s) = K_2^3(s) = 1 - s^2 + 2s^3 + s^4$$

$$K_2^2(s) = K_2^4(s) = 1 + s - s^2 + 2s^3 + s^4$$

The set $\mathbf{P}_E^1(s)$ therefore consists of the two plant segments

$$\frac{\lambda_1 K_1^1(s) + (1 - \lambda_1)K_1^3(s)}{K_2^1(s)} : \lambda_1 \in [0, 1]$$

$$\frac{\lambda_2 K_1^1(s) + (1 - \lambda_2)K_1^3(s)}{K_2^2(s)} : \lambda_2 \in [0, 1]$$

The set $\mathbf{P}_E^2(s)$ consist of the two plant segments

$$\frac{K_1^1(s)}{\lambda_3 K_2^1(s) + (1 - \lambda_3)K_2^2(s)} : \lambda_3 \in [0, 1]$$

$$\frac{K_1^3(s)}{\lambda_4 K_2^1(s) + (1 - \lambda_4)K_2^2(s)} : \lambda_4 \in [0, 1]$$

Thus, the extremal set $\mathbf{P}_E(s)$ consists of the following four plant segments:

$$\frac{0.5(1 + \lambda_1) - 2s - (1 + \lambda_1)s^2 + s^3}{1 - s^2 + 2s^3 + s^4} : \lambda_1 \in [0, 1]$$

$$\frac{0.5(1 + \lambda_2) - 2s - (1 + \lambda_2)s^2 + s^3}{1 + s - s^2 + 2s^3 + s^4} : \lambda_2 \in [0, 1]$$

$$\frac{0.5 - 2s - s^2 + s^3}{1 + \lambda_3 s - s^2 + 2s^3 + s^4} : \lambda_3 \in [0, 1],$$

$$\frac{1 - 2s - 2s^2 + s^3}{1 + \lambda_4 s - s^2 + 2s^3 + s^4} : \lambda_4 \in [0, 1]$$

Therefore, we can verify robust stability by checking the Hurwitz stability of the set $\mathbf{\Delta}_E(s)$, which consists of the following four polynomial segments:

$$F_1(s)(0.5(1 + \lambda_1) - 2s - (1 + \lambda_1)s^2 + s^3) + F_2(s)(1 - s^2 + 2s^3 + s^4)$$

$$F_1(s)(0.5(1 + \lambda_2) - 2s - (1 + \lambda_2)s^2 + s^3) + F_2(s)(1 + s - s^2 + 2s^3 + s^4)$$

$$F_1(s)(0.5 - 2s - s^2 + s^3) + F_2(s)(1 + \lambda_3 s - s^2 + 2s^3 + s^4)$$

$$F_1(s)(1 - 2s - 2s^2 + s^3) + F_2(s)(1 + \lambda_4 s - s^2 + 2s^3 + s^4)$$

$$\lambda_i \in [0, 1] : i = 1, 2, 3, 4$$

In other words, any compensator that stabilises the family of plants $\mathbf{P}(s)$ must stabilise the four one-parameter family of extremal plants $\mathbf{P}_E(s)$. If we had used the edge theorem it would have been necessary to check the 12 line segments of plants corresponding to the exposed edges of $\mathbf{\Delta}(s)$.

If the compensator polynomials $F_i(s)$ satisfy the "vertex conditions' in part II of GKT, it is enough to check that they stabilise the plants corresponding to the four Kharitonov vertices. This corresponds to checking the Hurwitz stability of the four *fixed* polynomials:

$$F_1(s)(1 - 2s - 2s^2 + s^3) + F_2(s)(1 - s^2 + 2s^3 + s^4)$$
$$F_1(s)(1 - 2s - 2s^2 + s^3) + F_2(s)(1 + s - s^2 + 2s^3 + s^4)$$
$$F_1(s)(0.5 - 2s - s^2 + s^3) + F_2(s)(1 + s - s^2 + 2s^3 + s^4)$$
$$F_1(s)(0.5 - 2s - 2s^2 + s^3) + F_2(s)(1 - s^2 + 2s^3 + s^4)$$

In the following section we show that the extremal set used in the GKT plays is crucial in frequency-domain design methods.

4.5 Frequency-domain results in RPRCT

The extremal sets introduced in the GKT play an important role in frequency-domain design methods. In particular they can be used to generate the uncertainty sets in the frequency domain needed to carry out robust design. We describe these properties in this section, again referring the reader to [2] for the proofs.

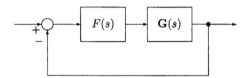

Figure 4.10   A unity feedback interval control system

Consider the feedback system shown in Figure 4.10 with

$$F(s) := \frac{F_1(s)}{F_2(s)}, \qquad G(s) := \frac{N(s)}{D(s)} \qquad (4.59)$$

We suppose that $F(s)$ is fixed but $G(s)$ contains uncertain real parameters which appear as the coefficients of $N(s)$ and $D(s)$. Write

$$D(s) := a_0 + a_1 s + a_2 s^2 + a_3 s^3 + \cdots + a_{n-1} s^{n-1} + a_n s^n$$
$$N(s) := b_0 + b_1 s + b_2 s^2 + b_3 s^3 + \cdots + b_{m-1} s^{m-1} + b_m s^m \qquad (4.60)$$

where $a_k \in [a_k^-, a_k^+]$, for $k \in \underline{n} := \{1, \ldots, n\}$ and $b_k \in [b_k^-, b_k^+]$, for $k \in \underline{m}$.

Let us define the interval polynomial sets:

$$\mathbf{D}(s) := \{D(s): a_0 + a_1 s + a_2 s^2 + \cdots + a_n s^n, a_k \in [a_k^-, a_k^+], \text{ for } k \in \underline{n}\}$$
$$\mathbf{N}(s) := \{N(s): b_0 + b_1 s + b_2 s^2 + \cdots + b_m s^m, b_k \in [b_k^-, b_k^+], \text{ for } k \in \underline{m}\}$$

and the corresponding set of *interval plants*:

$$\mathbf{G}(s) := \left\{ \frac{N(s)}{D(s)} : (N(s), D(s)) \in (\mathbf{N}(s) \times \mathbf{D}(s)) \right\} \qquad (4.61)$$

We refer to the unity feedback system in Figure 4.10 as an *interval control system*. For simplicity, we use the notational convention

$$\mathbf{G}(s) = \frac{\mathbf{N}(s)}{\mathbf{D}(s)} \tag{4.62}$$

to denote the family of eqns. 4.61. The characteristic polynomial of the system is

$$\delta(s) := F_1(s)N(s) + F_2(s)D(s) \tag{4.63}$$

and the set of system characteristic polynomials can be written as

$$\mathbf{\Delta}(s) := F_1(s)\mathbf{N}(s) + F_2(s)\mathbf{D}(s) \tag{4.64}$$

The control system is robustly stable if each polynomial in $\mathbf{\Delta}(s)$ is of the same degree and is Hurwitz. This is precisely the type of robust stability problem dealt with in the generalised Kharitonov theorem (GKT), where we showed that Hurwitz stability of the control system over the set $\mathbf{G}(s)$ could be reduced to testing over the much smaller extremal set of systems $\mathbf{G}_E(s)$.

Following the notation of the last section, let $\mathcal{K}_N(s)$ and $\mathcal{K}_D(s)$ denote Kharitonov polynomials associated with $\mathbf{N}(s)$ and $\mathbf{D}(s)$, and let $\mathcal{S}_N(s)$ and $\mathcal{S}_D(s)$ denote the corresponding sets of Kharitonov segments. Recall that these segments are pairwise convex combinations of Kharitonov polynomials sharing a common even or odd part. Define the *extremal subsets*, using the above notational convention:

$$\mathbf{G}_E(s) := \frac{\mathcal{K}_N(s)}{\mathcal{S}_D(s)} \cup \frac{\mathcal{S}_N(s)}{\mathcal{K}_D(s)} \quad \text{(extremal systems)} \tag{4.65}$$

$$\mathbf{G}_K(s) := \frac{\mathcal{K}_N(s)}{\mathcal{K}_D(s)} \quad \text{(Kharitonov systems)} \tag{4.66}$$

As before, we will say that $F(s)$ satisfies the *vertex condition* if the polynomials $F_i(s)$ are of the form

$$F_i(s) := s^{t_i}(a_i s + b_i)U_i(s)R_i(s) \qquad i = 1, 2 \tag{4.67}$$

where t_i are non-negative integers, a_i, b_i are arbitrary real numbers, $U_i(s)$ is an anti-Hurwitz polynomial, and $R_i(s)$ is an even or odd polynomial.

Let us first recall the result given by GKT: the control system of Figure 4.10 is robustly stable, i.e. stable for all $G(s) \in \mathbf{G}(s)$ if and only if it is stable for all $G(s) \in \mathbf{G}_E(s)$. If, in addition, $F(s)$ satisfies the vertex condition, robust stability holds if the system is stable for each $G(s) \in \mathbf{G}_K(s)$.

In the rest of this section we show that the systems $\mathbf{G}_E(s)$ and $\mathbf{G}_K(s)$ enjoy many useful boundary and extremal properties. They can be used constructively to carry out frequency response calculations in control system analysis and design. In fact, it will turn out that most of the important system properties such as worst-case stability and performance margins over the set of uncertain parameters can be determined by replacing $G(s) \in \mathbf{G}(s)$ by the elements of

$G(s) \in \mathbf{G}_E(s)$. In some special cases one may even replace $G(s)$ by the elements of $\mathbf{G}_K(s)$.

4.5.1 Frequency-domain properties

To carry out frequency response analysis and design incorporating robustness with respect to parameter uncertainty we need to be able to determine the complex plane images of various parametrised sets of polynomials and transfer functions as s runs along the imaginary axis, that is $s = j\omega, 0 \le \omega < \infty$.

In this section we develop some computationally efficient procedures to generate such sets. We first consider the complex plane images of $\mathbf{\Delta}(s)$ and $\mathbf{G}(s)$ at $s = j\omega$. These sets, called uncertainty templates, are denoted $\mathbf{\Delta}(j\omega)$ and $\mathbf{G}(j\omega)$. Since $\mathbf{N}(s)$ and $\mathbf{D}(s)$ are interval families, $\mathbf{N}(j\omega)$ and $\mathbf{D}(j\omega)$ are axis parallel rectangles in the complex plane. $F_1(j\omega)\mathbf{N}(j\omega)$ and $F_2(j\omega)\mathbf{D}(j\omega)$ are likewise rotated rectangles in the complex plane. Thus $\mathbf{\Delta}(j\omega)$ is the complex plane sum of two rectangles, whereas $\mathbf{G}(j\omega)$ is the quotient of two rectangles. We assume here that $0 \notin \mathbf{D}(j\omega)$. If this assumption fails to hold we can always 'indent' the $j\omega$ axis to exclude those values of ω which violate the assumption. Therefore, throughout this section we make the standing assumption that the denominator of any quotients exclude zero.

To determine $\mathbf{\Delta}(j\omega)$ and $\mathbf{G}(j\omega)$ we note that the vertices of $\mathbf{N}(j\omega)$ and $\mathbf{D}(j\omega)$ correspond to the Kharitonov polynomials, whereas the edges correspond to the Kharitonov segments. The set of points $\mathcal{K}_N(j\omega)$ are therefore the vertices of $\mathbf{N}(j\omega)$ and the four lines $\mathcal{S}_N(j\omega)$ are the edges of $\mathbf{N}(j\omega)$. $F_1(j\omega)\mathbf{N}(j\omega)$ is also a polygon with vertices $F_1(j\omega)\mathcal{K}_N(j\omega)$ and edges $F_1(j\omega)\mathcal{S}_N(j\omega)$. Similarly, $F_2(j\omega)\mathcal{K}_D(j\omega)$ and $F_2(j\omega)\mathcal{S}_D(j\omega)$ are the vertices and edges of the polygon $F_2(j\omega)\mathbf{D}(j\omega)$. The $j\omega$ image of the extremal systems $\mathbf{G}_E(s)$ defined earlier exactly coincides with these vertex-edge pairs. Let

$$(\mathbf{N}(s) \times \mathbf{D}(s))_E := (\mathcal{K}_N(s) \times \mathcal{S}_D(s)) \cup (\mathcal{S}_N(s) \times \mathcal{K}_D(s)) \qquad (4.68)$$

Recall that the *extremal systems* are

$$\mathbf{G}_E(s) := \left\{ \frac{N(s)}{D(s)} : (N(s), D(s)) \in (\mathbf{N}(s) \times \mathbf{D}(s))_E \right\} := \frac{\mathcal{K}_N(s)}{\mathcal{S}_D(s)} \cup \frac{\mathcal{S}_N(s)}{\mathcal{K}_D(s)} \qquad (4.69)$$

and define

$$\mathbf{\Delta}_E(s) := \{F_1(s)N(s) + F_2(s)D(s) : (N(s), D(s)) \in (\mathbf{N}(s) \times \mathbf{D}(s))_E\} \qquad (4.70)$$

We can now state an important result regarding the boundary of image sets.

Theorem 5 (Boundary generating property)

(a) $$\partial\mathbf{\Delta}(j\omega) \subset \mathbf{\Delta}_E(j\omega)$$

(b) $$\partial\mathbf{G}(j\omega) \subset \mathbf{G}_E(j\omega)$$

This boundary generating property of the extremal set is very important as it can reduce design and testing over the entire family to design and testing over this small subset. This is explored further below.

4.5.2 Closed-loop transfer functions

In this section we show that the boundary generating properties carry over to closed-loop transfer functions containing interval plants. Referring now to the control system in Figure 4.10, we consider the following transfer functions of interest in analysis and design problems:

$$\frac{y(s)}{u(s)} = G(s), \qquad \frac{u(s)}{e(s)} = F(s) \tag{4.71}$$

$$T^0(s) := \frac{y(s)}{e(s)} = F(s)G(s) \tag{4.72}$$

$$T^e(s) := \frac{e(s)}{r(s)} = \frac{1}{1 + F(s)G(s)} \tag{4.73}$$

$$T^u(s) := \frac{u(s)}{r(s)} = \frac{F(s)}{1 + F(s)G(s)} \tag{4.74}$$

$$T^y(s) := \frac{y(s)}{r(s)} = \frac{F(s)G(s)}{1 + F(s)G(s)} \tag{4.75}$$

As $G(s)$ ranges over the uncertainty set $\mathbf{G}(s)$ the transfer functions $T^0(s)$, $T^y(s)$, $T^u(s)$, $T^e(s)$ range over corresponding uncertainty sets $\mathbf{T}^0(s)$, $\mathbf{T}^y(s)$, $\mathbf{T}^u(s)$ and $\mathbf{T}^e(s)$, respectively. In other words,

$$\mathbf{T}^0(s) := \{F(s)G(s) : G(s) \in \mathbf{G}(s)\} \tag{4.76}$$

$$\mathbf{T}^e(s) := \left\{ \frac{1}{1 + F(s)G(s)} : G(s) \in \mathbf{G}(s) \right\} \tag{4.77}$$

$$\mathbf{T}^u(s) := \left\{ \frac{F(s)}{1 + F(s)G(s)} : G(s) \in \mathbf{G}(s) \right\} \tag{4.78}$$

$$\mathbf{T}^y(s) := \left\{ \frac{F(s)G(s)}{1 + F(s)G(s)} : G(s) \in \mathbf{G}(s) \right\} \tag{4.79}$$

We now show that the boundary generating property of the extremal subsets shown in Theorem 5 carries over to each of the system transfer functions listed above. In fact, we will show that the boundary of the image set at $s = j\omega$, the Nyquist plot and Bode plot boundaries of each of the above sets are all generated by the subset $\mathbf{G}_E(s)$. Introduce the subsets of eqns. 4.76−4.79 obtained by replacing $\mathbf{G}(s)$ by $\mathbf{G}_E(s)$:

$$\mathbf{T}_E^0(s) := \{F(s)G(s) : G(s) \in \mathbf{G}_E(s)\} \tag{4.80}$$

$$\mathbf{T}_E^e(s) := \left\{ \frac{1}{1 + F(s)G(s)} : G(s) \in \mathbf{G}_E(s) \right\} \tag{4.81}$$

$$\mathbf{T}_E^u(s) := \left\{ \frac{F(s)}{1 + F(s)G(s)} : G(s) \in \mathbf{G}_E(s) \right\} \tag{4.82}$$

$$\mathbf{T}_E^y(s) := \left\{ \frac{F(s)G(s)}{1 + F(s)G(s)} : G(s) \in \mathbf{G}_E(s) \right\} \tag{4.83}$$

The main results can now be stated.

Theorem 6   *For every $\omega \geq 0$,*

(a) $$\partial \mathbf{T}^0(j\omega) \subset \mathbf{T}_E^0(j\omega)$$

(b) $$\partial \mathbf{T}^e(j\omega) \subset \mathbf{T}_E^e(j\omega)$$

(c) $$\partial \mathbf{T}^u(j\omega) \subset \mathbf{T}_E^u(j\omega)$$

(d) $$\partial \mathbf{T}^y(j\omega) \subset \mathbf{T}_E^y(j\omega)$$

An important consequence of this is the following.

Theorem 7   (Nyquist, Bode and Nichols envelopes)   *The Nyquist, Bode magnitude and phase and Nichols plots of each of the transfer function sets $\mathbf{T}^0(s)$, $\mathbf{T}^y(s)$, $\mathbf{T}^u(s)$ and $\mathbf{T}^e(s)$ are bounded by their corresponding extremal subsets.*

Example 8

Consider the plant and controller

$$G(s) = \frac{b_1 s + b_0}{a_2 s^2 + a_1 s + a_0} \qquad C(s) = \frac{s^2 + 2s + 1}{s^4 + 2s^3 + 2s^2 + s}$$

where the plant parameters vary as

$$b_1 \in [0.1, 0.2], \qquad b_0 \in [0.9, 1.1]$$
$$a_2 \in [0.9, 1.0], \qquad a_1 \in [1.8, 2.0), \qquad a_0 \in [1.9, 2.1]$$

Figures 4.11−4.13 show the frequency domain plots for this example.

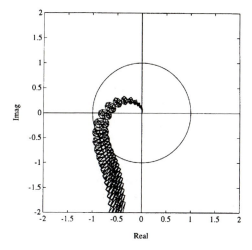

Figure 4.11   Nyquist templates (Example 8)

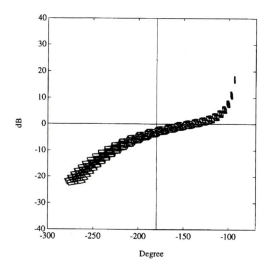

Figure 4.12   Nichols envelope (Example 8)

4.6 Concluding remarks

In this chapter we have described some theoretical results which are of great value in the analysis and design of control systems containing uncertain real parameters. In Chapter 8 we show, mainly through examples, how these can aid in the development of computer-aided tools for the design of robust control systems.

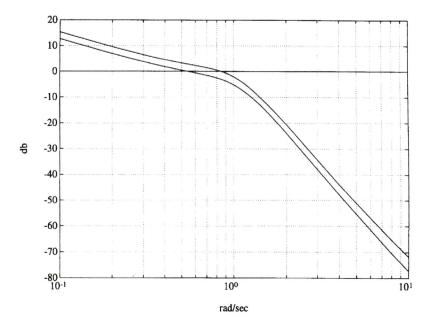

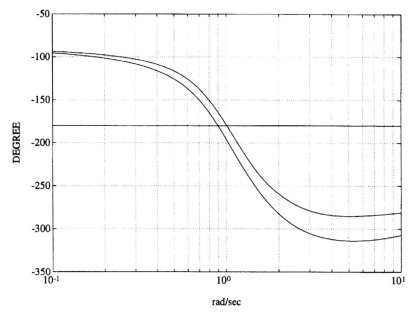

Figure 4.13   Bode envelopes (Example 8)

4.7 Acknowledgments

This research is supported in part by NASA grant NCC-5228 and NSF grants ECS-9417004 and HRD-9706268.

4.8 References

1 ŠILJAK, D. D.: 'Parameter space methods for robust control design: a guided tour', *IEEE Trans. Autom. Control*, **AC-34**, (7), 674−688 (1989)
2 BHATTACHARYYA, S. P., CHAPELLAT, H., and KEEL, L. H.: 'Robust control: the parametric approach', (Prentice Hall PTR, Upper Saddle River, NJ, 1995)
3 BARMISH, B. R.: 'New tools for robustness of linear systems', (Macmillan Publishing Co., New York, NY, 1994)
4 ACKERMANN, J.: 'Robust Control: systems with uncertain physical parameters' (Springer-Verlag, London, UK, 1993)

## Chapter 5

Using differential and real algebra in control

S. T. Glad and M. Jirstrand

5.1 Introduction

In the analysis of linear systems algebraic methods have been important since the early work of Routh and Hurwitz. Much of the work has been concerned with transfer functions or transfer matrices. For nonlinear systems the use of algebraic methods was for a long time made difficult by the complexity of the calculations involved. In recent years the availability of symbolic computer-based methods has made it possible to explore algebraic methods in the nonlinear context. Here we concentrate on two approaches: differential algebra, which can be used to eliminate variables in systems of differential equations, and real algebra, which gives computational tools to handle systems of polynomial equations and inequalities. We give very brief overviews of the computational algorithms and then present some control related applications.

To begin with we also note that the restriction to polynomial systems is not very severe. It can be shown that systems where the nonlinearities are not originally polynomial may be rewritten in polynomial form if the nonlinearities themselves are solutions to algebraic differential equations. For more details on this the reader is referred to [1, 2].

5.2 Differential algebra: motivating examples

Example 1

Consider a very simplified model for velocity control of an aircraft.

$$\dot{x}_1 = x_2 - f(x_1)$$
$$\dot{x}_2 = -x_2 + u \qquad (5.1)$$
$$y = x_1$$

Here x_1 is the velocity, x_2 is the engine thrust and $f(x_1) = x_1^2 - 2x_1 + 2$ is the aero-dynamic drag. The expression for the drag is typical for some fighter aircrafts operating at high angle of attack and low speed. The mass is normalised to 1 and we have assumed a first-order system response from pilot command to engine thrust.

Suppose we want an input-output relation for this nonlinear system, that is a differential equation directly relating u and y. Differentiating, we obtain

$$y = x_1 \tag{5.2}$$

$$\dot{y} = x_2 - x_1^2 + 2x_1 - 2 \tag{5.3}$$

$$\ddot{y} = -x_2 + u + (2 - 2x_1)\dot{y} \tag{5.4}$$

Eqn. 5.2 already has x_1 expressed in terms of y. Substituting this relation into eqn. 5.3, it is possible to express x_2 in terms of y and $\dot{y}$. Substituting these expressions for x_1 and x_2 into eqn. 5.4 gives the following three relations:

$$x_1 = y \tag{5.5}$$

$$x_2 = \dot{y} + y^2 - 2y + 2 \tag{5.6}$$

$$\ddot{y} + (2y - 1)\dot{y} + y^2 - 2y = u - 2 \tag{5.7}$$

The input-output relations is given by eqn. 5.7. In addition we have obtained the relations 5.5 and 5.6, which show that x_1 and x_2 can be uniquely computed from knowledge of u and y and hence are globally observable.

Example 2

Consider the system

$$\ddot{y} + 2\theta\dot{y} + \theta^2 y = 0 \tag{5.8}$$

We observe y and have a single unknown parameter θ. Is θ globally identifiable? In an ideal noise-free situation, knowledtge of y would enable us to calculate $\dot{y}$ and $\ddot{y}$. We can then regard eqn. 5.8 as a quadratic equation in θ with known coefficients. That equation would have two solutions, suggesting that θ is not globally identifiable. However, some fairly simple direct manipulations give the following result. Differentiate eqn. 5.8:

$$y^{(3)} + 2\theta\ddot{y} + \theta^2\dot{y} = 0 \tag{5.9}$$

Now multiply eqn. 5.8 by $\dot{y}$ and eqn. 5.9 by y and subtract the expressions

$$(\dot{y}\ddot{y} - yy^{(3)}) + 2\theta(\dot{y}^2 - y\ddot{y}) = 0 \tag{5.10}$$

This is a linear equation in θ that reveals that the structure is globally identifiable. (Further calculations show that this corresponds to one of the solutions of the quadratic eqn. 5.8.)

These examples show that it is possible to extract important information from nonlinear systems by doing algebraic manipulations together with differentiation. The manipulations we have made are similar to what is done for linear equations using Gaussian elimination, except that we also have nonlinear terms. It is also similar to the calculations involved in Gröbner bases, except that we also have to take into account differentiations of variables. There is a systematic way of doing calculations of this type, that goes back to the work by the American mathematician Ritt in the 1930s on what he called 'differential algebra' [3–5]. Similar algorithms were presented by Seidenberg [6]. Differential algebraic concepts were introduced into control theory by Fliess; for example [7].

5.3 The basic ideas of differential algebra

To perform calculations of the type that was shown in Examples 1 and 2 it is necessary to examine the pattern of variables and derivatives that are present in each equation. To begin with one decides on an ordering of all variables that are present in the problem and their derivatives. This is called a *ranking*. For a single variable y there is only one choice,

$$y < \dot{y} < \ddot{y} < y^{(3)} < y^{(4)} < \cdots \tag{5.11}$$

Here $<$ means 'is ranked lower than'.

If two variables u and y are involved, two rankings are natural. One could say that derivatives of u are always ranked lower than those of y. Then the ranking would be

$$u < \dot{u} < \ddot{u} < \cdots < y < \dot{y} < \ddot{y} < \cdots \tag{5.12}$$

One could also say that higher-order derivatives are always ranked higher, no matter what the varibles are. Then a ranking could be

$$u < y < \dot{u} < \dot{y} < \ddot{u} < \ddot{y} < \cdots \tag{5.13}$$

It turns out that any type of ranking can be used in the algorithms we are going to present, provided two conditions are satisfied. Let $u^{(\mu)}$ and $y^{(\nu)}$ be arbitrary derivatives of arbitrary variables. Then the ranking should be such that, for arbitrary positive σ, ν,

$$y^{(\nu)} < y^{(\nu+\sigma)} \tag{5.14}$$

$$u^{(\mu)} < y^{(\nu)} \Rightarrow u^{(\mu+\sigma)} < y^{(\nu+\sigma)} \tag{5.15}$$

Among three variables u, y and v there are a number of possible rankings; for example:

$$u < \dot{u} < \ddot{u} < \cdots < y < \dot{y} < \ddot{y} < \cdots < v < \dot{v} < \ddot{v} < \cdots \tag{5.16}$$

$$u < v < y < \dot{u} < \dot{v} < \dot{y} < \ddot{u} < \ddot{v} < \ddot{y} < \cdots \tag{5.17}$$

$$u < \dot{u} < \ddot{u} < \cdots < v < y < \dot{v} < \dot{y} < \ddot{v} < \ddot{y} < \cdots \tag{5.18}$$

5.3.1 Ranking of polynomials

The ranking of variables automatically carries over to a ranking of differential polynomials. When comparing two polynomials the procedure is as follows:

1. Find the highest ranked derivative in each polynomial. This is called the *leader* of the polynomial.
2. If the polynomials have different leaders, the one with the highest ranked leader is ranked highest.
3. If the polynomials have the same leader, the one with the highest power of the leader is ranked highest.

Example 3

Let the ranking be that of eqn. 5.16 and consider the polynomials

$$A = u\ddot{y}v + 1 \qquad B = \ddot{u} + (y^{(5)})^2$$

If the ranking is as eqn. 5.16 then A has the leader v and B has the leader $y^{(5)}$, so B is ranked lower than A. If instead the ranking eqn. 5.17 is used then A has the leader $\ddot{y}$ and B has the leader $y^{(5)}$, so that A is ranked lower.

Example 4

Consider again the ranking of eqn. 5.17 and the polynomials

$$A = v^2\dot{y} + \ddot{u}^2 \qquad B = \dot{v} + y + \ddot{u} + \ddot{u}^5$$

Both A and B have $\ddot{u}$ as leader but A is ranked lower since its leader is raised to a lower power (2) than in B (5).

5.3.2 Reducedness

Consider the polynomials

$$A = u\dot{y} + \dot{u} \qquad B = \ddot{y} + y \qquad (5.19)$$

under the ranking of eqn. 5.12. We can form a simpler polynomial C out of B by subtracting out the highest derivative

$$C = uB - \frac{d}{dt}A = uy - \dot{u}\dot{y} - \ddot{u} \qquad (5.20)$$

The operation was made possible by the fact that B contained a derivative of the leader of A. C can be further simplified to D by removing $\dot{y}$ using the computation

$$D = uC + \dot{u}A = u^2y - u\ddot{u} + \dot{u}^2 \qquad (5.21)$$

Now it is impossible to carry the calculations any further. The term in D containing y cannot be removed by subtracting a multiple of A or its derivative. To describe situations like these the notion of reducedness is introduced.

Suppose the differential polynomial F has leader $y^{(v)}$. The differential polynomial G is said to be:

1. *partially reduced* with respect to F, if it contains no higher derivative of y than $y^{(v)}$
2. *reduced* with respect to F, if it is partially reduced, and is a polynomial of lower degree than F in $y^{(v)}$.

Returning to the polynomials A and B of eqn. 5.19 we see that B is not partially reduced with respect to A, since it contains $\ddot{y}$, which is the derivative of the leader, $\dot{y}$, of A. It is precisely this fact which makes it possible to form C from eqn. 5.20. The polynomial C is partially reduced with respect to A so it is not possible to perform a similar calculation again. However, it is not reduced since it contains the leader of A. This makes it possible to form D, which does not contain $\dot{y}$. The polynomial D is reduced and so the process ends.

5.3.3 Autoreduced sets

Now consider the polynomials that are formed from eqn. 5.1 if all terms in the equations are moved to the left-hand sides:

$$A_1 = \dot{x}_1 - x_2 + x_1^2 - 2x_1 + 2, \qquad A_2 = \dot{x}_2 + x_2 - u, \qquad A_3 = y - x_1 \quad (5.22)$$

If the ranking

$$u < \dot{u} < \ddot{u} < \cdots < x_1 < x_2 < \dot{x}_1 < \dot{x}_2 < \cdots < y < \dot{y} < \ddot{y} < \cdots \quad (5.23)$$

is used, then the leaders of A_1, A_2 and A_3 are $\dot{x}_1$, $\dot{x}_2$ and y, respectively. Since the leaders are not present in any other polynomial, it follows that every polynomial is reduced with respect to every other polynomial. A set of polynomials, all of which are reduced with respect to each other, is called an *autoreduced set*. A state space description is thus an example of an autoreduced set. A further example is the set of eqns. 5.5–5.7 that were derived in Example 1. The corresponding polynomials,

$$B_1 = \ddot{y} + (2y - 1)\dot{y} + y^2 - 2y + 2 - u$$
$$B_2 = x_2 - \dot{y} - y^2 + 2y - 2 \qquad (5.24)$$
$$B_3 = x_1 - y$$

form an autoreduced set under ranking

$$u < \dot{u} < \cdots y < \dot{y} < \cdots < x_2 < \dot{x}_2 < \cdots < x_1 < \dot{x}_1 < \cdots \quad (5.25)$$

since the leaders are $\ddot{y}$, x_2 and x_1, respectively.

5.3.4 Remainders

Consider again the set of eqns. 5.22. Suppose we want to evaluate $\ddot{x}_1$ for soluitons to that set of equations. We can then form

$$\ddot{x}_1 - \frac{d}{dt}A_1 - A_2 + (2x_1 - 2)A_1$$

$$= u + x_2 - 2x_2x_2 + 2x_1^3 - 6x_1^2 + 8x_1 - 4 \qquad (5.26)$$

The result is a quantity which is reduced with respect to all the polynomials of eqn. 5.22. A corresponding calculation can be made for an arbitrary polynomial F and an arbitrary autoreduced set $A_1, \ldots, A_r$. It is then possible to find a polynomial R, the *remainder*, which is either zero or else reduced with respect to $A_1, \ldots, A_r$, and such that

$$S_1^{\nu_1} \cdots S_r^{\nu_r} I_1^{\mu_1} \cdots I_r^{\mu_r} F - R = \sum_{i,j} Q_{ij} \frac{d^j}{dt^j} A_i \qquad (5.27)$$

Here S_k denotes the derivative of A_k with respect to its leader, while I_k is the coefficient of the highest power of the leader of A_k. They are called *separants* and *initials*, respectively. The ν_k and μ_k are integers. In our Example, eqn. 5.26, all separants and initials were 1. Eqn. 5.27 is similar to the definition of remainder in polynomial division. It can in fact be seen as a natural extension of polynomial division to also handle formal differentiation.

5.3.5 Ritt's algorithm

The concepts of ranking, autoreduced sets and remainder form the basis of an algorithm developed by Ritt [3]. The algorithm starts with an arbitrary set of polynomials. By extending the concept of ranking also to autoreduced sets one can form a lowest autoreduced set among the given polynomials. By calculating remainders with respect to that set new polynomials are created. It is then possible to form a lower autoreduced set. New remainders are then calculated with respect to that set. The process continues until an autoreduced set is formed which has the property that the remainder of all original and created polynomials have remainder zero with respect to it. The autoreduced set which is created in this way can be viewed as representing the simplest set of equations (under a given ranking of variables) which describe the system defined by the original polynomials. The general way of calculating the input-output description discussed in Example 1 would be to use Ritt's algorithm with a ranking where u and its derivatives are ranked lowest and where y and its derivatives are ranked lower than the state variables. Since it is not possible to create an equation involving only u and its derivatives out of a state space description, the simplest possible polynomial under such a ranking would be one involving only u, y and their derivatives, that is an input-output description. In its search for a lowest autoreduced set the algorithm will therefore eventually produce such a polynomial. This application is discussed in [8–10].

5.3.6 Application of Ritt's algorithm to identifiability problems

It was shown in Example 2 how an identifiability question could be tackled. Using Ritt's algorithm the question of identifiability (or observability) can be answered by computing polynomials where all variables are eliminated except the parameter to be identified and the observable inputs and outputs. This approach is described in [11]. We give an example.

Example 5

The following example is taken from [12]:

$$\dot{x}(t) = -\frac{V_m x(t)}{k_m + x(t)} - k_{01} x(t)$$

$$x(0) = D \tag{5.28}$$

$$y(t) = cx(t)$$

D is here a known constant, while we shall investigate the identifiability of V_m, k_m, k_{01} and c. Using the ranking

$$y^{(\cdot)} < x^{(\cdot)} < c^{(\cdot)} < k_{01}^{(\cdot)} < k_m^{(\cdot)} < V_m^{(\cdot)}$$

the following equations are generated by Ritt's algorithm:

$$x\dot{y} - y\dot{x} = 0 \tag{5.29}$$

$$y - cx = 0 \tag{5.30}$$

$$(2\dot{y}^4 - y^2\ddot{y}^2 - 2y\dot{y}^2\ddot{y} + y^2\dot{y}y^{(3)})k_{01} - 3y\dot{y}\ddot{y}^2 + 2\dot{y}^3\ddot{y} + y\dot{y}^2 y^{(3)} = 0 \tag{5.31}$$

$$(y^2\dot{y}y^{(3)} - 2\dot{y}^2 y\ddot{y} + 2\dot{y}^4 - y^2\ddot{y}^2)k_m - y^2 x\ddot{y}^2 + y^2\dot{y}y^{(3)}x = 0 \tag{5.32}$$

$$\big(y^5\dot{y}^4 + 4y^4\dot{y}^3\ddot{y}^2 - 2y^5\dot{y}^2\ddot{y}y^{(3)} + 4y\dot{y}^8 - 8\dot{y}^6 y^2\ddot{y} + 4\dot{y}^5 y^3 y^{(3)} - 4y^4\dot{y}^3\ddot{y}y^{(3)}$$
$$+ y^5\dot{y}^2 y^{(3)^2}\big)V_m + 12x\dot{y}^5 y^2\ddot{y}^2 - 4x\dot{y}^3 y^3\ddot{y}^3 - 12x\dot{y}^7 y\ddot{y} + 4x\dot{y}^9 = 0 \tag{5.33}$$

We have omitted an equation containing only y and its derivatives. Since y and $x(0)$ are known, $x(t)$ can be considered known from eqn. 5.29 in all equations below. It then follows that the parameters k_{01}, c, k_m and V_m are all determined uniquely since they occur in equations of degree 1 where all other quantities are known.

5.4 Real algebra: motivating example

Example 6

Consider the problem of finding the equilibria of the system

$$\dot{x}_1 = x_1 + x_2 u$$

$$\dot{x}_2 = -x_2 + (1 + x_1^2)u + u^3 \tag{5.34}$$

subject to the constraints

$$-\tfrac{1}{2} \le u \le \tfrac{1}{2} \tag{5.35}$$

Putting the right-hand sides of the equations equal to zero and the side condition $-\tfrac{1}{2} \le u \le \tfrac{1}{2}$ gives a problem involving both equations and inequalities. The study of real solutions to such systems is part of the subject of real algebra. Formally the set of solutions is described by the formula

$$\exists u\left[-x_1 + x_2 u = 0 \wedge - x_2 + (1 + x_1^2)u + u^3 = 0 \wedge -\tfrac{1}{2} \le u \le \tfrac{1}{2}\right] \tag{5.36}$$

It would be nice to have a more explicit formula, not involving the existential quantifier $\exists$. This leads to a so-called quantifier elimination problem.

Many problems in control and analysis of dynamic systems can be formulated in a similar manner as in Example 6, i.e. as formulas involving polynomial equations, inequalities, quantifiers ($\exists, \forall$) and Boolean operators (e.g. $\vee, \wedge$). Formally these expressions constitute a sequence in the so-called first-order theory of real closed fields. The problem of *quantifier elimination* consists of finding an equivalent expression without any quantified variables. Many geometrical operations on sets defined by polynomial inequalities can be formulated as quantifier elimination problems: for example, projection, checking nonemptyness, and polynomial function optimisation.

In the late 1940s Tarski [13] gave a general algorithm for quantifier elimination in expressions of the above-mentioned type. However, this algorithm had a very high complexity and is impractical for most problems. In the mid-1970s Collins [14] presented a method, so-called cylindrical algebraic decomposition, that decreased the algorithmic complexity, and since then the algorithmic development has proceeded. Recently, more effective algorithms have made it possible to solve nontrivial quantifier elimination problems on workstations [15]. Introductions to quantifier elimination are given in, for example References 16–18 and an extensive bibliography over the early papers of the subject can be found in Reference 19.

In control theory, one of the first attempts to use quantifier elimination techniques was made by Anderson *et al.* [20] and recently, a few papers treating control related problems have been published [18, 21–25].

5.5 Constructive methods in real algebra

Solutions of systems (or Boolean combinations) of polynomial equations and inequalities correspond to subsets in $\mathbb{R}^n$ which are called *semi-algebraic sets*. Alternatively a set $S \subset \mathbb{R}^n$ is semi-algebraic if it can be constructed by a finite number of applications of union, intersection and complementation operations on sets of the form

$$\{x \in \mathbb{R}^n | f(x) \ge 0\}$$

where f is a polynomial with real coefficients. We use calligraphic letters such as $\mathcal{S}$ to denote semi-algebraic sets and $\mathcal{S}(x)$ to denote a defining formula.

There are several computational methods in real algebra such as real root isolation, algebraic number arithmetic, sign evaluation and Sturm sequences. Here we give a brief overview of two computational procedures and show how they can be used in control applications. These are cylindrical algebraic decomposition and quantifier elimination.

5.5.1 Cylindrical algebraic decompostion

Given a set $\mathcal{P}$ of multivariate polynomials a cylindrical algebraic decomposition (CAD) is a special partition of $\mathbb{R}^n$ into components, called *cells*, over which the polynomials have constant signs. The algorithm for computing a CAD also provides a point, called *sample point*, in each cell that can be used to determine the sign of the polynomials in the cell.

The algorithm can be divided into three phases: projection, base and extension. The projection phase consists of a number of steps, each in which new sets of polynomials is constructed. The zero set of the resulting polynomials of each step is the projection of 'significant' points of the zero set of the preceding polynomials, i.e. self-crossings, isolated points, vertical tangent points, etc. In each step the number of variables is decreased by one and hence the projection phase consists of $n-1$ steps. We denote the set of polynomials generated in step k of the projection phase for $\mathrm{proj}^k(\mathcal{P})$.

The base phase consists of isolation of real roots, $\alpha_i \in \mathbb{R}$, of the monovariate polynomials in $\mathrm{proj}^{n-1}(\mathcal{P})$. Each root and one point in each interval between two roots are chosen as sample points of a decomposition of $\mathbb{R}$. The cells of this decomposition consist of the real roots and the open intervals between them (the two infinite intervals are also cells). Hence, there will be $2m+1$ cells, where m is the number of distinct real roots computed in the base phase.

The purpose of the extension phase is to construct sample points of all cells of the CAD of $\mathbb{R}^n$. The extension phase consists of $n-1$ steps. In the first step a sample point, $(\alpha_i, \beta_j) \in \mathbb{R}^2$, of each cell of the 'cylinder' over the cells of the base phase is constructed. This is done by specialising the polynomials in $\mathrm{proj}^{n-2}(\mathcal{P})$ over each sample point from the base phase, i.e. $x_1 = \alpha_i$. Then the real root isolation can be performed as in the base phase and we get a number of sample points β_j for each α_i. These pairs of α_i and β_j are sample points of a decomposition of $\mathbb{R}^2$. Specialisation of a polynomial corresponds to evaluating it on a straight line over the sample point $x_1 = \alpha_i$. The real root isolation then gives the intersections of this line with the zero set of the polynomial. In the following extension steps the above procedure is repeated until we have sample points in all cells of the CAD of $\mathbb{R}^n$.

The sample points, together with all projection polynomials $\mathrm{proj}^k(\mathcal{P})$, $k = 0, \ldots, n-1$, can be used to give a defining formula for each cell of the decomposition (the sign of the polynomials over a sample point of the cell gives an inequality/equality description).

In the projection phase algebraic tools such as resultants and subresultants are used, which can be seen as generalisations of Sturm sequences for univariate polynomials. The base and extension phase utilise real root isolation and algebraic number arithmetics.

Observe that to determine if a system of polynomial inequalities has a solution, it is enough to determine the signs of $\mathcal{P}$ at a sample point of each cell since the polynomial in $\mathcal{P}$ has a constant sign for every point in each cell by construction.

Example 7

Consider the 3D-sphere of radius 1 centred at $(2, 2, 2)$ which is given by the set of real zeros of

$$f = (x_2 - 2)^2 + (x_2 - 2)^2 + (x_3 - 2)^2 - 1 \tag{5.37}$$

In this case the projection polynomial becomes

$$\text{proj}(f) = \{(x_1 - 2)^2 + (x_2 - 2)^2 - 1, (x_1 - 2)^2 + (x_2 - 2)^2 + 3\}$$

$$\text{proj}^2(f) = \{(x_1 - 1)(x_1 - 3)(x_1^2 - 4x_1 + 19), x_1^4 - 8x_1^3 + 30x_1^2 - 56 + 113,$$

$$x_1^2 - 4x_1 + 7, x_1^2 - 4x_1 + 11, (x_1 - 1)(x_1 - 3)\}$$

The only real zeros of the polynomials in $\text{proj}(f)$ correspond to the circle in the $x_1 x_2$-plane in Figure 5.1 and the only real zeros of the polynomials in $\text{proj}^2(f)$ are 1 and 3, i.e. the grey dots on the x_1-axis in Figure 5.1. The resulting CAD of $\mathbb{R}$ is $(-\infty, 1), 1, (1, 3), 3, (3, \infty)$. This corresponds to a partition of the x_1-axis into five cells and sample points for each cell (see the top left plot in Figure 5.2).

Specialising the polynomials of $\text{proj}(f)$ over each sample point gives ten univariate polynomials in x_2. For each specialised polynomial the same method as in the base phase can be used to construct sample points of the cells of $\mathbb{R}^2$. We may choose the following sample points:

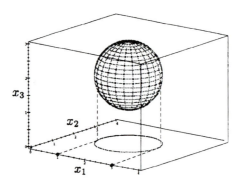

Figure 5.1   The set of real zeros of f and the projections of its 'significant' points (vertical tangents in this case)

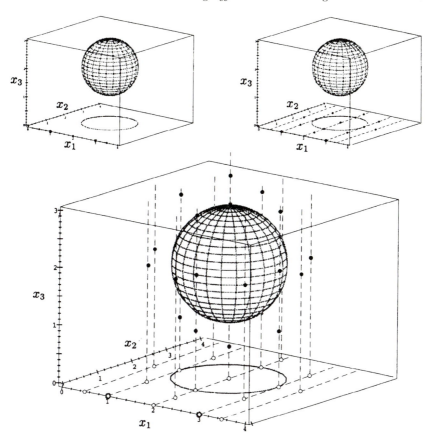

Figure 5.2   The sample points of each step of the extension phase

$$\left(\tfrac{1}{2}, 2\right),$$
$$(1, 1), (1, 2), (1, 3),$$
$$\left(2, \tfrac{1}{2}\right), (2, 1), (2, 2), (2, 3), \left(2, \tfrac{7}{2}\right),$$
$$(3, 1), (3, 2), (3, 3),$$
$$\left(\tfrac{7}{2}, 2\right)$$

(see the top right plot in Figure 5.2).

In the same way we specialise f over each sample point of $\mathbb{R}^2$ and construct the 25 sample points of the corresponding CAD of $\mathbb{R}^3$.

For intervals the sample points can always be chosen to be rational numbers but in general the sample points are real algebraic numbers, i.e. they are roots of algebraic equations with integer coefficients. Hence, an implementation of a CAD algorithm has to handle arithmetic operations with algebraic numbers.

In this example we chose very simple polynomials to avoid this kind of computation.

5.5.2 Quantifier elimination

By a quantifier-free Boolean formula we mean a formula consisting of polynomial equations $(f_i(x) = 0)$ and inequalities $(f_i(x) < 0)$ combined using the Boolean operators $\wedge$ (and), $\vee$ (or), and $\rightarrow$ (implies). A *formula* is an expression in the variables $x = (x_1, \ldots, x_n)$ of the following type:

$$Q_1 x_1 \ldots Q_s x_s \mathcal{F}(f_1(x), \ldots, f_r(x)) \tag{5.38}$$

where Q_i is one of the quantifiers $\forall$ (for all) or $\exists$ (there exists). Furthermore, $\mathcal{F}(f_1(x), \ldots, f_r(x))$ is a quantifier-free Boolean formula. In addition, there is always an implicit assumption that all variables are considered to be real. Eqn. 5.38 is said to be in prenex form, which means that all quantifiers appear in the beginning of the formula. There are rewriting rules for transforming a formula into prenex form (see Reference 16). In applications it is often more natural to formulate problems using a nonprenex form.

It is a nontrivial fact that to every formula including quantifiers there is an equivalent formula without any quantified variables. This was shown in the late 1940s by A. Tarski [13], who also presented an algorithm for quantifier elimination. The equivalence means that the use of quantifiers does not enlarge the set of geometrical objects that can be described. However, quantifiers are very convenient for formulating problems in applications.

Example 8

Consider the following formula:

$$\exists y \, \forall x [x^2 + y^2 > 1 \wedge (x - 1)^2 + (y - 1)^2 > 1]$$

which can be read as:

> There exists a real number y such that for all real numbers x we have that $x^2 + y^2 > 1$ and $(x - 1)^2 + (y - 1)^2 > 1$.

This particular formula is trivially seen to be TRUE since both inequalities are satisfied for all real x if $y = 3$.

The above problem is called a *decision problem* since all variables are bounded by quantifiers and the problem consists of deciding if the formula is TRUE or FALSE. If there are some *free* or unquantified variables as well one has a general *quantifier elimination* problem.

Example 9

The following is a general quanitifier elimination problem:

$$\forall x [x > 0 \rightarrow x^2 + ax + b > 0]$$

which can be read as:

For all real numbers x, if x > 0 then x² + ax + b > 0.

Performing quantifier elimination we obtain the following equivalent expression:

$$4b - a^2 > 0 \vee [a \geq 0 \wedge b \geq 0]$$

which are the conditions on the coefficients of a second-order polynomials such that it only takes positive values for $x > 0$.

To perform quantifier elimination in the nontrivial examples below we have used the software QEPCAD, presented in [15]. Quantifier elimination can be performed by sorting and simplifying polynomial inequalities defined by the projection polynomials from a cylindrical algebraic decomposition algorithm.

After performing quantifier elimination one obtains a Boolean combination of equations and inequalities. Solutions to such systems can be computed using CAD, which in many cases is a part of the quantifier elimination algorithm. Hence, a number of feasible solutions is often produced directly.

5.5.3 Applications

Many applications of quantifier elimination can be found both in linear and nonlinear control. However, for linear control system analysis and design there are already a huge amount of constructive methods based on linear algebra and matrix computations. Because of computational complexity of the quantifier elimination algorithms these methods should be preferred when applicable. Here we consider computations of stationarisable states, stability of these states and the ability to follow reference trajectories under actuator and state constraints for nonlinear systems.

5.5.3.1 Stationarisable points

Consider a dynamic system described by a nonlinear differential equation written in state space form:

$$\dot{x} = f(x, u)$$
$$y = h(x)$$

(5.39)

where $x \in \mathbb{R}^n$, $u \in \mathbb{R}^m$, $y \in \mathbb{R}^p$ and each component of f and h is a real polynomial, $f_i \in \mathbb{R}[x, u]$, $h_j \in \mathbb{R}[x]$. The x, u and y vectors will be referred to as the *state, control* and *output* of the system, respectively.

Suppose also that the system variables have to obey some additional constraints

$$x \in \mathcal{X} \quad \text{and} \quad u \in \mathcal{U}$$

(5.40)

where $\mathcal{X}$ and $\mathcal{U}$ are semi-algebraic sets which define the constraints on the state and control variables. We call $x \in \mathcal{X}$ the *admissible* states and $u \in \mathcal{U}$ the *admissible* controls.

Restrictions on amplitudes of control signals are common in real applications but hard to take into account in many classical design methods other than by simulation studies. A variety of constraints can be represented by semi-algebraic sets, e.g. amplitudes and direction constraints but also discontinuous and nonsmooth static input-output maps such as relays and saturations.

Example 10

Consider the following saturation function:

$$u_2 = \begin{cases} -1, & u_1 \leq -1 \\ u_1, & -1 < u_1 < 1 \\ 1, & u_1 \geq 1 \end{cases}$$

The set of admissible controls can then be described by the following semi-algebraic set:

$$\mathcal{U} = \{u \in \mathbb{R}^2 | u_1 \leq -1 \rightarrow u_2 = -1 \wedge$$
$$-1 < u_1 < 1 \rightarrow u_2 = u_1 \wedge u_1 \geq 1 \rightarrow u_2 = 1\}$$

Equilibrium or stationary points of dynamic system play an important role in control system design. A point, x_0, is stationary if $f(x_0, u_0) = 0$, where $x_0 \in \mathcal{X}$ and $u_0 \in \mathcal{U}$. For the class of dynamical systems considered here, this set of *stationarisable* states is a semi-algebraic set.

Definition 1: The stationarisable states of system 5.39 subject to constraints 5.40 is the set of states satisfying the formula

$$S(x) \overset{\Delta}{=} \exists u \big[f(x, u) = 0 \wedge \mathcal{X}(x) \wedge \mathcal{U}(u) \big] \tag{5.41}$$

The construction of a 'closed form' of the set of stationarisable states, i.e. an expression not including u, is a quantifier elimination problem and hence this set is semi-algebraic. Computation of extreme values of the states or outputs over this set can also be stated as quantifier elimination problems.

Example 11

Consider the system in Example 6 again and eqn. 5.36 describing the equilibria or stationarisable set. Performing quantifier elimination, we obtain

$$\big[x_2^4 - x_1^3 x_2^2 - x_1 x_2^2 - x_1^3 = 0 \wedge [x_2 + 2x_1 \leq 0 \vee x_2 - 2x_1 \geq 0] \big]$$

In this case the stationarisable set is easy to visualise (see Figure 5.3).

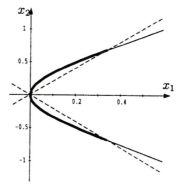

Figure 5.3 The stationarisable set (bold curve) of system 5.34 subjected to the control constraints $|u| \le \frac{1}{2}$

5.5.3.2 Stability

The subset of the stationarisable states that are asymptotically stable can be computed by augmenting the formula in Definition 1 by an additional number of inequalities. Let $f_x(x, u)$ be the Jacobian matrix corresponding to system 5.39. A stationary point is stable if the linearisation (given by the Jacobian) has all its eigenvalues in the open left half-plane.

Now, an asymptotic stability condition can be given by polynomial inequality constraints on the coefficients of the characteristic polynomial of $f_x(x, u)$. There are several such conditions due to Hurwitz, Routh or Lienard-Chipart [26, 27] where the Lienard-Chipart condition seems to be the least complex in terms of polynomial degrees. We do not state these conditions explicitly here, but denote this set of inequalities by $\mathrm{Re}(\mathrm{eig}(f_x(x, u))) < 0$.

Definition 2: The stationarisable states of system 5.39 subject to the constraints 5.40 that are asymptotically stable are given by the formula

$$\mathcal{AS}(x) \overset{\Delta}{=} \exists u \big[f(x, u) = 0 \wedge \mathcal{X}(x) \wedge \mathcal{U}(u) \wedge \mathrm{Re}(\mathrm{eig}(f_x(x, u))) < 0 \big] \qquad (5.42)$$

Example 12

Consider the following system:

$$\dot{x}_1 = -x_1^3 + x_2$$
$$\dot{x}_2 = -x_1^2 - x_2 - x_2^3 + u \qquad (5.43)$$

subject to the constraints $u^2 \leq 1$. We obtain the Jacobian

$$f_x(x, u) = \begin{bmatrix} -3x_1^2 & 1 \\ -2x_1 & -1 - 3x_2^2 \end{bmatrix}$$

and its corresponding characteristic polynomial

$$\lambda^2 + (3x_1^2 + 1 + 3x_2^2)\lambda + 3x_1^2 + 9x_1^2 x_2^3 + 2x_1$$

The zeros of a polynomial of degree two have negative real parts if and only if the coefficients all have the same sign. Hence, the inequalities $\mathrm{Re}(\mathrm{eig}(f_x(x, u))) < 0$ become

$$3x_1^2 + 1 + 3x_2^2 > 0, \qquad 3x_1^2 + 9x_1^2 x_2^2 + 2x_1 > 0$$

where the first inequality is trivially satisfied for all real x_1 and x_2. The asymptotically stable stationarisable points of system 5.43 are given by eqn. 5.42:

$$\mathcal{AS}(x) = \exists u \left[-x_1^3 + x_2 = 0 \wedge -x_1^2 - x_2 - x_2^3 + u = 0 \wedge \right.$$
$$\left. u^2 \leq 1 \wedge 2x_1 + 3x_1^2 + 9x_1^2 x_2^2 > 0 \right]$$

which after quantifier elimination becomes

$$\mathcal{AS}(x) = \left[-x_1^3 + x_2 = 0 \wedge [x_1 > 0 \vee 3x_1 + 9x_1 x_2^2 + 2 < 0] \wedge \right.$$
$$\left. -x_1^2 - x_2 - x_2^3 - 1 \leq 0 \wedge x_1^2 + x_2 + x_2^3 - 1 \leq 0 \right] \qquad (5.44)$$

(see Figure 5.4). Any point on the cubic which is not in the dark grey region is a stationarisable state. The part of the cubic in the light grey region corresponds to stationary points which are not asymptotically stable.

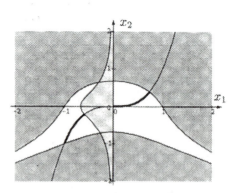

Figure 5.4   *The set of states of system (5.43) which is stationarisable and asymptotically stable (bold curve)*

5.5.3.3 Curve following

Consider system 5.39 subject to some semi-algebraic control constraints, $u \in \mathcal{U}$. Let Γ be a rationally parameterised curve in the output space $\mathbb{R}^p$, i.e.

$$\Gamma = \{y \in \mathbb{R}^p | y = g(s), g : \mathbb{R} \to \mathbb{R}^p, s \in [\alpha, \beta]\}$$

where the orientation of Γ is defined by increasing values of s.

To steer the system along the curve in the state space there has to be an admissible control u at each point on the curve such that the solution trajectory tangent vector $f(x, u)$ points in the same direction as a forward pointing tangent vector of the curve, i.e.

$$f(g(s), u) = \lambda \frac{\mathrm{d}}{\mathrm{d}s} g(s), \quad \lambda > 0, \quad \forall s \in [\alpha, \beta] \tag{5.45}$$

The more general case when a curve in the output space is given is treated in the following theorem. Note that if the number of outputs is lower than the number of states, a curve in the output space corresponds to a manifold of dimension > 1 in the state space.

Theorem 1 *There exists an admissible control $u \in \mathcal{U}$ such that the output y of system 5.39 follows the curve Γ if and only if the formula*

$$(\forall s \in [\alpha, \beta])(\exists x)(\exists u \in \mathcal{U})(\exists \lambda > 0)\left[h_x(x)f(x, u) = \lambda \frac{\mathrm{d}}{\mathrm{d}s} g(s) \wedge g(s) = h(x) \right] \tag{5.46}$$

is TRUE.

Here $h_x(x)$ denotes the Jacobian matrix $h(x)$ w.r.t. x.

Proof The curve Γ in the output space can be followed if and only if the output trajectory tangent $\dot{y}(t)$ can be chosen parallel to the tangent vector of Γ, at each point on Γ, by an admissible choice of u. Now

$$\dot{y}(t) = h_x(x)\dot{x} = h_x(x)f(x, u)$$

and a tangent of Γ is given by $(\mathrm{d}/\mathrm{d}s)g(s)$. Hence, the parallel condition becomes

$$h_x(x)f(x, u) = \lambda \frac{\mathrm{d}}{\mathrm{d}s} g(s)$$

for some $\lambda > 0$, $u \in \mathcal{U}$, and the theorem follows.

Now, quantifier elimination can be used to eliminate all quantified variables. In eqn. 5.46 all variables are quantified and the quantifier elimination algorithm gives TRUE or FALSE as a result, i.e. either there is an admissible control that steers the system along the curve or not. If the equations that specify the dynamic system or the curve contain additional parameters, then the quantifier elimination gives necessary and sufficient conditions on these *free* parameters such that an admissible control law exists. Solutions to these systems of polynomial equations and inequalities can, for example, be computed by cylindrical algebraic decomposition.

Example 13

Consider the following system:

$$\dot{x}_1 = ax_2 + \tfrac{1}{2}x_1 x_2$$

$$\dot{x}_2 = -x_1 - x_2 + (1 + x_1^2)u \qquad (5.47)$$

$$y = (x_1, x_2)^T$$

where $-2 \leq u \leq 2$ and Γ is the unit circle. Here we have one free design parameter a. The unit circle can be rationally parametrised according to

$$\Gamma = \left\{ x \in \mathbb{R}^2 \,|\, x_1 = \frac{s_2 - 1}{s^2 + 1}, \, x_2 = \frac{-2s}{s^2 + 1}, \, s \in \mathbb{R} \right\}$$

In this example the output map is the identity. Hence we can eliminate x directly by substitution. Clearing denominators the parallel condition in eqn. 5.46 becomes

$$s + 2bs - s^3 + 2bs^3 = 4\lambda s \wedge 1 + 2s + 2s^3 - s^4 + 2u + 2s^4 u = 2\lambda(s^2 - 1)$$

Next follows a few lines of Mathematica code:

```
In[1]:= EliminateQuantifiers[
        ∀s ∃u ∃λ
        (s + 2bs − s³ + 2bs³ = 4λs &&
        1 + 2s + 2s³ − s⁴ + 2u + 2s⁴u = 2λ(s² − 1)
        && − 2 ≤ u && u ≤ 2)]
Out[1]= − 1 + 2a ≥ 0 && − 4409 + 2548a + 790a² −
        452a³ − 73a⁴ + 24a⁵ + 4a⁶ ≤ 0
In[2]:= N[InequalitySolve[%, a]]
Out[2]= 0.5 ≤ a ≤ 1.8891
```

Here the function `EliminateQuantifiers` calls the external program QEPCAD and returns the output from this program. The second line of input is a built-in Mathematica function for simplification of Boolean combinations of univariate inequalities followed by a numerical evaluation of the result.

Hence, there is an admissible control such that the states of system 5.47 follow the unit circle if and only if $\tfrac{1}{2} \leq a \leq \alpha$, where $\alpha \simeq 1.8891$ is given by a real zero of a sixth-order polynomial.

5.5.4 *Software*

At the moment there is no complete implementation of quantifier elimination in any of the major computer algebra systems. However, implementations of parts of the algorithm are available as standard add-on packages in Mathematica.

Algorithms for differential algebra are being implemented by Dongming Wang of RISC-LINZ (see also [28]).

5.6 Conclusions

We have presented some examples from control theory where symbolic methods from differential and real algebra are used. The strength of the methods lies in their ability to handle general problems without any special structure (except for the restriction to polynomials). The main obstacle to widespread use is that some of the calculations have high complexity. This is to some degree offset by the development of algorithms and computers. There are also many interesting possibilities in the combination of numerical and symbolic techniques.

5.7 References

1 RUBEL, L. A. and SINGER, M. F.: 'A differentially algebraic elimination theorem with applications to analog computability in the calculus of variations'. *Proc. Amer. Math. Soc.*, vol. 94, (Springer, 1985), pp. 653–658

2 LINDSKOG, P.: 'Methods, algorithms and tools for system identification based on prior knowledge'. PhD thesis 436, Department of Electrical Engineering, Linkping University, Sweden, May 1996

3 RITT, J. F.: 'Differential algebra' (American Mathematical Society, Providence, RI, 1950)

4 KOLCHIN, E. R.: 'Differential algebra and algebraic groups' (Academic Press, New York, 1973)

5 FLIESS, M. and GLAD, S. T: 'An algebraic approach to linear and nonlinear control', in TRENGELMAN, H. L. and WILLEMS, J. C. (Eds.). 'Essays on control: Perspectives in the theory and its applications', (Birkhäuser, 1993), pp. 2233–267

6 SEIDENBERG, A.: 'An elimination theory for differential algebra', in WOLF, F., HODGES, J. L., and SEIDENBERG, A. (Eds), *University of California Publications in Mathematics: New Series*, (University of California Press, Berkeley and Los Angeles, CA, 1956), pp. 31–66

7 FLIESS, M.: 'Generalized controller canonical forms for linear and nonlinear dynamics', *IEEE Trans. Autom. Control*, 1990, **AC-35**, pp. 994–1001.

8 GLAD, S. T.: 'Nonlinear state space and input output descriptions using differential polynomials', in DESCUSSE, J., FLIESS, M., ISIDORI, A., and LEBORGNE, D., (Eds.), 'New trends in nonlinear control theory', *Lect. Notes Control Inform. Sci.*, *122*, (Springer, 1989), pp. 182–189

9 DIOP, S.: 'A state elimination procedure for nonlinear systems', *in* DESCUSSE, J. FLIESS, M., ISIDORI, A., and LEBORGNE, D., (Eds.), 'New trends in nonlinear control theory', *Lect. Notes Control Inform. Sci.*, *122*, (Springer, 1989), pp. 190–198

10 DIOP, S.: 'Elimination in control theory', *Math. Control Signals Syst.*, **4**, 17–32 (1991)

11 LJUNG, L. and GLAD, S. T.: 'On global identifiability of arbitrary model parameterizations', *Automatica*, 1994, **30**(2), pp. 265–276

12 CHAPELL, M., GODFREY, K., and VAJDA, S.: 'Global identifiability of non-linear systems with specified inputs: A comparison of methods', *Math. Biosci.*, **102**, 41–73 (1990)

13 TARSKI, A.: 'A decision method for elementary algebra and geometry' (University of California Press, 2nd edn., 1948)

14 COLLINS, G. E.: 'Quantifier elimination for real closed fields by cylindrical algebraic decompositioin'. 2nd GI Conf. *Automata Theory and Formal Languages* (Springer, 1975)

15 COLLINS, G. E., and HONG, H.: 'Partial cylindrical algebraic decomposition for quantifier elimination', *J. Symbol. Comput.*, 1991, **12**, (3), pp. 299–328

16 MISHRA, B.: 'Algorithmic algebra, 'Texts and monographs in computer science' (Springer-Verlag, 1993)

17 DAVENPORT, J. H., SIRET, Y., and TOURNIER, E.: 'Computer algebra Systems and algorithms for algebraic computation' (Academic Press, 1988)

18 JIRSTRAND, M.: 'Constructive methods for inequality constraints in control,' PhD thesis no. 527, Department of Electrical Engineering, Linköping University, Sweden, May 1998

19 ARNON, D. S.: 'A bibliography of quantifier elimination for real closed fields', *J. Symbol. Comput.*, 1988, 5(1–2), pp. 267–274

20 ANDERSON B., BOSE, N., and JURY, E.: 'Output feedback stabilization and related problems–solution via decision methods', *IEEE Trans. Autom. Control.*, 1975, **20**, pp. 53–65

21 DORATO, P., YANG, W., and ABDALLAH, C.: 'Robust multi-objective feedback design by quantifier elimination', *J. Symbol. Comput.*, 1997, **24**, 153–159

22 GLAD, S. T.: 'Algebraic approach to bang-bang control'. *Proc. ECC'95*, 1995.

23 JIRSTRAND, M. and GLAD, S. T.: 'Computational questions of equilibrium calculation with application to nonlinear aircraft dynamics.' *Proc. MTNS'96*, St. Louis, USA, 1996

24 NESIC, D., and MAREELS, I. M. Y.: 'Deciding dead beat controllability using QEP-CAD'. *Proc. MTNS'96*, St. Louis, USA, 1996

25 JIRSTRAND, M.: 'Nonlinear control system design by quantifier elimination', *J. Symbol. Comput.*, 1997, **24**(2), pp. 137–152

26 PARKS, P. C. and HAHN, V.: 'Stability theory' (Prentice Hall, 1993)

27 GANTMACHER, F. R.: 'Matrix theory', Vol. II (Chelsea, 1971)

28 WANG, D. M.: 'Characteristic sets and zero structure of polynomial sets'. Technical report, RISC-LINZ, 1989, Lecture Notes

Chapter 6

Approximate algebraic computations of algebraic invariants

N. Karcanias and M. Mitrouli

6.1 Introduction

Algebraic and geometric invariants are instrumental in describing system properties and characterising the solvability of control problems. This chapter deals with the computation of certain types and values of invariants, the presence of which on a family of linear models is nongeneric. The computation of such invariants on models with numerical inaccuracies requires special methods, which may lead to approximate meaningful results to the computation problem. A classification of the algebraic computations according to their behaviour on numerically uncertain models is given, and then two of the key problems underlying the computation of a number of system invariants are considered; these are the problems of approximate computations of the greatest common divisor (GCD) and least common multiple (LCM) of polynomials. Some fundamental issues in the transformation of the GCD and LCM algebraic computations in an analytic, 'approximate' sense are considered, and methodologies yielding approximate solutions to GCD, LCM problems are examined. The results are illustrated in terms of a number of examples.

The theory of algebraic and geometric invariants in linear systems is instrumental in describing system properties and it is linked to solvability of fundamental control theory problems [1–3]. These invariants are defined on rational, polynomial matrices and matrix pencils under different transformation groups (co-ordinate, compensation, feedback type) and their computation relies on algebraic algorithms. The use of symbolic tools may thus be considered as natural in developing algorithms for their computation. The underlying assumption behind the use of symbolic computations is that the mathematical model has fixed parameters. The distinguishing feature of all engineering applications is that mathematical models always have numerical inaccuracies, and this has a significant effect on the selection and development of the computational tools. The existence of certain types and/or values of invariants and system properties may be either generic or nongeneric [1] on a family of linear models. Computing or evaluating nongeneric types or values of invariant, and thus associated system properties on models with numerical inaccuracies, is crucial for applications. For such cases, symbolic tools fail, since they 'almost always' lead to a generic solution, which does not represent the 'approximate presence' of the value-property on the set of models under consideration. A methodology that allows the computation of nongeneric types or values of invariants in a meaningful way is of numerical nature, and this is what we refer to here as nongeneric computations. The subject of this chapter is the discussion of the fundamentals of this important area of algebraic computations and presentation of an overview of some of the key problems.

The development of a methodology for robust computation of nongeneric algebraic invariants, or nongeneric values of generic ones, has as prerequisites: (*a*) the development of a numerical linear algebra characterisation of the invariants, which may thus allow the measurement of degree of presence of the property on every point of the parameter set; (*b*) the development of special numerical tools, which avoid the introduction of additional errors; (*c*) the formulation of appropriate criteria which will allow the termination of algorithms at certain steps and the definition of meaningful approximate solutions to the algebraic computation problem. It is clear that the formulation of the algebraic problem as an equivalent numerical linear algebra problem is essential in transforming concepts of an algebraic nature to equivalent concepts of analytic character; this property is referred to as *numerical reducibility* (NR) of the algebraic computation and it depends on the nature of the particular invariant. The last two prerequisities are referred to in brief as numerical tools for nongeneric computations (NGC).

The work described here goes back to the attempt to introduce the notion of almost zero of a set of polynomials [4] and study the properties of such zeros from the feedback viewpoint. This work was subsequently developed to a methodology for computing the approximate GCD of polynomials using numerical linear algebra methods, such as the ERES [5, 6] and matrix pencil methods [7, 8]. The numerical methods in References 5–8 have used a variety of procedures for NGC, and a richer set of such tools is given in Reference 9. The current chapter provides an overview of issues and a summary of results on some crucial problems. The classification of types of computational problems on numerically uncertain linear models is first considered and then a set of fundamental tools for NGC are examined. The notion of almost zero is then reviewed and two methods for approximate greatest common divisor (GCD) evaluation are examined. Some recent results on the approximate factorisation of polynomials and evaluation of the approximate least common multiple (LCM) of a set of polynomials are considered. The methodologies are illustrated in terms of examples. The study of GCD and LCM of a set of polynomials is central in many algebraic synthesis problems and thus the derivation of methodologies for their approximate definition is crucial for the development of the field of approximate algebraic computations.

Throughout the chapter, $\mathcal{R}^{m \times n}$ denotes the set of all $m \times n$ real matrices and $\mathcal{R}[s]$ denotes the ring of real polynomials. If $A \in \mathcal{R}^{m \times n}$, then $\rho(A)$ denotes its rank, $n_r(A)$ denotes it right nullity and $\mathcal{N}_r(A), \mathcal{N}_l(A)$ denote its right, left nullspace, respectively. Capital letters denote matrices and small underlined letters denote vectors. $C_r(A)$ denotes the rth-order compound matrix of A [10]. The symbol $\triangleq$ means equal by definition, whereas the symbol $\partial[a(s)]$ denotes the degree of a given polynomial $a(s)$.

6.2 Models with numerical inaccuracies and classification of algebraic computation problems

The development of robust algebraic computation procedures for engineering type models always has to take into account that the models have a certain accuracy and that it is meaningless to continue computations beyond the accuracy of the original data set. In fact, engineering computations are defined not on a single model of a system S, but on a ball of system models $\Sigma(S_0, r(\varepsilon))$, where S_0 is a nominal system and $r(e)$ is some radius defined by the data error order ε. The result of computations has thus to be representative for the family $\Sigma(S_0, r(\varepsilon))$ and not just the particular element of this family. From this viewpoint, symbolic computations carried out on an element of the $\Sigma(S_0, r(\varepsilon))$ family may lead to results which do reveal the desired properties of the family. Numerical computations have to stop when we reach the original data accuracy, and an approximate solution to the computational task has to be given. We consider algebraic computation problems which are numerically reducible (i.e. have a numerical linear algebra equivalent formulation). The classification of such computational tasks according to their behaviour under numerical errors is important, since it reveals those requiring special attention; this classification is undertaken here and reveals an important class—that of nongeneric computations which form the main subject of this chapter. To motivate the basic ideas we consider first a simple example.

Example 1 [4]

Consider the polynomials $t_1(s) = s + 1.1$ and $t_2(s) = s(s + 1)$. Clearly, the polynomials are coprime and any symbolic procedure for the computation of GCD will lead to this result. It is worth pointing out that the two polynomials have roots $-1, -1.1$, which are very close to each other and thus the notion of an 'approximate common zero' emerges. For such a notion to make sense, there must be a continuity with some important properties of the exact zero. A key property of the common zero z of two polynomials $t_1(s), t_2(s)$ is that any combination $k_1 t_1(s) + k_2 t_2(s)$ for all $k_1, k_2 \in \mathcal{R}$ will also have z as a common zero. Thus exact zeros correspond to fixed zeros of the polynomial combinants of $(t_1(s), t_2(s))$. For our example the zeros of $k_1(s + 1.1) + k_2s(s + 1)$ are defined by the root locus diagram of the single-input single-output system with transfer function

$$g(s) = \frac{k(s + 1.1)}{s(s + 1)}, \qquad k = k_1/k_2$$

The corresponding root locus is indicated in Figure 6.1, and this reveals the presence of a circle that contains at least one of the zeros of the combinant for all values of k. The fixed point interpretation of the exact zero now becomes a circle with the same property, as far as distribution of zeros of the corresponding combinants; the size of disc depends on the proximity of the zeros of the two polynomials. The continuity of properties between the exact and the approximate case (root distribution of combinants) allows us to say that an 'almost zero' exists in the neighbourhood of -1. This notion of 'almost zero' is not an algebraic property any more, but an analytical property, and it depends on the distance between root sets of polynomials. *The above example demonstrates that the fundamental notion of common divisor may be extended to an 'approximate' sense. Although the existence of GCD of polynomials is a nongeneric property [1, 11], 'almost common divisors' may be introduced under certain conditions. This example demonstrates that the notions of genericity and nongenericity are fundamental in the classification of different types of algebraic computation.*

To make the idea of genericity precise [1], we borrow some terminology from algebraic geometry. Consider polynomials $\phi(\lambda_1, \ldots, \lambda_n)$ with coefficients in $\mathcal{R}$. A variety $V \subset \mathcal{R}^N$ is defined to be the locus of common zeros of a finite number of polynomials $\phi_1, \ldots, \phi_k$:

$$V = \{P \in \mathcal{R}^N : \phi_i(P_1, \ldots, P_N) = 0, i \in k\}$$

For example, one can prove that the set of all (A, B, C, D) of fixed dimensions modulo co-ordinate state transformations is a variety. A property Π on V is

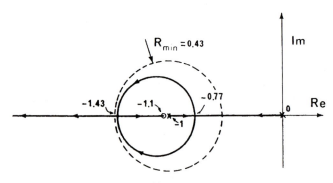

Figure 6.1   Root locus associated with g(s)

merely a function $\Pi\colon V \to \{0, 1\}$, where $\Pi(P) = 1$(or 0) means Π holds (or fails) at P. Let V be a proper variety; then we shall say that Π is *generic relative* to V provided $\Pi(P) = 0$ only for points $P \in V' \subset V$, where V' is a proper subvariety of V, and that Π is generic provided such a V' exists. If Π is generic, we sometimes write $\Pi = 1(g)$. As V' is a locus of zeros of polynomials in V, the subset of V such that the property is not true is a negligible set (measure zero). On the basis of the above we are led to the following classification of algebraic computations.

Definition 1 Numerical computations dealing with the derivation of an approximate value of a property, function, which is nongeneric on a given model set, will be called *nongeneric computations* (NGC). If the value of a function always exists on every element of the model set and depends continuously on the model parameters, then the computations leading to the determination of such values will be called *normal numerical* (NNC). Computational procedures aimed at defining the generic value of a property function on a given model set (if such values exist), will be called *generic* (GC). □

On a set of polynomials with coefficients taking values from a certain parameter set, the existence of GCD in nongeneric numerical procedures that aim to produce an approximate nontrivial value by exploring the numerical properties of the parameter set are typical examples of NG computations, and two approximate GCD procedures will be considered subsequently. NG computations refer to both continuous and discrete type system invariants. On the other hand, the eigenvalues of a square matrix, or the zeros of a square polynomial matrix, are always defined on any model set and their numerical values continuously depend on the numerical values of the parameter set; such cases are examples of NNC computations, covered well in numerical analysis books [12, 13] and not considered here. For unstructured model sets the computation of the generic value of discrete invariants is an issue that is usually simple and follows by the dimensionality of the matrices involved and genericity arguments [14]. For model sets characterised by some underlying graph structure, either explicitly or implicitly, the computation of the generic values can benefit by exploiting the genericity argument. In the case of structured state space models with well identified graphs, computations based on structured (Boolean) matrices may frequently benefit from numerical procedures and genericity arguments. For large-scale transfer functions corresponding to composite systems and characterised by model uncertainty, the computation of discrete invariants such as McMillan degree, and the order of infinite zeros, are problems within the class of generic computations; these latter problems have been recently considered in Reference 15 and are referred to as problems of structural identification. In this Chapter we consider two fundamental problems of nongeneric computations, that of almost zeros, almost GCD and the associated problem of the least common multiple; such problems are central in the development of algebraic control synthesis methodologies. We consider first some fundamental problems that arise in the development of algorithms for non-generic computations.

6.3 Numerical tools for approximate computations

The key algebraic computation problems we consider are those of GCD, LCM and problems of factorisation of polynomials. The development of the numerical algorithms requires some special tools, and these are considered below.

6.3.1 *Issues related to numerical dependence and independence of vectors*

Let $A = \{\underline{r}_1, \underline{r}_2, \ldots, \underline{r}_m\}$ be a set of m given vectors $\underline{r}_i \in \mathcal{R}^n$, $i = 1, 2, \ldots, m$. This set can be expressed in terms of a matrix $A = [\underline{r}_1, \underline{r}_2, \ldots, \underline{r}_m]^t \in \mathcal{R}^{m \times n}$. The notion of dependence or independence of the set A is tested in terms of the rank $(\rho(A))$ of the associated matrix. The numerical version of this property is considered next.

6.3.1.1 *The SVD criterion*

If $\sigma_1 \geq \sigma_2 \geq \cdots \geq \sigma_r, r = \min(m, n)$, are the singular values of A [12], we may define the following.

Definition 2   For a given tolerance ε we say that:

(i)   The set A is ε-independent if $\sigma_i > \varepsilon, i = 1, 2, \ldots, r$, i.e. all the singular values are greater than ε.

(ii)   The set A is numerically ε-dependent if $\sigma_i > \varepsilon$, and $\sigma_j \leq \varepsilon$, for some i, j, i.e. some singular values are greater than ε and others are smaller than ε.

(iii)   The set A is strongly ε-dependent if $\sigma_1 > \varepsilon$, $\sigma_i \leq \varepsilon, i = 2, 3, \ldots, r$, i.e. the maximal singular value is greater than ε and all the others are less than ε.

(iv)   The set A is fuzzy ε-dependent if $\sigma_i \leq \varepsilon$, $i = 1, 2, \ldots, r$, i.e. all the singular values are less than ε.   □

Since scaling affects the singular values of a matrix, the above definition will be more suitable when it is applied in a normalised set of vectors. If $A = [\underline{a}_1, \underline{a}_2, \ldots, \underline{a}_m]^t \in \mathcal{R}^{m \times n}$, then the normalisation of A is a matrix $A_N = [\underline{v}_1, \underline{v}_2, \ldots, \underline{v}_m]^t \in \mathcal{R}^{m \times n}$ with the property: $\underline{v}_i = \underline{a}_i / \|\underline{a}_i\|_2, i = 1, 2, \ldots, m$; it is obvious that $\underline{v}_i \in \mathcal{R}^{n \times 1}$, $i = 1, 2, \ldots, m$, are unit length vectors ($\|\underline{v}_i\|_2 = 1$). Doing the above normalisation, we may avoid the strange situations of fuzzy ε-dependence, which is mostly encountered when we are dealing with extremely low value data. The normalisation process of a given set of vectors is numerically stable [13], and thus the set A may always be assumed to be normalised. The above notions of dependence, and independence are strongly connected to the notions of numerical ε-rank $(\rho_\varepsilon(A))$ [12] and numerical ε-nullity $(n_\varepsilon(A))$ [16] of a matrix. A simplified condition for their determination is given next [16].

Theorem 1 *For a matrix $A \in \mathcal{R}^{m \times n}$ and a specified tolerance ε we have:*

(i) $\rho_\varepsilon(A) = \{$*number of singular values of A that* $> \varepsilon\}$
(ii) $n_\varepsilon(A) = \{$*number of singular values of A that are* $\leq \varepsilon\}$
(iii) $\rho_\varepsilon(A) = n - n_\varepsilon(A)$. □

The above result suggests one method for calculating the numerical ε-rank and nullity of a matrix via the singular value decomposition. Based on the above result and Definition 1, the following remark applies.

Remark 1

(i) The set A is ε-independent if and only if $\rho_\varepsilon(A) = r$.
(ii) The set A is numerically ε-dependent if and only if $\rho_\varepsilon(A) < r$.
(iii) The set A is strongly ε-dependent if and only if $\rho_\varepsilon(A) = 1$. □

The numerical computation of right and left null spaces of matrices is also a problem appearing frequently in several applications. Based on the singular value decomposition of the given matrix, we can directly compute bases for its null spaces.

6.3.1.2 The Gramian criterion [27]

Definition 3 Let $A = \{\underline{a}_1, \underline{a}_2, \ldots, \underline{a}_m\}$ be a set of m given vectors $\underline{r}_i \in \mathcal{R}^n$, $i = 1, 2, \ldots, m$. The matrix defined by

$$
G_A = \begin{bmatrix}
(\underline{a}_1 \cdot \underline{a}_1) & (\underline{a}_1 \cdot \underline{a}_2) & \cdots & (\underline{a}_1 \cdot \underline{a}_m) \\
(\underline{a}_2 \cdot \underline{a}_1) & (\underline{a}_2 \cdot \underline{a}_2) & \cdots & (\underline{a}_2 \cdot \underline{a}_m) \\
\cdots & \cdots & \cdots & \cdots \\
(\underline{a}_m \cdot \underline{a}_1) & (\underline{a}_m \cdot \underline{a}_2) & \cdots & (\underline{a}_m \cdot \underline{a}_m)
\end{bmatrix} \in \mathcal{R}^{m \times m}
$$

where $(\underline{a} \cdot \underline{b})$ denotes the inner product of vectors $\underline{a}, \underline{b}$, is called the Gram matrix of A, and the determinant $|G_A|$ is called the Gramian of A. □

One of the most important abilities of Gramian is that it provides us with an important criterion about the linear dependency of vectors.

Theorem 2 *For any normalised set A we have:* $0 \leq |G_A| \leq 1$, *where the left equality holds when the set is linearly dependent and the right holds when the set is orthogonal.* □

6.3.2 Compound matrices

A few aspects concerning the notation of compound matrices [10] are described next.

Notation 1

(a) $Q_{p,n}$ denotes the set of strictly increasing sequences of p integers $(1 \leq p \leq n)$ chosen from $1, \ldots, n$, e.g. $Q_{2,3} = \{(1, 2)\}, (1, 3), (2, 3)\}$. Thus, the number of sequences which belong to $Q_{p,n}$ is $\binom{n}{p}$.

    If $\alpha, \beta \in Q_{p,n}$ we say that α precedes β $(\alpha < \beta)$ if there exists an integer t $(1 \leq t \leq p)$ for which $\alpha_1 = \beta_1, \ldots, \alpha_{t-1} = \beta_{t-1}, \alpha_t < \beta_t$, where α_i, β_i denote the elements of α, β, respectively, e.g. in the set $Q_{3,8}(3, 5, 8) < (4, 5, 6)$. This describes the lexicographic ordering of the elements of $Q_{p,n}$. The set of sequences $Q_{p,n}$ from now on will be assumed with its sequences lexicographically ordered, and the elements of the ordered set $Q_{p,n}$ will be denoted by ω.

(b) Suppose $A = [a_{ij}] \in R^{m \times n}$; let k, p be positive integers satisfying $1 \leq k \leq m$, $1 \leq p \leq n$, and let $\alpha = (i_1, \ldots, i_k) \in Q_{k,m}$ and $\beta = (j_1, \ldots, j_p) \in Q_{p,n}$. Then $A[\alpha|\beta] \in F^{k \times p}$ denotes the submatrix of A which contains the rows $i_1, \ldots, i_k$ and the columns $j_1, \ldots, j_p$.

    Let $A \in R^{m \times n}$ and $1 \leq p \leq \min\{m, n\}$, then the pth compound matrix or pth adjugate of A is the $\binom{m}{p} \times \binom{n}{p}$ matrix whose entries are $\det(A[\alpha|\beta])$ (for convenience, A_β^α), $\alpha \in Q_{p,m}, \beta \in Q_{p,n}$, arranged lexicographically in α and β. This matrix will be designated by $C_p(A)$.

For the numerical computation of the pth compound matrix of a given matrix $A \in R^{m \times n}$ or $\in R^{m \times n}[s], 1 \leq p \leq \min\{m, n\}$ [7,8].

6.3.3 Selection of a best uncorrupted base for a numerically dependent set

In algebraic control theory and other applications very frequently the problem of selecting a base out of a given set of data is encountered. In many cases when a numerically ε-dependent set of data is given a selection of a base out of the original data without transforming them is required. The already known methods for finding bases, orthogonal or not, for given sets of vectors are based on the fact that they virtually transform the original data by using mostly Gaussian or orthogonal techniques. Evidently, they obtain new sets, and amongst the new vectors they choose the required ones that span the original set. Thus, the base will consist of vectors completely different from the given ones.

    The following problem is considered next. For a given set of vectors we would like to choose a 'best uncorrupted' base, the 'best' in a sense to be made precise later. By the term 'uncorrupted' we mean that we want to find a base for this set without transforming the original data and evidently introducing roundoff error even before the method starts. Of course, especially when orthogonal techniques are used for the transformation of the given data the roundoff error is not actually remarkable. But when we are dealing with nongeneric computations and the evaluation of a base out of a given set of vectors is required, it is important to select this base from the original set of data for the following reasons:

1. If we are given, for example, a numerically ε-dependent set of vectors, by choosing an uncorrupted base we keep the original structure of the given set, which is particularly required in starting nongeneric computations. On the other hand, by selecting an orthogonal base we completely corrupt the status of the given data and this may lead us to wrong results if we want to start nongeneric computations using this set.

2. When we are interested in evaluating the GCD of a given set, it is extremely important to begin the calculating process using the concrete set of data or a subset of it.

3. The singular values of the original set will be altered when orthogonal techniques are used; therefore the computations will start with a different set of singular values.

Let $A = \{r_1, r_2, \ldots, r_m\}$ be a set of m given vectors $r_i \in \mathcal{R}^n$, $i = 1, 2, \ldots, m$. This set can be expressed in terms of a matrix $A = [r_1, r_2, \ldots, r_m]^t \in \mathcal{R}^{m \times n}$. Then, the problem of finding a 'best uncorrupted' base for the set A is transferred into finding a 'best uncorrupted' base for the row space of matrix A. For the evaluation of an uncorrupted base without the restriction of being in a sense 'best', the row-searching algorithm [17] can be used. A better stable algorithm that applies Gaussian elimination with partial pivoting or Householder transformations to the columns of the given matrix can also be found in References 13 and 17. In the sequel, a method for the evaluation of a 'most orthogonal uncorrupted' base is developed. The whole process is based on compound matrices and Gram's criterion.

Proposition 1 (see [11]) Let $A = [r_1, r_2, \ldots, r_m]^t \in \mathcal{R}^{m \times n}$, $\rho(A) = r \leq \min\{m, n\}$, and $A_N = [v_1, v_2, \ldots, v_m]^t \in \mathcal{R}^{m \times n}$, the normalisation of A. Suppose $G \in \mathcal{R}^{m \times n}$, the Gram matrix of the vectors $v_1, v_2, \ldots, v_m$, and $C_r(G) = [c_{ij}] \in \mathcal{R}^{\binom{m}{r} \times \binom{m}{r}}$, the rth compound matrix of G. If $c_{ii} = \det(G[a/a])$, $a = (i_1, i_2, \ldots, i_r) \in Q_{r,m}$, is the maximum diagonal element of $C_r(G)$, then a most orthogonal uncorrupted base for the row space of A consists of the vectors $\{r_{i_1}, r_{i_2}, \ldots, r_{i_r}\}$. □

Remark 2

- Proposition 1 provides an uncorrupted base for the row space of a given matrix. This base contains vectors that are mostly orthogonal; therefore, if we define the notion of a 'best base' as a base consisting of vectors that are mostly orthogonal, the previous defined base satisfies this definition. Any such base will be referred to as a 'best uncorrupted base' and may be uniquely defined.

- The advantage of using the Gramian-based approach rather than that of an equivalent condition number is established by Proposition 2, which provides a procedure of selection of a most orthogonal set out of a given one. A corresponding procedure based on the condition number will necessitate testing for all possible permutations of selected subsets.   □

The above technique is illustrated with the following example.

Example 2

Let $A = \{\underline{a}_1 = (3, 1, 0)', \ \underline{a}_2 = (-3, 2, 1)', \ \underline{a}_3 = (6, 5, 1)'\}$ be a given set of vectors. We want to find a best uncorrupted basis for this set. The corresponding matrix is

$$A = [\underline{a}_1, \underline{a}_2, \underline{a}_3]' = \begin{bmatrix} 3 & 1 & 0 \\ -3 & 2 & 1 \\ 6 & 5 & 1 \end{bmatrix} \in \mathcal{R}^{3 \times 3}, \qquad \rho(A) = 2 \leq 3$$

The normalisation of A is

$$A_N = [\underline{v}_1, \underline{v}_2, \underline{v}_3]' = \begin{bmatrix} 3/\sqrt{10} & 1/\sqrt{10} & 0 \\ -3/\sqrt{14} & 2/\sqrt{14} & 1/\sqrt{14} \\ 6/\sqrt{62} & 5/\sqrt{62} & 1/\sqrt{62} \end{bmatrix}$$

The Gram matrix of vectors $\underline{v}_1, \underline{v}_2, \underline{v}_3$ and its 2-compound are:

$$G = A_N \cdot A_N^T = \begin{bmatrix} 1 & -7/(\sqrt{10}\sqrt{14}) & 23/(\sqrt{10}\sqrt{62}) \\ -7/(\sqrt{10}\sqrt{14}) & 1 & -7/(\sqrt{14}\sqrt{62}) \\ 23/(\sqrt{10}\sqrt{62}) & -7/(\sqrt{14}\sqrt{62}) & 1 \end{bmatrix}$$

$$C = C_2(G) \simeq \begin{bmatrix} 0.65 & 0.30914 & -0.78 \\ 0.30914 & 0.15 & -0.37 \\ -0.78 & -0.37 & 0.94 \end{bmatrix} \in \mathcal{R}^{\binom{3}{2} \times \binom{3}{2}}$$

The maximum diagonal element of $C_2(G)$ is c_{33} and is given by $c_{33} = \det(G[a/a])$, $a = (2, 3) \in Q_{2,3}$. Therefore, a best uncorrupted base for the set A consists of the vectors $\underline{a}_2, \underline{a}_3$, and is analytically, $B_{u_1} = \{\underline{b}_1 = (-3, 2, 1)', \ \underline{b}_2 = (6, 5, 1)'\}$. From the angle θ formed by the normalised form of the two vectors of the base, $\cos \theta = 0.238$.

If the method of Gaussian elimination with partial pivoting [13] or the method of Householder transformations [12, 13] is applied to matrix A^T, then we take an uncorrupted base for the set A, which consists of the vectors $\underline{a}_1, \underline{a}_2$ (top to bottom selection, as expected) and is analytically, $B_{u_2} = \{\underline{b}_1 = (3, 1, 0)', \ \underline{b}_2 = (-3, 2, 1)'\}$. From the angle θ formed by the normalised form of the two vectors of the above base, $\cos \theta = 0.592$. Comparing the angles of the vectors of bases B_{u_1} and B_{u_2} we deduce that B_{u_1} consists of vectors that are mostly orthogonal.   □

Remark 3

- The available stable numerical methods for selecting an uncorrupted base for the row space of a matrix provide us with a random base only based on a top to bottom technique, without searching for any 'best' combinations of the selected linearly independent vectors. On the other hand, the compound matrix-based method selects a base consisting of linearly independent original vectors having mostly orthogonal normalised forms.
- Let A_N be the normalisation of the set A given in Example 1. The condition number of the row-independent matrix A_N consisting of rows 2 and 3 determined from B_{u_1} is equal to 1.274, and is much smaller than the condition number of the row-independent matrix A_N consisting of rows 1 and 2 determined from B_{u_2}, which equals 1.974.   □

Example 3

Let $A = \{\underline{a}_1 = (2, -3, 1)', \ \underline{a}_2 = (1.99, -2.99, 1)', \ \underline{a}_3 = (3.99, -5.99, 2)'\}$ be a given set of numerically ε-dependent vectors, from the matrix

$$A = \begin{bmatrix} 2 & -3 & 1 \\ 1.99 & -2.99 & 1 \\ 3.99 & -5.99 & 2 \end{bmatrix}$$

with singular values $\sigma_1 \cong 9.1488$, $\sigma_2 \cong 0.0033$ and $\sigma_3 \cong 0$.

Using the compound matrix method we compute an uncorrupted base consisting of vectors $B_{unc} = \{\underline{a}_1 = (2, -3, 1)'$ and $\underline{a}_2 = (1.99, -2.99, 1)'\}$. The matrix that these vectors form has singular values $\sigma'_1 \cong 5.2821$ and $\sigma'_2 \cong 0.0033$. However, if we select an orthogonal base for the set, this consists of the vectors $B_{orth} = \{\underline{a}_1 \cong (0.534, -0.802, 0.268)', \quad \underline{a}_2 \cong (-0.617, -0.154, 0.771)'\}$. The matrix that these vectors form has singular values $\sigma'_1 \cong 1$ and $\sigma'_2 \cong 1$.

By choosing an orthogonal base, the structure of the original set is totally changed (all the singular values of the new linearly independent set are completely different from the values of the original set). On the other hand, the compound matrix method preserves the structure of the original set. This fact is extremely useful when we use the selected set for the application of nongeneric computations such as evaluation of the GCD of polynomials. Furthermore, starting a numerical method involving nongeneric evaluations with a set similar to that of B_{orth} there is great danger of the method converging to the generic solution [5], whereas if it starts with the set of B_{unc} an approximate solution will be calculated. The following example demonstrates this effect.   □

Example 4

Find the GCD of the following set of polynomials $\{p_1(s) = s^2 - 3s + 2,$ $p_2(s) = s^2 - 2.998s + 1.997001, \ p_3(s) = 4s^2 - 11.996s + 7.994001\}$.

For $\varepsilon = 10^{-3}$, the above set has numerical ε-rank $= 2$. Thus, the selection of a base for the row space is required. We select a most orthogonal uncorrupted base of the above set $B_{unc} = \{p_1(s) = s^2 - 3s + 2, p_2(s) = s^2 - 2.998s + 1.997001\}$. Then, for $\varepsilon = 10^{-6}$, using a method for estimating an approximate GCD [5, 6], we compute $s^2 - 3s + 2$ as the estimated GCD.

On the contrary, if we select an orthonormal base $B_{orth} = \{p_1(s) = 0.9428s^2 + 0.2356s + 0.2358, \ p_2(s) = -0.000199s^2 - 0.7069s + 0.707269\}$ for every value of ε, the above method will always end up with a coprime set of polynomials. □

The numerical algorithm of the method, its analysis and several examples concerning its application can be found in Reference 11.

6.4 Almost zeros of a set of polynomials

The subject of nongeneric computations has as one of its most important topics the study of almost zeros. A summary of the basics related to the notion is given next. The computational issues and the feedback significance of the notion (trapping discs for multiparameter root locus) is given in References 4 and 28.

Let

$$P_{m,d} = \left\{ p_i(s) : p_i(s) = a_0^i + a_1^i s + \cdots + a_{d_i}^i s^{d_i} \in \mathcal{R}[s], \quad a_{d_i}^i \neq 0, \qquad i = 1, 2, \ldots, m \right\}$$

be a set of polynomials of maximum degree $d = \max\{d_i, i = 1, 2, \ldots, m\}$. A polynomial vector $\underline{p}(s) = [p_1(s), \ldots, p_m(s)]$ may always be associated with the set $P_{m,d}$. This vector can be written in the form

$$\underline{p}(s) = [p_1(s), p_2(s), \ldots, p_m(s)]^t = P_d \cdot \underline{e}_d(s), \quad \underline{e}_d(s) = [1, s, \ldots, s^d]^t, \quad P_d \in \mathcal{R}^{m \times (d+1)}$$

$$(6.1)$$

For the set $P_{m,d}$ the polynomial vector $\underline{p}(s)$ will be referred to as a *vector representative* (v.r.) and the matrix P_d as a *basis matrix* (b.m.) of $P_{m,d}$.

When $s \in C$, $\underline{p}(s)$ defines a vector valued analytic function with domain C and codomain C^m; the norm of $\underline{p}(s)$ is defined as a positive definite real function with domain C as

$$\|\underline{p}(s)\| = \sqrt{\underline{p}'(s^*)\underline{p}(s)} = \sqrt{\underline{e}_d'(s^*)P_d^T P_d \underline{e}_d(s)} \qquad (6.2)$$

where s^* is the complex conjugate of s. Note that if $q(s) = s + a$ is a common factor of the polynomials $p_i(s), i = 1, 2, \ldots, m$, then for all $i = 1, 2, \ldots, m$, $p_i(-a) = 0$, $\underline{p}(-a) = \underline{0}$ and thus $\|\underline{p}(-a)\| = 0$. This observation leads to the following definition.

Definition 4   Let $\mathcal{P}$ be a set of polynomials of $R[s]$, $\underline{p}(s)$ be the v.r. and let $\phi(\sigma, w) = ||\underline{p}(s)||$, where $s = \sigma + wj \in C$.

An ordered pair (z_k, ε_k), $z_k \in C$, $\varepsilon_k \in R$ and $\varepsilon_k \geq 0$ defines an almost zero (a.z.) of $\mathcal{P}$ at $s = z_k$ and of order ε_k, if $\phi(\sigma, w)$ has a minimum at $s = z_k$ with value ε_k. From the set $\mathcal{Z} = \{(z_k, \varepsilon_k), k = 1, \ldots, r\}$ of almost zeros of $\mathcal{P}$ the element (z^*, ε^*) for which $\varepsilon^* = \min\{\varepsilon_k, k = 1, \ldots, r\}$ is defined as the prime almost zero of $\mathcal{P}$.   □

It is clear that if $\mathcal{P}$ has an exact zero, then the corresponding ε is zero. Clearly, Definition 1 is an extension of the concept of exact zero to that of the almost zero. The magnitude of ε at an almost zero $s = z$ provides an indication of how well z may be considered as an approximate zero of p_i; we should note, however, that ε depends on the scaling of the polynomials $p_i(s)$ in $\mathcal{P}$ by a constant c, $c \in R - \{0\}$. The general properties of the distribution of the almost zeros of a set of polynomials $\mathcal{P}$ on the complex plane were considered in Reference 11 and are summarised below.

Theorem 3   *The prime almost zero of $\mathcal{P}$ is always within the circle centred at the origin of the complex plane and with radius p^*, defined as the unique positive solution of the equation*

$$1 + r^2 + \cdots + r^{2d} = \overline{\gamma}^2/\underline{\gamma}^2 = \theta^2 \tag{6.3}$$

where $\overline{\gamma}$, $\underline{\gamma}$ are the maximum, minimum singular values of P_d, respectively.   □

The disk $[0, p^*]$ within which the prime almost zero lies is referred to as the *prime disk* of p. The following general results may be stated for the radius p^*.

Proposition 2   If d is the degree and θ is the condition number of $\mathcal{P}$, then the radius $p^* = g(d, \theta)$ of the prime disc is a uniquely defined function of d and θ and it has the following properties:

1.   The radius p^* is invariant under the scaling of the polynomial of $\mathcal{P}$ by the same nonzero constant c.
2.   The radius p^* is a monotonically decreasing function of d and $1/\theta$.
3.   The radius p^* is within the following intervals:
     (a)   If $d + 1 > \theta^2$, then $0 < p^* < 1$.
     (b)   If $d + 1 < \theta^2$, then $2 < p^* < \theta^{1/2}$.
     (c)   If $d + 1 = \theta^2$, then $p^* = 1$.   □

The conditioning of the polynomials plays an important role in determining the position of the prime almost zero. In fact, the prime almost zero is always in the vicinity of the origin of the complex plane. The uncertainty in its exact position is measured by the radius of the prime disc. Well conditioned sets of polynomials $\mathcal{P}$ (i.e. $\theta \simeq 1$) have a very small radius prime disc even for very small values of the degree d. Badly conditioned sets of polynomials $\mathcal{P}$ (i.e. $\theta \gg 1$) have very large radius discs, even for large values of the degree d. The computation of almost zeros may be achieved by deriving the necessary conditions for the minimum [4], or using standard numerical hill climbing techniques [18]. The latter approach is used in the following example.

Example 5

Let the set of polynomials be defined by the polynomial vector $\underline{p}(s)$:

$$
\underline{p}(s) = \begin{bmatrix} s^3 + 5.5s^2 + 11s + 7.5 \\ s^2 - 1 \\ s - 2 \end{bmatrix} = \begin{bmatrix} 7.5 & 11 & 5.5 & 1 \\ -1 & 0 & 1 & 0 \\ -2 & 1 & 0 & 0 \end{bmatrix} \begin{bmatrix} 1 \\ s \\ s^2 \\ s^3 \end{bmatrix} = P_3 \cdot \underline{e}_3(s)
$$

We want to evaluate the almost zero z of the above set of polynomials. Applying the optimisation algorithm developed in Reference 18 we obtain the following results (recorded to an accuracy of five decimal digits). The singular values of matrix P_3 are

$$
\sigma_1 = 14.44295, \qquad \sigma_2 = 2.42926, \qquad \sigma_3 = 1.00000, \, \theta = \frac{\sigma_1}{\sigma_3}
$$

The radius p^* is the unique positive solution of the equation $1 + r^2 + r^4 + r^6 = (14.44295)^2$. We obtain $p^* = r = 2.35644$; the almost zero is located at the point $z = (-0.99999, 0)$ and the corresponding function norm has the value $\varepsilon = 10.0$. The size of ε indicates that the set of polynomials contains no common zeros.   □

Example 6

Let the set of polynomials be defined by the polynomial vector $\underline{p}(s)$:

$$
\underline{p}(s) = \begin{bmatrix} p_1(s) \\ p_2(s) \end{bmatrix} = \begin{bmatrix} 2s^3 + 3s + 2s + 3 \\ 10s^3 + 15s^2 + 14s + 21 \end{bmatrix} = \begin{bmatrix} 3 & 2 & 3 & 2 \\ 21 & 14 & 15 & 10 \end{bmatrix} \begin{bmatrix} 1 \\ s \\ s^2 \\ s^3 \end{bmatrix}
$$

$$
= P_2 \cdot \underline{e}_3(s), \qquad P_2 \in \mathcal{R}^{2 \times 4}
$$

It is known that the above two polynomials $p_1(s)$, $p_2(s)$ are not coprime and they have an exact zero at -1.5. Consequently their almost zero $z = \sigma + wj$ must be located somewhere in the area of -1.5. Once more we use optimisations for the computation of almost zeros and select different types of initial conditions to demonstrate the effect on the minimum and its location. The singular values of P_2 and the condition numbers are

$$
\sigma_1 = 31.42157, \qquad \sigma_2 = 0.82746, \qquad \theta = 37.97367
$$

The radius of the prime disk is $p^* = 3.30798$ and the results are summarised in Table 6.1. We observe that when the initial point is somewhere in the area of the exact zero, then the final value of the almost zero is very close to the exact one. On the other hand, if we select an initial point that does not belong to the same

Table 6.1

Initial point	Almost zero (σ, w)	Norm
$\left(\dfrac{p^*}{3}, -\dfrac{p^*}{3}\right)$	$(-0.00044, -1.17635)$	2.236
$(-1, -0.2)$	$(-1.5, 0)$	0.26177×10^{-14}
$(-1.4, -0.0001)$	$(-1.5, 0)$	-0.93213×10^{-11}
$(-2, 0.5)$	$(-1.5, 0)$	0.13644×10^{-14}
$(1, 1)$	$(-0.00044, 1.17635)$	2.236

half-plane with the exact zero, then the final computed value of the almost zero varies significantly from that of the exact zero. □

The position of the almost zero varies according to the scaling which is used. This suggests that there is a need for a better definition of the concept which is independent of scaling. Instead of looking for approximate common roots we can search for approximate common factors, and this is considered next.

6.5 GCD method based on ERES

The problem of finding the greatest common divisor (GCD) of a polynomial set has been a subject of interest for a very long time and has widespread applications. Since the existence of a common divisor of polynomials is a property that holds for specific sets and is not true generically, extra care is needed in the development of efficient numerical algorithms correctly calculating the required GCD. Several numerical methods for the computation of the GCD of a set $\mathcal{P}_{m,d}$ of m polynomials of $\mathcal{R}[s]$ of maximal degree d have been proposed [8, 19–21]. All of them can be classified as follows: (i) numerical methods based on Euclid's algorithm and its generalisations; (ii) numerical methods based on procedures involving matrices (matrix-based). The matrix-based methods usually perform specific transformations to a matrix formed directly from the coefficients of the given polynomials. Their main characteristics are: (i) good performance in the case of large sets of polynomials; (ii) guaranteed numerical stability. Such methods are used exclusively when the given $\mathcal{P}_{m,d}$ has $m \geq 3$. According to the process applied to the consequent coefficient matrix, these methods can be further classified into the following two subclasses: (i) iterative matrix-based methods; (ii) direct matrix-based methods. Iterative methods perform an iterative process finally converging to the required result, whereas direct methods apply a specific formula leading directly to the desired computation.

The ERES method is a new iterative matrix-based method developed in References 5, 6 and 22 and is based on the properties of the GCD as an invariant of $\mathcal{P}_{m,d}$ under extended row equivalence and shifting (ERES) operations. In the present Chapter, we describe analytically the numerical algorithm produced from the ERES method.

6.5.1   The ERES method: theoretical and numerical issues

Let $\mathcal{P}_{m,d} = \{p_i(s) : p_i(s) \in \mathcal{R}[s], \ i = 1, 2, \ldots, m, \ d_i = \deg\{p_i(s)\}, \ d = \max\{d_i, \ i = 1, 2, \ldots, m\}$ be the set of m polynomials of $\mathcal{R}[s]$ of maximal degree d.

For any $\mathcal{P}_{m,d}$ set we define a vector representative $\underline{p}_m(s)$ and a basis matrix P_m by $\underline{p}_m(s) = [p_1(s), \ldots, p_m(s)]^t = [\underline{p}_0, \underline{p}_1, \ldots, \underline{p}_d] \underline{e}_d(s) = P_m \underline{e}_d(s)$, where $P_m \in \mathcal{R}^{m \times (d+1)}$, $\underline{e}_d(s) = [1, s, \ldots, s^d]^t$.

By GCD $\{\mathcal{P}_{m,d}\} = \phi(s)$ we shall denote the GCD of the set. If c is the integer for which $\underline{p}_0 = \cdots = \underline{p}_{c-1} = 0, \underline{p}_c \neq 0$, then $c = w(P_{m,d})$ is called the order of $\mathcal{P}_{m,d}$ and s^c is an elementary divisor of the GCD. The set $\mathcal{P}_{m,d}$ will be called proper if $c = 0$, and nonproper if $c \geq 1$.

In the ERES (extended row equivalence and shifting) method for a given $\mathcal{P}_{m,d}$ with a basis matrix P_m the following operations are defined:

(i)   elementary row operations with scalars from $\mathcal{R}$ on P_m
(ii)   addition or elimination of zero rows on P_m
(iii) if $\underline{a}^t = [0, \ldots, 0, a_\varepsilon, \ldots, a_{d+1}] \in \mathcal{R}^{1 \times (d+1)}$, $a_\varepsilon \neq 0$ is a row of P_m, then we define the shifting operation shf : shf$(\underline{a}^t) = (\underline{a}^*)^t = [a_\varepsilon, \ldots, a_{d+1}, 0, \ldots, 0]$.

shf$(\mathcal{P}_{m,d}) = \mathcal{P}^*_{m,d}$ denotes the set obtained from $\mathcal{P}_{m,d}$ by applying shifting on every row of P_m. Type (i), (ii) and (iii) operations are referred to as extended row equivalence and shifting (ERES) operations. The following theorem describes the properties characterising the GCD of any given $\mathcal{P}_{m,d}$, and it is proved in Reference 22. If we are given a nonproper set $\mathcal{P}_{m,d}$ with $w(\mathcal{P}_{m,d}) = c$, then we can always consider the corresponding proper one $\mathcal{P}_{m,d'}$ by dismissing the c leading zero columns. Then GCD$\{\mathcal{P}_{m,d}\} = s^c \cdot$ GCD$\{\mathcal{P}_{m,d'}\}$.

Theorem 4   *For any set $\mathcal{P}_{m,d}$, with a basis matrix P_m, $r(P_m) = r$ and GCD $\{\mathcal{P}_{m,d}\} = \phi(s)$, we have the following properties:*

(i)   *If R is the row space of P_m, then $\phi(s)$ is an invariant of R (e.g. $\phi(s)$ remains invariant after the execution of elementary row operations on P_m). Furthermore, if $r = \dim(R) = d + 1$, then $\phi(s) = 1$.*

(ii)   *If $w(\mathcal{P}_{m,d}) = c \geq 1$ and shf$(\mathcal{P}_{m,d}) = \mathcal{P}^*_{m,d}$, then $\phi(s) = $ GCD$\{\mathcal{P}_{m,d}\} = s^c \cdot$ GCD$\{\mathcal{P}^*_{m,d}\}$.*

(iii) *If $\mathcal{P}_{m,d}$ is proper, then $\phi(s)$ is invariant under the combined ERES set of operations.*

From Theorem 4 it is evident that ERES operations preserve the GCD of any $\mathcal{P}_{m,d}$ and thus can be easily applied to obtain a modified basis matrix with a much simpler structure.

Example 7

Let $\mathcal{P}_{3,4} = \{p_1(s) = -s^4 - s^3 + s + 1, p_2(s) = s^3 + 3s^2 - s - 3, p_3(2) = s^4 - 1\}$ be a proper set. We form the basis matrix of the set

$$P_3 = [\underline{r}_1, \underline{r}_2, \underline{r}_3]^t = \begin{bmatrix} 1 & 1 & 0 & -1 & -1 \\ -3 & -1 & 3 & 1 & 0 \\ -1 & 0 & 0 & 0 & 1 \end{bmatrix} \in \mathcal{R}^{3 \times 4}$$

$$\rho(P_3) = 3, \qquad GCD\{\mathcal{P}_{3,4}\} = s^2 - 1, \qquad d = 4$$

We perform the following ERES operations to the basis matrix P_3.

Step 1 $\quad \vec{r}_3' := \underline{r}_3' + \underline{r}_1', \underline{r}_2' := \underline{r}_2' + 3\underline{r}_1'$, shifting

$$P_3 = \begin{bmatrix} 1 & 1 & 0 & -1 & -1 \\ 2 & 3 & -2 & -3 & 0 \\ 1 & 0 & -1 & 0 & 0 \end{bmatrix}, \qquad d = 2$$

Step 2 $\quad \vec{r}_1' := \underline{r}_1' - \underline{r}_3', \underline{r}_2' := \underline{r}_2' - 2\underline{r}_3'$, shifting

$$P_3 = \begin{bmatrix} 1 & 1 & -1 & -1 & 0 \\ 3 & 0 & -3 & 0 & 0 \\ 1 & 0 & -1 & 0 & 0 \end{bmatrix}, \qquad d = 2$$

Step 3 $\quad \vec{r}_1' := \underline{r}_1' - \underline{r}_3'$, shifting

$$P_3 = \begin{bmatrix} 1 & 0 & -1 & 0 & 0 \\ 3 & 0 & -3 & 0 & 0 \\ 1 & 0 & -1 & 0 & 0 \end{bmatrix}, \qquad \text{and} \quad \rho(P_3) = 1, \qquad GCD = -s^2 + 1$$

Thus, after successive applications of ERES operations on an initial basis matrix, the maximal degree of the resulting set of polynomials is reduced and after a finite number of steps the resulting basis matrix has rank 1. In that stage, any row of the matrix specifies the coefficients of the required GCD of the set. A useful criterion deciding when the iterative process applying ERES operations on a specific basis matrix will be terminated is Theorem 5 [5].

Theorem 5 *Let* $A = [\underline{r}_1, \underline{r}_2, \ldots, \underline{r}_m]^t \in \mathcal{R}^{m \times n}$, $\underline{r}_i \neq \underline{0}$, $i = 1, 2, \ldots, m$. *Then* $\rho(A) = 1$, *if and only if the singular values* $\sigma_m \leq \sigma_{m-1} \leq \cdots \leq \sigma_1$ *of the normalisation* $A_N = [\underline{v}_1, \underline{v}_2, \ldots, \underline{v}_m]^t \in \mathcal{R}^{m \times n}$, $\underline{v}_i = \underline{r}_i / \|\underline{r}_i\|_2$, *of* A *satisfies the conditions* $\sigma_1 = \sqrt{m}$, $\sigma_i = 0$, $i = 2, 3, \ldots, m$. $\qquad \square$

To derive an effective numerical algorithm for the ERES method we have to resolve the following numerical problems:

I: *Numerical application of the ERES operations*

The most reliable and stable numerical method performing ERES operations is the method of Gaussian elimination with partial pivoting. Thus, after each iteration an upper triangular or trapezoidal form of the basis matrix will be computed. This form of the matrix can be changed so that its first row corresponds to the least degree polynomial. Simultaneously, the matrix is scaled appropriately to retain as the first row the least degree polynomial, followed by the application of partial pivoting. Since Gaussian elimination preserves the structure of the first row of the matrix, after a finite number of iterations it can achieve a quick reduction to the maximal degree of the initial polynomial set. During the elimination process an appropriate accuracy ε_G is selected such that all elements of the matrix with values less than ε_G will be set equal to zero throughout the ERES operations. The specified accuracy ε_G will be defined as the Gaussian accuracy.

II: *Numerical interpretation of the termination criterion*

The successive implementation of Gaussian elimination and shifting will be terminated when the resulting matrix has rank 1. This property can be detected numerically according to the following theorem.

Theorem 6   *Let* $A = [\underline{r}_1, \underline{r}_2, \dots, \underline{r}_m]^t \in \mathcal{R}^{m \times n}$, $m \leq n$, $\underline{r}_i \neq \underline{0}$, $i = 1, 2, \dots, m$. *Then for an appropriate accuracy* $\varepsilon_t > 0$ *the numerical* ε_t *rank of A equals one* $(\rho_{\varepsilon_t}(A) = 1)$ *if and only if the singular values* $\sigma_m \leq \sigma_{m-1} \leq \cdots \leq \sigma_1$ *of the normalisation* $A_N = [\underline{v}, \dots, \underline{v}_m]^t \in \mathcal{R}^{m \times n}$, $\underline{v}_i = \underline{r}_i / \|\underline{r}_i\|_2$ *of A satify the conditions:* $|\sigma_1 - \sqrt{m}| \leq \varepsilon_t$, $\sigma_i \leq \varepsilon_t$, $i = 2, 3, \dots, m$.   □

The specified accuracy ε_t will be defined as the *termination accuracy*.

III: *Selection of the representative row containing the coefficients of the* GCD

When we achieve the computation of the resulting rank-1 matrix an important question arising is the following: can we define an appropriate criterion according to which we can select a specific vector containing the coefficients of the required GCD? The above issue is connected with the case of approximating a given matrix $A \in \mathcal{R}^{m \times n}$ by a rank-1 matrix A_1 [23]. The following proposition can be applied.

Proposition 3   Let $A = V \cdot \Sigma \cdot W^t$ be the singular value decomposition of a given matrix $A \in \mathcal{R}^{m \times n}$, $\rho(A) = 1$. Then a 'best' rank-1 approximation to A in the Frobenius norm is given by $A_1 = \sigma_1 \cdot \underline{v} \cdot \underline{w}^t$, where σ_1 is the largest singular value of A and $\underline{v}$ and $\underline{w}$ are the first columns of the orthogonal matrices V and W of the singular value decomposition of A, respectively. The vector $\underline{w}$ is then the 'best' representative of the rows of matrix A in the sense of the rank-1 approximation.   □

The numerical algorithm of the method and its analysis can be found in References 5 and 6.

6.6 GCD method based on matrix pencils

The computation of GCD may be achieved alternatively by using concepts from systems theory, which lead to an alternative formulation using matrix pencils. The process involves the following steps [7].

6.6.1 Connection of $\mathcal{P}_{m,d}$ with a linear system

For any set $\mathcal{P}_{m,d}$ we may define an associated linear system such that the GCD of the set is the output decoupling zero polynomial [7] of the system.

Theorem 7 Let $\mathcal{P}_{m,d} \in \{\mathcal{P}_d\}$, P_m be a basis matrix, $\rho(P_m) = r < d + 1$, $M \in R^{(d+1) \times \mu}$, $\mu = d - r + 1$ be a basis matrix for $\mathcal{N}_r(P_m) = \mathcal{M}$ and $M_1 \in R^{d \times \mu}$ be the submatrix of M obtained by deleting the last row of M. If $p(s) \in \mathcal{P}_{m,d}$ is any monic polynomial of degree d, $\widehat{A} \in R^{d \times d}$ is the associated companion matrix and $\widehat{C} \in R^{(r-1) \times d}$, $\rho(\widehat{C}) = r - 1$, is such that $\widehat{C}M_1 = 0$, then the unobservable modes of the system: $S(\widehat{A}, \widehat{C})$, $\dot{x} = \widehat{A}x$, $y = \widehat{C}x$ with multiplicities included, define the roots of the GCD of $\mathcal{P}_{m,d}$. □

$S(\widehat{A}, \widehat{C})$ will be called the *associated system* of $\mathcal{P}_{m,d}$ and the observability matrix

$$Q\,(\widehat{A}, \widehat{C}) = [\widehat{C}^t, \widehat{A}^t\widehat{C}, \ldots, (\widehat{A}^t)^{d-1}\widehat{C}^t]^t \in R^{d(r-1) \times d}$$

will be referred to as a *reduced resultant* of $\mathcal{P}_{m,d}$. From the above we have the following.

Remark 4 If $\mathcal{N}_r\{P_m\} = \{0\}$, then the set $\mathcal{P}_{m,d}$ is coprime. If $\rho(P_m) = 1$, then $\mu = d$ and any polynomial in $\mathcal{P}_{m,d}$ defines the GCD. □

Remark 5 With any set $\mathcal{P}_{m,d}$ there is a family of associated systems $S(\widehat{A}, \widehat{C})$. Furthermore, if any $S(\widehat{A}, \widehat{C})$ is observable, then $\mathcal{P}_{m,d}$ is coprime. □

6.6.2 Reduction of the original set

From the last interpretation of the GCD we have some further results, which provide the basis for the numerical algorithm.

Corollary 1 [7] Let $\mathcal{P}_{m,d} \in \{\mathcal{P}_d\}, \rho(\mathcal{P}_m) = r < d + 1, S(\widehat{A}, \widehat{C})$ be the associated system of $\mathcal{P}_{m,d}$ and let $\widehat{\mathcal{P}}_{r-1,d'}$ be the set of $r-1$ polynomials of degree $d' \leq d - 1$ defined by $\widehat{\mathcal{P}}_{r-1}(s) = \widehat{C}e_{d-1}(s)$. Then the sets $\mathcal{P}_{m,d}$ and $\widehat{\mathcal{P}}_{r-1,d'}$ have the same GCD. □

The set $\widehat{\mathcal{P}}_{r-1,d'}$ defined above is equivalent to $\mathcal{P}_{m,d}$ as far as the GCD and there are clear advantages in deploying $\widehat{\mathcal{P}}_{r-1,d'}$ instead of $\mathcal{P}_{m,d}$ for computing the GCD. Successive application of the above result leads to equivalent sets of smaller or equal numbers of elements and degree.

6.6.3 Determination of the associated pencil

Let $\mathcal{P}_{m,d} \in \{\mathcal{P}_d\}, \rho(\mathcal{P}_m) = r < d + 1$ and let $Q(\widehat{A}, \widehat{C})$ be the corresponding reduced resultant. Let $\rho\{Q(\widehat{A}, \widehat{C})\} < d$ (when it is equal to d it can be proved that the set is coprime) and $W \equiv \mathcal{N}_r\{Q(\widehat{A}, \widehat{C})\} \neq \{0\}, k = \dim\{W\}$. The pencil $T(s) = sW - \widehat{A}W$ characterises the set $\mathcal{P}_{m,d}$ and it is called the *associated pencil* of the set. The following result forms a basis for the numerical computation of the GCD.

Corollary 2 [7] Let $T(s) = sW - \widehat{A}W \in R^{d \times k}[s]$ be the associated pencil of $\mathcal{P}_{m,d}$. If $\phi(s)$ is the GCD of $\mathcal{P}_{m,d}$, then $C_k(T(s)) = \phi(s) \cdot C_k(W)$. □

The above result leads to the following procedure for computing the GCD.

Remark 6   Let $T(s) = sW - \widehat{A}W = sW - \widetilde{W}$ and let $sW_\alpha - \widetilde{W}_\alpha \alpha \in Q_{k,d}$ be any minor of maximal order such that $|W_\alpha| \neq 0$. Then the GCD of $\mathcal{P}_{m,d}$ is defined by $|sW_\alpha - \widetilde{W}_\alpha|$. □

To derive an effective numerical algorithm for the **MP** method based on the above theoretical results we have to resolve the following numerical problems: numerical interpretation of the notion of nullity; numerical computation of right and left null spaces of matrices; computation of compound matrices. Section 6.3 has shown how these problems can be handled.

Example 8

Find the GCD of the set $\mathcal{P}_{4,4} = \{p_1(s) = 2s^3 + 5s^2 + 2s + 5,\quad p_2(s) = s^4 + s^3 + 12s^2 + s + 11,\ p_3(s) = 3s^4 + 7s^3 + 4s^2 + 7s + 1,\ p_4(s) = 4s^4 + 5s^3 - s^2 + 5s + 3\}$ according to the developed MP algorithm.

Using four decimal digits and by setting the nullity accuracy $\varepsilon = 10^{-16}$ we have:

$$P_4 = \begin{bmatrix} 5 & 2 & 5 & 2 & 0 \\ 11 & 1 & 12 & 1 & 1 \\ 1 & 7 & 4 & 7 & 3 \\ -4 & 5 & -1 & 5 & 3 \end{bmatrix}$$

$$\hat{C} = \begin{bmatrix} 0.6967 & 0.1211 & 0.6967 & 0.1211 \\ -0.1211 & 0.6967 & -0.1211 & 0.6967 \end{bmatrix}$$

$$\hat{A} = \begin{bmatrix} 0 & 1 & 0 & 0 \\ 0 & 0 & 1 & 0 \\ 0 & 0 & 0 & 1 \\ -11 & -1 & -12 & -1 \end{bmatrix}$$

$$Q(\hat{A}, \hat{C}) = \begin{bmatrix} 0.6967 & 0.1211 & 0.6967 & 0.1211 \\ -0.1211 & 0.6967 & -0.1211 & 0.6967 \\ -1.3318 & 0.5756 & -1.3318 & -0.5756 \\ -7.6633 & -0.8177 & -7.6633 & -0.8177 \\ -6.3315 & -1.9074 & -6.3315 & -1.9074 \\ 8.9951 & -6.8456 & 8.9951 & -6.8456 \\ 20.9812 & -4.4241 & 20.9812 & -4.4241 \\ 75.3013 & 15.8407 & 75.3013 & 15.8407 \end{bmatrix}$$

$$W = \begin{bmatrix} 0.5741 & -0.4128 \\ -0.4128 & -0.5741 \\ -0.5741 & 0.4128 \\ 0.4128 & 0.5741 \end{bmatrix}$$

$$T(s) = sW - \hat{A}W, \qquad c = (1, 2), \qquad \alpha = (1, 2)$$

$$\text{GCD} = |sW_\alpha - \tilde{W}_\alpha| = s^2 + 1 \qquad \qquad \square$$

The numerical algorithm of the method and its analysis can be found in References 7 and 8.

6.7 LCM computation and approximate factorisation of polynomials

The problem of computing the LCM of polynomials has widespread applications and requires implementation of algorithms computing the GCD. It is the aim of this section to investigate the problem of computing the LCM, in a numerical robust way that avoids the compution of roots of the corresponding polynomials, and the associated factorisation problem. The main problem addressed here is as follows: given a set of polynomials $t_i(s) \in \mathcal{R}[s]$, $i = 1, 2, \ldots, \tau$, define a numerical procedure for the computation of their LCM by avoiding root finding. The current objective is to transform the LCM computations problem to real matrix computations and thus also introduce a notion of 'almost LCM'.

6.7.1 Fundamental definitions and properties of LCM of polynomials

For a given set of polynomials $\mathcal{P} = \{t_i(s) : t_i(s) \in \mathcal{R}[s], i = 1, 2, \ldots, \tau\}$ we shall denote by $z_{\mathcal{P}}(s) \stackrel{\triangle}{=} \langle \mathcal{P} \rangle$ a greatest common divisor (GCD) and by $p_{\mathcal{P}}(s) \stackrel{\triangle}{=} [\mathcal{P}]$ a least common multiple (LCM) of $\mathcal{P}$. In fact, $\langle \cdot \rangle, [\cdot]$ denote the operations of extracting the GCD, LCM, respectively, from the set. With this notation for the set $\mathcal{P} = \{t_1(s), t_2(s)\}$ we have: $p_{1,2}(s) \stackrel{\triangle}{=} [t_1(s), t_2(s)]$, $z_{1,2}(s) \stackrel{\triangle}{=} \langle t_1(s), t_2(s) \rangle$. For the LCM we have some important properties. One of them is stated below and proved in Reference 24.

Proposition 4   Let $\mathcal{P} = \{t_i(s) : t_i(s) \in R[s], i = 1, 2, \ldots, \mu\}$ and assume any ordering of its elements. If we denote by $\mathcal{P}_\mu = \{t_1(s), \ldots, t_{\mu-1}(s), t_\mu(s)\} = \mathcal{P}$, $\mathcal{P}_{\mu-1} = \{t_1(s), \ldots, t_{\mu-1}(s)\}, \ldots, \mathcal{P}_2 = \{t_1(s), t_2(s)\}$ then we may define $[\mathcal{P}_\mu] \stackrel{\triangle}{=} p_\mu(s)$ as

$$
\begin{aligned}
[\mathcal{P}_\mu] &= p_\mu(s) &&= \big[[\mathcal{P}_{\mu-1}], t_\mu(s)\big] = \big[p_{\mu-1}(s), t_\mu(s)\big] \\
[\mathcal{P}_{\mu-1}] &= p_{\mu-1}(s) &&= \big[[\mathcal{P}_{\mu-2}], t_{\mu-1}(s)\big] = \big[p_{\mu-2}(s), t_{\mu-1}(s)\big] \\
& \vdots \\
[\mathcal{P}_3] &= p_3(s) &&= \big[[t_1(s), t_2(s)], t_3(s)\big] = \big[p_2(s), t_3(s)\big] \\
[\mathcal{P}_2] &= p_2(s) &&= [t_1(s), t_2(s)]
\end{aligned}
\tag{6.4}
$$

The above proposition establishes the *associativity* property for the LCM. This property can reduce the computation of LCM of a general set $\mathcal{P}$ to the computation of LCM of two polynomials. This computation is now considered with no root finding procedures. In Reference 24 the following result is proved.

Theorem 8 *Let $t_1(s)$, $t_2(s) \in R[s]$ and let $z(s) = \langle t_1(s), t_2(s) \rangle$ be their GCD and express $t_1(s)$, $t_2(s)$ as $t_1(s) = z(s) \cdot \tilde{t}_1(s)$, $t_2(s) = z(s) \cdot \tilde{t}_2(s)$. The LCM $p(s) = [t_1(s), t_2(s)]$ is then defined by*

$$p(s) = z(s)\tilde{t}_1(s)\tilde{t}_2(s) \tag{6.5}$$

Remark 7 The associativity property established above is fundamental in the computation of LCM. Such a methodology involves the following fundamental steps: (i) computation of GCD; (ii) factorisation of polynomials into two factors when one is given. □

An alternative way of computing the LCM of two polynomials is in terms of the following result [24].

Theorem 9 *Let $t_1(s)$, $t_2(s) \in R[s]$ and let $z(s) = \langle t_1(s), t_2(s) \rangle$ be their GCD. Then the LCM is defined as $p(s) = [t_1(s), t_2(s)]$ and satisfies the relationship*

$$t_1(s)t_2(s) = p(s)z(s) \tag{6.6}$$

6.7.2 Factorisation and almost factorisation of polynomials

The numerical computation of GCD of many polynomials has been well addressed and algorithms producing an evaluation of the 'numerical' or 'almost' GCD have been previously derived based on the ERES method [6], as well as the matrix pencil method [8]. The evaluation of the 'approximate GCD' also has implications on factorisation of polynomials, which is an integral part of the LCM algorithm as stated by Theorem 9. Such issues are examined next. We first examine the problem of factorisation of polynomials.

Let $a(s)$, $b(s)$, $c(s) \in R[s]$ and assume that

$$a(s) = b(s)c(s) \tag{6.7a}$$

where

$$
\begin{aligned}
a(s) &= a_0 + sa_1 + \cdots + s^k a_k, & a_k &\neq 0 \\
b(s) &= b_0 + sb_1 + \cdots + s^\mu b_\mu, & b_\mu &\neq 0 \\
c(s) &= c_0 + sc_1 + \cdots + s^\nu c_\nu, & c_\nu &\neq 0
\end{aligned} \tag{6.7b}
$$

Condition eqn. 6.7a implies that $k = \mu + \nu$ and

$$
\begin{bmatrix}
b_0 & 0 & \cdots & & 0 \\
b_1 & b_0 & \cdots & & 0 \\
\vdots & b_1 & \ddots & & b_0 \\
\vdots & \vdots & \ddots & & b_1 \\
b_\mu & \vdots & & & \vdots \\
0 & b_\mu & & & \vdots \\
\vdots & & \ddots & & \vdots \\
0 & \cdots & & 0 & b_\mu
\end{bmatrix}
\begin{bmatrix}
c_0 \\
c_1 \\
\vdots \\
c_\nu
\end{bmatrix}
\underset{\triangleq\,\underline{c}^\nu}{} =
\begin{bmatrix}
a_0 \\
a_1 \\
\vdots \\
\vdots \\
\vdots \\
a_k
\end{bmatrix}
\underset{\triangleq\,\underline{a}^k}{} , \qquad
\underline{b}^\mu =
\begin{bmatrix}
b_0 \\
b_1 \\
\vdots \\
b_\mu
\end{bmatrix}
\tag{6.7c}
$$

$$
\triangleq T_{\nu+1}(\underline{b}^\mu) \in \mathcal{R}^{[[(\mu+1)+\nu] \times [(\nu+1)]]}
$$

Given that eqn. 6.7a holds true and assuming that $a(s)$ and $b(s)$ are given, then eqn. 6.7c always has a solution which implies [24].

Proposition 5　If $a(s), b(s), c(s) \in R[s]$ and satisfy eqn. 6.7b with $k = \mu + \nu$, then the following conditions are equivalent: (i) $a(s) = b(s)c(s)$; (ii) $T_{\nu+1}(\underline{b}^\mu)\underline{c}^\nu = \underline{a}^k$; (iii) $\underline{a}^k \in col\ sp\{T_{\nu+1}(\underline{b}^\mu)\}$. □

Assuming that $b_0 \neq 0$, we may partition $T_{\nu+1}(\underline{b}^\mu)$ as shown below and express the solution of the factorisation problem as

$$
T_{\nu+1}(\underline{b}^\mu) =
\begin{bmatrix}
b_0 & 0 & \cdots & 0 \\
b_1 & b_0 & \ddots & \vdots \\
\vdots & & \ddots & 0 \\
b_\nu & & & b_0 \\
\cdots & \cdots & \cdots & \cdots \\
\vdots & & & b_1 \\
b_\mu & & & \vdots \\
0 & \ddots & & \vdots \\
\vdots & \ddots & \ddots & \vdots \\
0 & \cdots & 0 & b_\mu
\end{bmatrix}
=
\begin{bmatrix}
T_1 \\
\cdots \\
T_2
\end{bmatrix} , \qquad
T_1 \in \mathcal{R}^{(\nu+1) \times (\nu+1)}
\tag{6.8}
$$

Proposition 6   If $a(s) = b(s)c(s)$, where $a(s)$, $b(s)$ are given as in eqn. 6.7b and $c(s)$ is also as described in (6.7b), then the polynomial $c(s)$ is defined by

$$
\begin{bmatrix} c_0 \\ c_1 \\ \vdots \\ c_\nu \end{bmatrix} = \begin{bmatrix} b_0 & 0 & \cdots & 0 \\ b_1 & \ddots & & \vdots \\ \vdots & \ddots & \ddots & 0 \\ b_\nu & \cdots & b_1 & b_0 \end{bmatrix}^{-1} \begin{bmatrix} a_0 \\ a_1 \\ \vdots \\ a_\nu \end{bmatrix}
\tag{6.9}
$$

whereas the parameters of $a(s)$, $b(s)$ satisfy the condition

$$
\begin{bmatrix} b_{\nu+1} & \cdots & \cdots & b_1 \\ \vdots & & & \vdots \\ b_\mu & & & \vdots \\ 0 & \ddots & & \vdots \\ \vdots & \ddots & \ddots & \vdots \\ 0 & \cdots & 0 & b_\mu \end{bmatrix} \begin{bmatrix} b_0 & 0 & \cdots & 0 \\ \vdots & \ddots & & \vdots \\ \vdots & & \ddots & 0 \\ b_\nu & \cdots & & \cdots b_0 \end{bmatrix}^{-1} \begin{bmatrix} a_0 \\ a_1 \\ \vdots \\ a_\nu \end{bmatrix} = \begin{bmatrix} a_{\nu+1} \\ \vdots \\ \vdots \\ \vdots \\ \vdots \\ a_k \end{bmatrix}
\tag{6.10}
$$

Given that $|T_1| \neq 0$ (since we assume $b_0 \neq 0$), the result readily follows.   □

Remark 8   If $b_0 = 0$, $b_1 = 0, \ldots, b_{\sigma-1} = 0$, $b_\sigma \neq 0$, then we define as T_1 the first square submatrix of $T_{\nu+1}(\underline{b}^\mu)$ and we repeat the process. In fact, then also we have $a_0 = a_1 = \cdots = a_{\sigma-1} = 0$, $a_\sigma \neq 0$, and the upper zero block of $T_{\nu+1}(\underline{b}^\mu)$ leads to an identity.   □

The linear algebra formulation of the factorisation problem allows the introduction of the notion of approximate factorisation of polynomials introduced below.

Definition 5   Let $a(s), b(s) \in R[s]$, $\partial[b(s)] \leq \partial[a(s)]$. With the notation introduced by eqns. 6.7c and 6.7b, $b(s)$ may be defined as an **almost factor** of $a(s)$ of order $\phi = \min \phi(\underline{c}^\nu)$ for $\forall \underline{c}^\nu \in R^{\nu+1}$, where

$$
\phi(\underline{c}^\nu) = \| T_{\nu+1}(\underline{b}^\mu)\underline{c}^\nu - \underline{a}^k \|
\tag{6.11}
$$ □

Clearly, when $\phi = 0$, then $b(s)$ is an exact factor of $a(s)$; otherwise, it is an almost factor and the vector $\underline{c}^\nu$ associated with the minimum of eqn. 6.11 defines the **complement** to the $b(s)$ almost factor, $c(s)$. The problem of finding the minimum of eqn. 6.11 when $\underline{b}^\mu$, $\underline{a}^k$ are given will be referred to as the **approximate factorisation problem (AFP)**, and it is central in the development of any robust numerical method for computing LCM. The study of AFP may be reduced to a problem of computing the *almost null space* of an operator, and it is discussed briefly below. The study of AFP involves two distinct issues: (i) measuring the closeness of $b(s)$ to an exact factor of $a(s)$; and (ii) given an almost factor $b(s)$, define the 'best' complementary factor $c(s)$.

Measuring the closeness of a given factor from an exact factor will be referred to as an **order of approximation problem (OAP)**, whereas given an almost factor determining the 'best' complement will be called the **optimal completion problem (OCP)**.

Let us now assume that $b(s)$ is an almost factor of $a(s)$. Then eqn. 6.7c can be written in the form

$$T_{\nu+1}(\underline{b}^\mu)\underline{c}^\nu = \underline{a}^k - \underline{\varepsilon} \tag{6.12}$$

where $\underline{\varepsilon}$ expresses the error. The exact factorisation corresponds to the case where $\underline{\varepsilon} = 0$, and this is when condition eqn. 6.7c holds true. Note that if $T_{\nu+1}(\underline{b}^\mu) \triangleq T_b$ and $T_b^\perp$ denotes a left annihilator of T_b, then condition 6.12 implies that

$$\underbrace{T_b^\perp T_b}_{0}\underline{c}^\nu = T_b^\perp \underline{a}^k - \underbrace{T_b^\perp \underline{\varepsilon}}_{\triangleq \varepsilon'}$$

or

$$\underline{\varepsilon}' = T_b^\perp \underline{a}^k \tag{6.13}$$

It is clear that when $b(s)$ is an exact divisor, then $\underline{\varepsilon}' = \underline{0}$ and thus a good measure for the closeness of an almost divisor to that of an exact factor is the function defined by

$$\phi = \left\| T_b^\perp \underline{a}^k \right\|_2. \tag{6.14}$$

Clearly the value of ϕ is a measure of the proximity of $b(s)$ to be a true factor of $a(s)$; the smaller the value, the better the approximation, and when $\phi = 0$, then $b(s)$ becomes a true factor.

Of course, given an almost factor $b(s)$, the error $\underline{\varepsilon}$ in eqn. 6.12 also depends on the choice of $c(s)$. If we denote by $T_a^\perp$ a left annihilator of $\underline{a}^k$ (i.e. $T_a^\perp \in \mathcal{R}^{k \times (k+1)}$, $T_a^\perp \underline{a}^k = 0$, rank$(T_a^\perp) = k$), then eqn. 6.12 implies that

$$T_a^\perp T_{\nu+1}(\underline{b}^\mu)\underline{c}^\nu = \underbrace{T_a^\perp \underline{a}^k}_{0} - \underbrace{T_a^\perp \underline{\varepsilon}}_{\triangleq \underline{\varepsilon}''}$$

or

$$\underline{\varepsilon}'' = T_a^\perp T_b \underline{c}^\nu \tag{6.15}$$

From the above, it is clear that $\underline{\varepsilon}''$ provides an expression for measuring the 'best' selection $c(s)$ when $b(s)$ and $a(s)$ are given. The solution of the **optimal completion problem** is thus reduced to finding $\underline{c}^\nu$ such that $\|\underline{\varepsilon}''\|$ is minimised; such a minimum is denoted by ψ. We may summarise the analysis as follows.

Proposition 7:   Let $a(s) \in R[s]$, $\partial[a(s)] = k$, and let $b(s) \in R[s]$, $\partial[b(s)] = \mu$, be an almost factor with order ϕ as defined by eqn. 6.14. The solution of the optimal completion problem is equivalent to finding the vector $\underline{c}^\nu = \underline{c}$, where $\underline{c}$ is the singular vector of $T_a^\perp T_b$ that corresponds to the smallest singular value. $\square$

6.7.3 *Algorithm for computing the LCM of many polynomials*

Given a set of polynomials $\mathcal{P} = \{t_i(s), i = 1, 2, \ldots, \tau\}$ the computation of LCM is based on the following steps:

Step 1   Take two polynomials $t_1(s), t_2(s) \in R[s]$ and compute their GCD using any of the available numerical procedures. Let $z(s) = [t_1(s), t_2(s)]$ be this GCD. Then

(a) If $z(s) = 1$, i.e. the polynomials are coprime, then the LCM $p(s)$ is defined by $p(s) = [t_1(s), t_2(s)] = t_1(s) \cdot t_2(s)$.

(b) If $z(s) \neq 1$, i.e. the polynomials are not coprime, then evaluate $p(s)$ as the factor in the factorisation $a(s) = t_1(s)t_2(s) = p(s) \cdot z(s)$ by carrying out the following steps:

   (i)   Define the order ϕ of $z(s)$ in the approximate factorisation using eqn. 6.14.

   (ii) Define the solution of the numerical factorisation problem by defining the coefficient vector p of the LCM $p(s)$ as the singular vector corresponding to the smallest singular value of the matrix $Q = T_a^{\perp} T_z$, where $a(s) = t_1(s)t_2(s)$, $z(s)$ is the GCD, T_z is the appropriate Toeplitz matrix and $T_a^{\perp}$ is the left annihilator of $\underline{a}$. The corresponding singular value is the order ψ of the approximation.

The result of step 1 is the definition of the LCM of $t_1(s), t_2(s)$, i.e. $p_{1,2}(s) = [t_1(s), t_2(s)]$, defined as $p(s)$ from the above process.   □

Step 2   Using the associative property of LCM, consider now the pair $(p_{1,2}(s), t_3(s))$ and repeat step 1 to define their LCM $p_{1,2,3}(s)$.

The process will terminate when all polynomials of the set have been considered.   □

The numerical algorithm of the above method and its analysis can be found in [24].

6.8 Applications, examples and numerical performance of methods

ERES, MP methods and the LCM algorithm were programmed in the MATLAB environment and tested on a Pentium machine over several sets of polynomials $P_{m,d}$ characterised by various properties. Next, for each set of data we present the exact or approximate GCD and a table summarising the achieved results (Table 6.2). In the first column the applied method is given, and in the second column the obtained relative error in the final result is written; the third column shows the required accuracy for several intermediate calculations performed by each method, and in the fourth column the number of required iterations is recorded. For each set of data we also compute the exact or approximate LCM and the achieved results. The parameter accur1 expresses the required accuracy for GCD computation, whereas values lower than accur2 are set equal to zero. 'Rel. error' expresses the relative error.

Example 9

The following polynomial set contains randomly chosen real polynomials

$$\mathcal{P}_{3,3} = \{p_1(s) = 2.9s^2 + 14.85s + 15.75,$$

$$p_2(s) = 6.1s^3 + 11.65s^2 + 11.85s + 12.15,$$

$$p_3(s) = 3.7s^3 + 17.05s^2 + 30.35s + 19.65\}$$

Initial basis matrix: $P_3 \in \mathcal{R}^{3 \times 4}$
   Exact GCD $= s + 1.5$

Table 6.2

Method	Rel. Error	Accuracy	Iter.
ERES	$\leq 10^{-16}$	$\varepsilon_G = 10^{-15}$, $\varepsilon_t = 10^{-16}$	3
MP	$\leq 10^{-16}$	$\varepsilon = 10^{-15}$	—

$$\text{Exact LCM} = 25.5333 + 54.3697s + 66.6242s^2 + 56.1367s^3 + 29.5876s^4$$
$$+ 8.6386s^5 + s^6$$
$$\phi = O(1.0 \times 10^{-16}), \psi = O(1.0 \times 10^{-15})$$
$$accur1 = 10^{-12}, accur2 = 10^{-12}, rel = 1.8858 \times 10^{-15}$$

Example 10

The given polynomial set contains a large number of high degree polynomials:

$\mathcal{P}_{11,20} =$

$\{p_1(s) = s^{20} + 4s^{19} + 7s^{18} + 21s^{17} + 54s^{16} + 82s^{15} + 61s^{14} + 29s^{13} + 36s^{12}$
$\qquad + 47s^{11} + 26s^{10} + 7s^9 + 15s^8 + 20s^7 + 12s^6 + 6s^5 + 27s^4 + 131s^3$
$\qquad + 286s^2 + 318s + 140$

$p_2(s) = s^{20} + 3s^{19} + 4s^{18} + 2s^{17} + 3s^{14} + 9s^{13} + 12s^{12} + 6s^{11} + 5s^{10} + 15s^9$
$\qquad + 22s^8 + 16s^7 + 9s^6 + 7s^5 + 4s^4 + 2s^3$

$p_3(s) = s^{20} + 3s^{19} + 4s^{18} + 2s^{17} + 3s^{13} + 3s^{12} + 4s^{11} + 2s^{10} + s^6 + 3s^5 + 15s^4$
$\qquad + 35s^3 + 44s^2 + 22s$

$p_4(s) = 5s^{20} + 15s^{19} + 20s^{18} + 10s^{17} + 4s^{13} + 12s^{12} + 16s^{11} + 8s^{10} + 2s^8 + 6s^7$
$\qquad + 8s^6 + 4s^5 + 10s^3 + 30s^2 + 40s + 20$

$p_5(s) = -s^{20} - 3s^{19} - 4s^{18} - 2s^{17} - s^8 - 3s^7 - 4s^6 - 2s^5 + 30s^3 + 90s^2 + 120s$
$\qquad + 60$

$p_6(s) = s^{20} + 3s^{19} + 4s^{18} + 2s^{17} - 2s^{16} - 6s^{15} - 8s^{14} - 4s^{13} + s^{12} + 3s^{11} + 4s^{10}$
$\qquad - s^9 - 9s^8 - 12s^7 - 6s^6 + 11s^3 + 33s^2 + 44s + 22$

$p_7(s) = s^{20} + 3s^{19} + 4s^{18} + 2s^{17} + 11s^{10} + 33s^9 + 44s^8 + 22s^7 + 20s^3 + 60s^2$
$\qquad + 80s + 40$

$p_8(s) = s^{20} + 3s^{19} + 7s^{18} + 11s^{17} + 12s^{16} + 8s^{15} + 6s^{14} + 8s^{13} + 4s^{12} + 5s^9$
$\qquad + 15s^8 + 20s^7 + 10s^6 + 9s^3 + 27s^2 + 36s + 18$

$p_9(s) = s^{20} + 3s^{19} + 4s^{18} + 3s^{17} + 3s^{16} + 4s^{15} + 5s^{14} + 9s^{13} + 9s^{11} + 9s^{10}$
$\qquad + 17s^9 + 20s^8 + 10s^7 + s^6 + 3s^5 + 4s^4 + 5s^3 + 9s^2 + 12s + 6$

$p_{10}(s) = s^{20} + 2s^{19} + s^{18} - 2s^{17} - 2s^{16} + s^{12} + 3s^{11} + 4s^{10} + 2s^9 - s^8 - 3s^7$
$\qquad - 4s^6 + 2s^5 - 4s^3 - 12s^2 - 16s - 8$

$p_{11}(s) = s^{20} + 3s^{19} + 15s^{18} + 35s^{17} + 44s^{16} + 22s^{15} + 3s^{14} + 9s^{13} + 9s^{11}$
$\qquad + 4s^{10} + 2s^9 + 30s^3 + 90s^2 + 120s + 60 \}$

Initial basis matrix: $P_{11} \in \mathcal{Z}^{11 \times 21}$
  Exact GCD $= s^3 + s^2 + 4s + 2$

Table 6.3

Method	Rel. error	Accuracy	Iter.
ERES	$\leq 10^{-6}$	$\varepsilon_G = 10^{-8}, \varepsilon_t = 10^{-6}$	7
MP	$\leq 10^{-16}$	$\varepsilon = 10^{-15}$	—

Applying the LCM algorithm to the three first polynomials of the set we have:

$$\text{Exact LCM} = 140s^4 + 318s^5 + 286s^6 + 131s^7 + 27s^8 + 6s^9 + 12s^{10} + 20s^{11}$$
$$+ 15s^{12} + 7s^{13} + 26s^{14} + 47s^{15} + 36s^{16} + 29s^{17} + 61s^{18} + 82s^{19}$$
$$+ 54s^{20} + 21s^{21} + 7s^{22} + 4s^{23} + s^{24}$$
$$\phi = O(1.0 \times 10^{-15}), \psi = O(1.0e - 13)$$
$$\text{accur1} = 10^{-15}, \text{accur2} = 10^{-15}, \text{rel} = 2.3207 \times 10^{-13} \qquad \square$$

Example 11

The polynomials belonging to the next polynomial set have approximately equal root clusters [25]:

$$\mathcal{P}_{5,5} = \{p_1(s) = (s - 0.5)(s - 0.502)(s + 1)(s - 2)(s - 1.5),$$
$$p_2(s) = (s - 0.501)(s - 0.503)(s - 1)(s + 2)(s + 1.5),$$
$$p_3(s) = -4.998s^4 - 5.013997s^3 + 4.7414925s^2 - 6.0174985s + 1.509009$$
$$(\text{roots}: -1.09553, 0.50066, 0.502393, 1.09568),$$
$$p_4(s) = 0.697880794s^3 - 0.701037162s^2 + 0.178930204s - 0.0014439101$$
$$(\text{roots}: 0.00833993, 0.494572, 0.501611),$$
$$p_5(s) = 0.840067492s^2 - 0.841442765s + 0.210704053$$
$$(\text{roots}: 0.499717, 0.50192)\}$$

Thus, theoretically the polynomials of the above set are coprime. However, in several engineering computations it is useful to define an approximate GCD or LCM of the set within a specified accuracy. This approximate GCD can be used in several applications such as the definition of the almost zeros of a given polynomial set [4].

Initial basis matrix: $P_3 \in \mathcal{Z}^{3 \times 3}$
Exact GCD $= 1$ (computed with the MP method)
Approximate GCD (computed with the ERES method)

Table 6.4

ε_t	ε_G	Approximate GCD
2.0×10^{-7}	1.0×10^{-5}	$s - 0.5013$
1.4×10^{-2}	1.0×10^{-5}	$s^2 - 0.9909s + 0.2454$
		(roots: 0.5039,0.4870)

Approximate LCM
$$= s^{14} - 2.5047s^{13} - 5.9366s^{12} + 19.8998s^{11} + 3.0323s^{10} - 49.3315s^{9}$$
$$+ 30.1875s^{8} + 38.3610s^{7} - 50.1373s^{6} + 6.6894s^{5}$$
$$+ 18.3898s^{4} - 12.7507s^{3} + 3.4615s^{2} - 0.3632s + 0.0028$$

$\phi = O(1.0e - 011)$, $\psi = O(1.0e - 011)$

accur1 $= 2.0 \times 10^{-7}$, accur2 $= 1.0 \times 10^{-5}$, rel $= 5.3651 \times 10^{-14}$

Approximate LCM
$$= s^{13} - 0.8041s^{12} - 7.4025s^{11} + 7.3504s^{10} + 16.3555s^{9} - 21.9220s^{8}$$
$$- 9.1182s^{7} + 24.2975s^{6} - 7.0628s^{5} - 7.1821s^{4} + 6.0355s^{3}$$
$$- 1.7428s^{2} + 0.1883s - 0.0016$$

$\phi = O(1.0e - 004)$, $\psi = O(1.0e - 004)$

accur1 $= 1.0 \times 10^{-3}$, accur2 $= 1.0 \times 10^{-5}$, rel $= 0.0020$ □

6.8.1 *Performance of the ERES and MP methods*

Both methods start their computations with a matrix $P_m \in \mathcal{R}^{m \times (d+1)}$ of reasonable dimensions formed directly from the coefficients of the given polynomial set $\mathcal{P}_{m,d}$. They work satisfactorily with real polynomials, and in all the examples the achieved time for the execution of their algorithms was very satisfactory. If the rank of the original basis matrix P_m is less than m then both methods attain a reduction to the number of polynomials used by actually computing a basis of the original set. The property of reducing the original polynomial set is very important because you can also reduce the dimensions of the original matrix required for the method. For example, if we are given a $\mathcal{P}_{15,20}$ set with $\rho(P_{15}) = 5$ the methods will find a basis for the row space of P_{15} and will start their evaluations using only five polynomials. A main difference between the two methods is in the way that this basis is selected. The ERES method selects a most orthogonal 'uncorrupted' basis of the initial P_m (described in Section 6.3) and thus ERES can estimate an approximate GCD of a given set according to specified accuracies (see Example 5 above). On the other hand, the MP method finds an orthogonal basis for the row space of the given matrix. As far as accuracy is concerned, both methods give satisfactory results. The MP method always attains high accuracy in the final GCD approximation due to the application of a direct formula achieving the required computation, whereas in the ERES method we have to take into account the number of required iterations. When we encounter polynomial sets $\mathcal{P}_{m,d}$ characterised by large values of m, d the convergence of the ERES method does not require a large number of iterations since the increase of m causes a quicker reduction to the degree d.

The problem of selecting the accuracies is crucial for both methods. In ERES, the appropriate choice of ε_G and ε_t influences the correctness of the results. In all the examples the values of ε_G remained in the range $\{10^{-16}, \ldots, 10^{-7}\}$, whereas the values of ε_t remained in the range $\{10^{-16}, \ldots, 10^{-3}\}$. Generally the values of the Gaussian accuracy ε_G and the termination accuracy ε_t are different. In the MP method, the values of the nullity accuracy ε were always $\leq 10^{-15}$ for almost all the tested examples.

The MP algorithm can successfully separate nearly identical factors. For example, if the MP algorithm is applied to the following two polynomials: $p_1(s) = (s-1)^2(s-2)(s-2-n)$, $p_2(s) = p_1{}'(s)$ for values of $n = 10^{-2}$, 10^{-3}, 10^{-4}, 10^{-5} and $\varepsilon = 10^{-16}$, the method computes the factor $s - 1$ exactly as the GCD of the polynomials. For $n = 10^{-6}$, $\varepsilon = 10^{-16}$ the method distinguishes the above factor with three significant digits. For values of $n \geq 10^{-7}$ and for $\varepsilon = 10^{-16}$ the method calculates the polynomial $(s-1)(s-2)$ as the GCD.

It has been observed empirically that, if the elements of the initial basis matrix P_m vary greatly in size, then it is likely that loss of significance errors will be introduced and that the rounding error propagation will be worse. To avoid this problem, especially when an iterative process such as the ERES method is applied, it is appropriate in the beginning of the evaluations to scale matrix P_m so that its elements vary less. A specific row-scaling named B-scaling (see [26]), which uses the machine's arithmetic base B and does not cause any rounding errors, is applied.

Estimating the performance of ERES and MP methods for the computation of GCD of a given $\mathcal{P}_{m,d}$ it is difficult to select a 'best' method according to specified criteria and suitable for any given polynomial set. Each method combines advantages and disadvantages, and the important thing is to select the method combining the most advantages with respect to the given set of data. According to the previous analysis, Table 6.5 proposes the recommended method as a function of the given data.

A comparison of the ERES and MP methods with other matrix-based processes can be found in References 6 and 8.

6.8.2 *Performance of the LCM method*

The proposed LCM method computes the LCM of several polynomials, avoiding the computation of the roots of the polynomials. The method works satisfactorily for various polynomial sets, real or integer. According to the

Table 6.5

Data	Method
$P_{m,d}$, $r(P_m) = r \ll m$, $m = \min\{m, d\}$ $= r \ll d$, $d = \min\{m, d\}$	MP,ERES
$P_{m,d}$, m large	MP,ERES
$P_{m,d}$ with approximate GCD	ERES
$P_{2,d}$ or other cases	MP

method that will be selected for the required GCD computations (such as the ERES), an approximate LCM can be computed whenever it is needed.

6.9 Conclusions

The paper has provided an overview of the main issues in the emerging area of 'approximate' algebraic computations, and it has been focused on the nongeneric computations. We have considered the problem of almost zeros and the computation of the GCD and LCM of many polynomials. Although the current definition of almost zeros serves the interpretation of certain phenomena of the multiparameter root locus [4] well, it is sensitive to scaling and better definitions are needed. The GCD algorithms which have been presented perform extremely well; these algorithms, together with appropriate tools from exterior algebra computation, provide the means for the computation of the Smith form of polynomial matrices. This approach, however, suffers from computational explosion as the size of the matrices increases. Alternative numerical methods for computing Smith forms are under investigation. The LCM problem uses the GCD methodology and associativity. Improved methods are required here to avoid the computational explosion that may occur when many polynomials are involved. The overall area is in its early stages of development, and there are many open issues. One that needs special attention is the merging of symbolic and numerical procedures.

6.10 References

1 WONHAM, W. M.: 'Linear multivariable control: a geometric approach' (Springer-Verlag, New York, 1984)

2 ROSENBROCK, H.: 'State space and multivariable theory' (Nelson, London, 1970)

3 KAILATH, T.: 'Linear systems' (Prentice Hall, Inc., Englewood Cliffs, NJ, 1980)

4 KARCANIAS, N., and GIANNAKOPOULOS, C.: 'Almost zeros of a set of polynomials of $\mathcal{R}[s]$', *Int. J. Control*, 1983, **38**, pp. 1213–1238

5 MITROULI, M., and KARCANIAS, N.: 'Computation of the GCD of polynomials using Gaussian transformation and shifting', *Int. J. Control*, 1993, **58**, pp. 211–228

6 MITROULI, M., KARCANIAS, N., and KOUKOUVINOS, C.: 'Further numerical aspects of the ERES algorithm for the computation of the greatest common divisor of polynomials and comparison with other existing methodologies', *Util. Math.*, 1996, **50**, pp. 65–84

7 KARCANIAS, N., and MITROULI, M.: 'A matrix pencil based numerical method for the computation of the GCD of polynomials', *IEEE Trans. Autom. Control*, 1994, **39**, pp. 977–981

8 MITROULI, M., KARCANIAS, N., and KOUKOUVINOS, C.: 'Numerical performance of the matrix pencil algorithm computing the greatest common divisor of polynomials and comparison with other matrix-based methodologies', *J. Comput. Appl. Math.*, 1996, **76**, pp. 89–112

9 MITROULI, M.: 'Numerical issues and computational problems in algebraic control theory'. PhD thesis, City University, London, UK, 1991

10 MARCUS, M., and MINC, M.: 'A survey of matrix theory and matrix inequalities' (Allyn and Bacon, Boston, 1964)
11 MITROULI, M., KARCANIAS, N., and KOUKOUVINOS, C.: 'Numerical aspects for nongeneric computations in control problems and related applications', *Congressus Numerantium*, 1997, **126**, pp. 5–9
12 GOLUB, G. H., and VAN LOAN, C. F.: 'Matrix computations' (The Johns Hopkins University Press, Baltimore, London, 1989, 2nd edn.)
13 WILKINSON, J. H.: 'Rounding errors in algebraic processes' (Her Majesty's Stationary Office, London, 1963)
14 LEVENTIDES, J., and KARCANIAS, N.: 'Genericity, generic system properties and generic values of invariants'. Research Report SDCU 047, 20 May 1996, Control Engineering Centre, City University
15 KARCANIAS, N., SHAN, X. Y., and MILONIDES, E.: 'Structural identification problem in early process design: the generic McMillan degree problem'. Proceedings of the 4th IEEE Med. Symp. MSCA, 96
16 FOSTER, L.: 'Rank and null space calculations using matrix decomposition without column interchanges', *Lin. Alg. Appl.*, 1986, **74**, pp. 47–71
17 CHEN, C. T.: 'Linear Systems Theory and Design', (Holt-Saunders International Edition, 1984)
18 MITROULI, M., and KARCANIAS, N.: 'Computational aspects of almost zeros and related properties'. Proceedings of the 3rd European Control Conference, Rome, Italy, September 1995, pp. 2094–2099
19 BARNETT, S.: 'Greatest common divisor of several polynomials', *Proc. Camb. Philos. Soc.*, 1970, **70**, pp. 263–268
20 PACE, I., and BARNETT, S.: 'Comparison of algorithms for calculation of GCD of polynomials', *Int. J. Syst. Sci.*, 1973, **4**, pp. 211–226
21 BLANKISHIP, W.: 'A new version of Euclid algorithm', *Am. Math. Mon.*, 1963, **70**, pp. 742–744
22 KARCANIAS, N.: 'Invariance properties and characterisation of the greatest common divisor of a set of polynomials', *Int. J. Control*, 1987, **40**, pp. 1751–1760
23 HORN, R., and JOHNSON, C.: 'Matrix analysis' (Cambridge University Press, 1985)
24 KARCANIAS, N., and MITROULI, M.: 'Numerical computation of the least common multiple of a set of polynomials', Control Engineering Centre, City University Research Report, February 1996, CEC/NK-MM/165
25 NODA, M., and SASAKI, T.: 'Approximate GCD and its applications to ill-conditioned algebraic equations', J. Comput. Appl. Math., 1991, **38**, pp. 335–351
26 ATKINSON, K.: 'An introduction to numerical analysis' (John Wiley and Sons, New York, 1978)
27 GANTMACHER, F.: 'The theory of matrices', Vol. I (Chelsea Publishing Company, New York, NY, 1959)
28 KARCANIAS, N., GIANNAKOPOULOS, C.: 'Grassmann invariants, almost zeros and the determinantal zero, pole assignment problems of linear systems', *Int. J. Control*, 1984, **40**, pp. 673–609

## Chapter 7

Robust stability conditions for MIMO systems with parametric uncertainty

E. Kontogiannis and N. Munro

7.1 Introduction

This chapter examines the robust stability of multi-input/multi-output (MIMO) systems with parametric uncertainty, based on Rosenbrock's Direct Nyquist Array stability theorem. Two classes of structured uncertainty are considered, namely the interval and the affine linear types. Various measures of the diagonal dominance of a transfer function matrix are reviewed, and the original matrix row/column dominance concept, the subsequent generalised dominance concept, and more recent fundamental diagonal dominance measure are all extended to incorporate uncertainty in the system parameters. The fundamental dominance condition is also used as a nominal system approximation measure, generalising the diagonal system approximation assumption inherent in all diagonal dominance measures. Illustrative examples are presented.

The H_∞ approach to robust control system design is now well established [1–3]. However, in recent years another approach to robust stability and performance has emerged, based on the theory of interval polynomials [4–6], that can result in much simpler controllers. The familiar Nyquist diagram, Bode plots, Nichols chart, and Evans root-locus have all been generalised, in this framework, to systems where the coefficients of the polynomials describing the elements of the model are uncertain, in the sense that they are not known exactly, but are known to lie within certain limits. This situation arises frequently in practice, due to our inability to exactly model the dynamics of a system, parameter variations occurring as the system moves within its operating envelope, or can arise due to the effects of manufacturing tolerances on the performance of a system or subsystem. An introduction to this theory of uncertain systems is given.

Although some work [7] has also recently been carried out to extend this approach to the case of multivariable systems, the approach to be explained here is based on extensions to Rosenbrock's Direct Nyquist Array stability theorem [8] for multivariable systems described by a square transfer-function matrix, which is diagonally dominant. The diagonal dominance concept, and Rosenbrock's stability theorem for systems that are diagonally dominant, is first briefly reviewed, and several subsequent further refinements of this approach are also described. This lays the foundation for the recent extensions of these results to uncertain multivariable systems presented here.

In the sequel, italics are used to denote uncertain functions, such as transfer functions, polynomials, etc., and bold type is used to denote vectors and/or matrices. Following this notation, bold italics will be used to denote vectors or matrices whose elements are uncertain.

7.2 Direct Nyquist array design method

For the feedback configuration shown in Figure 7.1, let $G(s) = [g_{rc}(s)]_{r,c=1,\ldots,n}$ denote an $n \times n$ transfer function matrix of a MIMO plant, and $\mathbf{F} = \mathrm{diag}[f_1, \ldots, f_n]$ denote a constant diagonal matrix of feedback gains $(f_1 \neq 0)$.

In 1974, Rosenbrock [8] showed that a MIMO system design can be simplified into a number of single-input/single-output (SISO) designs if the interaction within the system is small. As a measure of interaction Rosenbrock introduced the concept of row/column diagonal dominance, which is based on the well known Gershgorin theorem [8, 9], and is defined as follows.

Definition 1 A transfer function $\mathbf{Z}(s) = [z_{rc}(s)]_{r,c=1,\ldots,n}$ is said to be column diagonally dominant on the Nyquist D-contour if $z_{ii}(s)$ has no pole on D; $i = 1, \ldots, n$; and the column diagonal dominance measure function $\phi_c(j\omega)$, defined as

$$\phi_c(\omega) = |z_{cc}(j\omega)| - \sum_{\substack{r=1 \\ r \neq c}}^{n} |z_{rc}(j\omega)| \qquad c = 1, \ldots, n \qquad (7.1)$$

where $|\cdot|$ denotes the magnitude, is positive $\forall s$ on D. The corresponding row diagonal dominance measure can be defined similarly. Finally, $\mathbf{Z}(s)$ is called diagonally dominant if, for each s on D, $\mathbf{Z}(s)$ is either row or column diagonally dominant.

An alternative definition of the column diagonal dominance measure function is given below:

$$\phi_c(\omega) = \frac{\sum_{\substack{r=1 \\ r \neq c}}^{n} |z_{rc}(j\omega)|}{|z_{cc}(j\omega)|} \qquad (7.2)$$

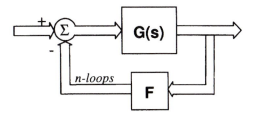

Figure 7.1   Basic MIMO feedback configuration

in which case the necessary condition for diagonal dominance becomes $\phi_c(\omega) < 1$.

The graphical interpretation of diagonal dominance at each frequency ω_k over the bandwidth of interest is a disc centred on the diagonal element $z_{cc}(j\omega_k)$, with radius equal to the sum of the magnitudes of the off-diagonal elements. Consequently, as s traverses the Nyquist D-contour, the corresponding discs on each diagonal term sweep out a band, which is called the Gershgorin band. If the bands so produced exclude the origin in all the loops, then the transfer-function matrix $\mathbf{Z}(s)$ is diagonally dominant on D.

Let $\mathbf{R}(s)$ denote the modified return difference matrix,[1] where

$$\mathbf{R}(s) = \big[r_{ij}(s)\big]_{i,j=1,\dots,n} = \mathbf{F}^{-1} + \mathbf{G}(s) \tag{7.3}$$

and define the direct Nyquist array (DNA) of the transfer-function matrix $\mathbf{G}(s)$ as the Nyquist plots of its individual elements. Then, for the feedback configuration of Figure 7.1, the following stability theorem exists [8], which is called the DNA stability theorem.

Theorem 1   *Consider the feedback configuration of Figure 7.1, and suppose that the modified return difference matrix $\mathbf{R}(s)$ is diagonally dominant. Let the transfer function $g_{cc}(s)$, corresponding to the cth diagonal element of $\mathbf{G}(s)$, map the Nyquist D-contour into Γ_c. Suppose that Γ_c encircles the critical point $(-f_c^{-1}, 0)$ N_c times anti-clockwise, where $c = 1, \dots, n$. Then, the closed-loop system is asymptotically stable if*

$$\sum_{i=1}^{n} N_i + p_0 = 0 \tag{7.4}$$

where p_0 denotes the number of open-loop unstable poles.

---

[1]The actual return difference transfer function matrix is defined as $\mathbf{R}_a(s) = \mathbf{I} + \mathbf{G}(s)\mathbf{F}$. However,

$$\det[\mathbf{I} + \mathbf{G}(s)\mathbf{F}] = 0 \Leftrightarrow \det\big[(\mathbf{F}^{-1} + \mathbf{G}(s))\mathbf{F}\big] = 0 \Leftrightarrow \det\big[\mathbf{F}^{-1} + \mathbf{G}(s)\big]\det[\mathbf{F}] = 0$$
$$\Leftrightarrow \det\big[\mathbf{F}^{-1} + \mathbf{G}(s)\big] = 0$$

since $\mathbf{F}$ is diagonal and nonsingular. Hence, for the feedback configuration in Figure 7.1, the diagonal dominance of the modified return difference matrix, defined here, implies the diagonal dominance of the normally used form.

Note that exclusion of the origin by a band based on $f_c^{-1} + g_{cc}(s)$ is the same as exclusion of the critical point $(-f_c^{-1}, 0)$ by the band based on $g_{cc}(s)$, since $\mathbf{F}$ is assumed to be constant. The latter graphical interpretation of the diagonal dominance of $\mathbf{R}(s)$ will be used in the sequel.

The DNA stability theorem has been used extensively in recent years. However, a main drawback of this method is that the row/column diagonal dominance measure is conservative, and the diagonal dominance condition as defined by Rosenbrock can sometimes be quite hard to achieve. However, alternative definitions, such as the generalised diagonal dominance measure [10], the **M**-matrix condition [11], the fundamental dominance [12] and the L-dominance [12] measures were subsequently introduced, which improved the applicability of the above theorem. In the sequel, the generalised diagonal dominance and the fundamental dominance measures are presented.

The generalised dominance measure of a transfer-function matrix $\mathbf{Z}(s)$ is based on the Perron-Frobenius (PF) eigenvalue of a non-negative matrix[2] [13], and is defined as follows.

Definition 2   Let $\mathbf{T}(\omega)$ be the comparison matrix of a transfer-function matrix $\mathbf{Z}(s)$, defined as

$$\mathbf{T}(\omega) = |\mathbf{Z}(j\omega)| \left(\text{diag}\left[|\mathbf{Z}(j\omega)|\right]\right)^{-1} \tag{7.5}$$

where $|\mathbf{A}|$ denotes a matrix $\mathbf{A}$ whose elements are replaced by their moduli. If $\lambda[\mathbf{T}(\omega)] < 2$, where $\lambda[\cdot]$ denotes the PF eigenvalue of $[\cdot]$, for all frequencies ω around the Nyquist D-contour, then $\mathbf{Z}(s)$ is said to be generalised diagonally dominant.

Based on this definition, Mees [14] defined an optimal constant diagonal transformation (scaling), which can improve the row/column diagonal dominance of a transfer-function matrix at some frequency ω_k and in the nearby neighbourhood, in the sense that the corresponding rows/columns of the resulting matrix $\mathbf{S}\,\mathbf{Z}(j\omega_k)\,\mathbf{S}^{-1}$, where $\mathbf{S} = \text{diag}[x_i]$, and the x_i are the components of the PF left eigenvector of $\mathbf{T}(\omega_k)$, have balanced row/column diagonal dominance measures. However, in the general context of improving the column (row) diagonal dominance of $\mathbf{Z}(j\omega)$, the application of the post- (pre-) transformation matrix only is sufficient. Moreover, by allowing $\mathbf{S}$ to be dynamic, $\mathbf{S}(j\omega)$, Munro [15] showed that the optimal column (row) diagonal dominance can be achieved over a wider range of frequencies, by approximating the magnitude variation of each component of the PF left (right) eigenvector by a suitable simple transfer function.

As in the row/column diagonal dominance case, the generalised dominance of the cth loop of the transfer-function matrix $\mathbf{R}(s)$ for each frequency can be

---

[2] The maximum eigenvalue of a non-negative matrix is called the PF eigenvalue. For such matrices, it has been proven that the PF eigenvalue is always real and non-negative.

visualised graphically in a DNA plot as a disc centred on the diagonal element $g_{cc}(s)$, with radius

$$\phi(j\omega) = (\lambda[\mathbf{T}(j\omega)] - 1) \left| g_{cc}(j\omega) + f_c^{-1} \right| \tag{7.6}$$

Substituting now for the terms 'row/column dominance' of $\mathbf{R}(s)$ by the 'generalised diagonal dominance' of the same matrix in Theorem 1, the DNA stability theorem is transformed to the generalised diagonal dominance stability theorem [10].

An alternative type of dominance that exploits the concept of system approximation is the fundamental dominance concept. This dominance criterion includes the diagonal dominance measure, considered above, as a special case.

Definition 3 Given a regular splitting of a transfer-function matrix $\mathbf{Z}(s)$, defined as

$$\mathbf{Z}(s) = \mathbf{D}(s) + \mathbf{B}(s) \tag{7.7}$$

such that $\mathbf{D}(s)$ is nonsingular, the transfer-function matrix $\mathbf{Z}(s)$ is said to be fundamentally dominant if

$$\rho[\mathbf{B}(j\omega)\mathbf{D}^{-1}(j\omega)] < 1, \qquad \forall\omega \tag{7.8}$$

where $\rho[\cdot]$ denotes the spectral radius of $[\cdot]$.

The fundamental dominance (FD) stability theorem can then be defined as follows.

Theorem 2 *Consider the feedback configuration shown in Figure 7.1, and let the modified return difference matrix have a regular splitting, defined as $\mathbf{R}(s) = \mathbf{T}_D(s) + \mathbf{B}(s)$, over the Nyquist D-contour, where $\mathbf{T}_D(s) \triangleq \mathbf{F}^{-1} + \mathbf{D}(s)$ is nonsingular for all $s \in D$. Also, let Γ_D be the graph of $\det[\mathbf{T}_D(s)]$ which maps the Nyquist contour onto the complex plane. Then, if*

(i) $$\rho[\mathbf{B}(j\omega)\mathbf{T}_D^{-1}(j\omega)] < 1, \qquad \forall\omega$$

and

(ii) $$n_D + p_0 = 0$$

then the closed-loop system is asymptotically stable, where n_D is the total winding number of Γ_D about the origin.

In the case where the matrix $\mathbf{T}_D(s)$ is chosen to be the diagonal part of the matrix $\mathbf{R}(s)$, i.e.

$$\mathbf{T}_D(s) = \text{diag}[f_c^{-1} + g_{cc}(s)], \qquad c = 1, \dots, n \tag{7.9}$$

then condition (i) above corresponds to the fundamental diagonal dominance condition, and condition (ii) can be substituted for eqn. 7.4. On a DNA plot, the fundamental diagonal dominance discs in the cth loop are defined as the discs centred on the diagonal element with radius

$$\phi(\omega) = \rho[\mathbf{B}(j\omega)\mathbf{T}_D^{-1}(j\omega)] \left| f_c^{-1} + g_{cc}(j\omega) \right| \tag{7.10}$$

However, depending on the problem being addressed, different choices of splitting can be considered. For instance, if the open-loop system $\mathbf{G}(s)$ is uncertain, an alternative choice might be to consider $\mathbf{T}_D(s)$ as representing the nominal system, and $\mathbf{B}(s)$ as representing the error term due to modelling errors, parameter variations, etc. It could also be the case that in order to simplify the design task, a specific splitting has to be considered. Such a case is the diagonal splitting given by eqn. 7.9, when a decentralised control strategy is the desired solution. For a complete elaboration on this subject, the reader is referred to Reference 16.

Note now that the column diagonal dominance of $\mathbf{R}(j\omega)$ can be expressed in terms of the regular diagonal splitting of eqn. 7.9, as $\|\mathbf{B}(j\omega)\mathbf{T}_D^{-1}(j\omega)\|_1$, where $\|\cdot\|_1$ denotes the 1-norm of the matrix, the row diagonal dominance can be expressed as the corresponding ∞-norm, and the generalised diagonal dominance as the W-(optimally) scaled norm. Based now on the fact that all eigenvalues of a matrix are bounded above by any induced norm, it is then obvious that the fundamental diagonal dominance condition is the best dominance measure, i.e. the least conservative; and also that all previous measures are special cases of this one.

It is also interesting to note that the FD stability theorem provides sufficient conditions for the approximation of the open-loop system $\mathbf{G}(s)$ by a subsystem $\mathbf{D}(s)$, the nonsingular term of the regular splitting chosen. In other words, the stabilisation of $\mathbf{D}(s)$ would result in the stabilisation of the original system $\mathbf{G}(s)$, provided that the FD condition is satisfied. However, the FD condition is affected by the addition of any compensator (even diagonal) in the forward path, or any change in existing compensators (previously implemented), as well as any change in the diagonal feedback matrix $\mathbf{F}$. Hence, each time such a change is made, re-calculation of the FD condition should be carried out. Also, the fundamental dominance test is carried out for all the loops simultaneously.[3] For these reasons, the FD stability theorem is mainly considered as an analytical tool rather than a design tool, as opposed to the DNA stability theorem, which is more conservative as an analysis tool, but allows each control-loop to be designed and tuned independently.

7.3 Uncertain parametric systems and polynomial families

Prior to the seminal theorem of Kharitonov [17] on the robust stability analysis of uncertain systems, very little attention had been given by control researchers to the real parameter uncertainty case. The problem of real parametric uncertainty was computationally very complex, if not impossible, depending on the dimensionality of the problem and the computing power available. The only

---

[3] This is also a drawback of the generalised diagonal dominance condition.

possible solution in most of the cases was the substitution of the parameter uncertainty by overbounding complex plane discs, which is obviously conservative. Kharitonov's work revealed to the control community the potential of a completely new approach to robust control. Many significant developments have been reported since, and the interested reader is referred to the books by Barmish [4], Bhattacharyya [5] and Ackermann [6] for an extended review of the most notable ones. For the purpose of this chapter, a few stability testing results of polynomial families and the frequency response of uncertain parametric systems are considered in the subsequent Sections. Before doing this, a classification of polynomials with uncertain coefficients (polynomial families), according to the way the uncertainty enters into the coefficients, as well as an equivalent classification for uncertain parametric systems, is given below.

An uncertain polynomial $p(s) = \sum_{i=0}^{n} \alpha_i s^i$ is called an *interval* polynomial family if its coefficients $\alpha_i \in [\alpha^-, \alpha^+]$ are independent intervals. If the coefficients of the polynomial family are affine functions with respect to the vector of uncertainties $\mathbf{q} = [q_1, q_2, \ldots, q_\ell] \in Q \subseteq \mathcal{R}^\ell$ (i.e. $\alpha_i(\mathbf{q}) = \beta_{i,0} + \sum_{k=1}^{\ell} \beta_{i,k} q_k$, where $\beta_{i,k} \in \mathcal{R}$; $i = 1, \ldots, n$; $k = 0, \ldots, \ell$), then the polynomial family is called *affine linear*. An uncertain polynomial $p(s, \mathbf{q}) = \sum_{i=0}^{n} \alpha_i(\mathbf{q}) s^i$ is called a *multilinear (multiaffine)* polynomial if its coefficients $\alpha_i(\mathbf{q})$ are multilinear (multiaffine) functions with respect to the elements of the uncertainty vector $\mathbf{q}$. That is, if all but one component of $\mathbf{q}$ is fixed, then $\alpha_i(\mathbf{q})$ is (affine) linear in the remaining components, for all $i = 1, \ldots, n$. More generally, $p(s, \mathbf{q})$ is said to have a *polynomic* uncertainty structure, if each of the coefficient functions $\alpha_i(\mathbf{q})$ is a multivariate polynomial with respect to the elements of the uncertainty vector $\mathbf{q}$.

It is well known that the stability analysis problem of linear time-invariant systems is equivalent to locating the roots of the corresponding characteristic polynomial. Thus, it is evident that the solution of the problem of calculating the root space accurately (as opposed to roots) of a characteristic polynomial family in the complex plane would be a breakthrough in the study of systems 'suffering' from real parametric uncertainty. The following Section is devoted to this problem. In the following, a classification equivalent to that of polynomial families is given for uncertain linear time-invariant systems.

Consider a general MIMO system described by the transfer-function matrix $\mathbf{G}(s) = [g_{rc}(s)]_{r,c=1,\ldots,n}$ and assume that the coefficients of the rational polynomial functions defining each of the $g_{rc}(s)$ vary within prescribed limits. Two types of uncertain parametric systems are considered here, namely the interval and the affine linear systems.

A transfer function $g_{rc}(s, \mathbf{a} = [a_0, a_1, \ldots, a_m], \mathbf{b} = [b_0, b_1, \ldots, b_p])$, defined as a ratio of two independent interval polynomial families, is called an interval transfer function, i.e.

$$g_{rc}(s, \mathbf{a}, \mathbf{b}) = \frac{n_{rc}(s, \mathbf{a})}{d_{rc}(s, \mathbf{b})} = \frac{a_m s^m + a_{m-1} s^{m-1} + \cdots + a_0}{b_p s^p + b_{p-1} s^{p-1} + \cdots + b_0} \tag{7.11}$$

where

$$a_i \in \left[a_i^-, a_i^+\right] \subseteq \mathcal{R}, \qquad i = 0, \ldots, m$$

$$b_i \in \left[b_i^-, b_i^+\right] \subseteq \mathcal{R}, \qquad i = 0, \ldots, p \tag{7.12}$$

and $\underline{\mathbf{a}} \in Q_A$, $\underline{\mathbf{b}} \in Q_B$ are independent perturbation vectors.

The uncertain transfer function $g_{rc}(s, \underline{\mathbf{q}})$ is said to be affine linear if it can be written as

$$g_{rc}(s, \underline{\mathbf{q}}) = \frac{n_{rc}(s, \underline{\mathbf{q}})}{d_{rc}(s, \underline{\mathbf{q}})} = \frac{n_{0,rc}(s) + \displaystyle\sum_{k=1}^{\ell} q_k n_{k,rc}(s)}{d_{0,rc}(s) + \displaystyle\sum_{k=1}^{\ell} q_k d_{k,rc}(s)} \tag{7.13}$$

where

$$\underline{\mathbf{q}} = [q_1, q_2, \ldots, q_\ell] \in Q \subseteq \mathcal{R}^\ell$$

$$q_i \in \left[q_i^-, q_i^+\right], \qquad i = 1, \ldots, \ell \tag{7.14}$$

and ℓ is the number of uncertain parameters. Again, the elements of $\underline{\mathbf{q}}$ are assumed to be independent.

Classical control system analysis and design tools like the Nyquist and the Bode plot, as well as the Nichols chart, are heavily dependent on the ability to calculate accurately the frequency response of the system under consideration. In the following Section, such algorithms are given for the case of interval and affine linear systems.

7.4 From classical to robust control

Associated with an interval polynomial family $P = \left\{ p(s, \underline{\mathbf{q}}), \ \underline{\mathbf{q}} \in Q \subset \mathcal{R}^n \right\} = \sum_{i=1}^{n} q_i s^i$, where $q_i \in [q_i^-, q_i^+]$, are its four fixed Kharitonov polynomials, defined as:

$$K_1(s) = q_0^- + q_1^- s + q_2^+ s^2 + q_3^+ s^3 + q_4^- s^4 + q_5^- s^5 + q_6^+ s^6 + \cdots$$

$$K_2(s) = q_0^+ + q_1^+ s + q_2^- s^2 + q_3^- s^3 + q_4^+ s^4 + q_5^+ s^5 + q_6^- s^6 + \cdots$$

$$K_3(s) = q_0^+ + q_1^- s + q_2^- s^2 + q_3^+ s^3 + q_4^+ s^4 + q_5^- s^5 + q_6^- s^6 + \cdots \tag{7.15}$$

$$K_4(s) = q_0^- + q_1^+ s + q_2^+ s^2 + q_3^- s^3 + q_4^- s^4 + q_5^+ s^5 + q_6^+ s^6 + \cdots$$

Kharitonov [17] showed that the interval polynomial family P is robustly stable if and only if the four polynomials defined in eqn. 7.15, which are called the Kharitonov polynomials, are stable. The computational savings introduced by this theorem are evident; namely, the problem of verifying the stability of an infinite number of polynomials is transformed to the stability analysis of four fixed polynomials. Dasgupta [18] also showed that the value (image) set of P on the complex plane at any given frequency s $= j\omega$, denoted by $P(j\omega)$, is an axis-parallel rectangle called the Kharitonov rectangle, whose four vertices correspond to the four Kharitonov polynomials. Based on the latter, the zero exclusion condition, a simple graphical test for robust stability of interval polynomials is defined below [4].

Lemma 1 (zero exclusion condition)   Suppose that an interval polynomial family $P = \{p(s, \mathbf{q}), \mathbf{q} \in Q\}$ has invariant degree and at least one stable member $p(s, \underline{\mathbf{q}}^0)$. Then, P is robustly stable if and only if the origin is excluded from the Kharitonov rectangle calculated at all non-negative frequencies.

Lemma 1 can be generalised easily to tackle the robust D-stability problem; i.e. robust stability with respect to any open subset of the complex plane; as well as more complicated uncertainty structures, such as affine linear and multilinear. To do so, it is only necessary to substitute the Kharitonov rectangle at a fixed $z \in C$ by the plot of the corresponding value set of the polynomial family. For the case of affine linear polynomial families with uncertainty bounding set $Q = \text{conv}\{\mathbf{q}^i\}$, where $\mathbf{q}^i, i = 1, \ldots, 2^\ell$, denotes the ith vertex of the parameter space, and 'conv' denotes the convex hull function, it has been proved that the value set of the polynomial family for a fixed $z \in C$ is a polygon defined as $P(z) = \text{conv}\{p(z, \mathbf{q}^i)\}$. If P is a multilinear polynomial family, then, using the mapping theorem [19], it can be proved that the polygon $P(z) = \text{conv}\{p(z, \mathbf{q}^i)\}$ is the convex hull approximation of the actual value set at the given $z \in C$. This means that the zero exclusion condition associated with these polygons provides sufficient (but not necessary) conditions for the robust stability of multilinear polynomial families. The tightness of the convex hull approximation of the actual value set can be improved using a subdivision procedure of the parameter space ($Q = \cup Q_i$), a procedure initially proposed by Gaston [20]. Using this procedure, the value set of the polynomial family is defined as the union of the value sets at each subspace Q_i, i.e. $P(s) = \cup_i p(s, Q_i)$.

Another well-known robust stability condition for affine linear polynomial families is the edge theorem [21], which guarantees that the boundary of the root space of an affine linear polynomial family is a subset of the root space of the polynomials corresponding to the edges of the parameter space. The generalised Kharitonov theorem [5] provides necessary and sufficient conditions for the stability of linear interval polynomials, i.e. polynomial families, which can be written in the form $P = \sum_{i=1}^m F_i(s)P_i$, where the $F_i(s)$ are fixed polynomials, and the P_i are interval polynomials. As illustrated in [6], the value set of a multilinear polynomial family may include a hole. This situation cannot be depicted with the mapping theorem-based approach described earlier. Necessary and sufficient conditions for the stability of multilinear polynomial families can be found in [22], which showed that real unstable roots are always attained at vertices of the parameter space, whereas checking for complex unstable roots involves examining the real solutions of up to $\ell + 1$ simultaneous equations, where ℓ is the number of uncertain parameters. Polyak *et al.* [23] introduced the notion of principal points, which also provide necessary and sufficient conditions for the robust stability of multilinear polynomials, but as in the previous case the computational complexity of the algorithm may be enormous.

Up to this point robust stability analysis results of uncertain parametric systems based on the root location of the corresponding characteristic polynomial have been given. In the sequel, the robust frequency response of interval and affine linear systems is given, based on the classical frequency-domain control techniques, such as the Nyquist diagram, the Bode plots and the Nichols chart; they are easily extended for such uncertain systems. The robust frequency response of uncertain parametric systems is defined as follows:

'Determine the boundary of the value set of $g(j\omega, \underline{\mathbf{q}}) \ \forall \underline{\mathbf{q}} \in Q$, denoted as $\partial(g(j\omega, Q)), \forall \omega \in D$'.

For the case of interval systems, the boundary of the frequency response is obtained by a one-parameter scan of the frequency response of 32 (at most) edges, obtained from the definition of the Kharitonov polynomials and the Kharitonov segments (generalised Kharitonov theorem [5]). More precisely, the boundary of the frequency response of the interval transfer function, $g(s, \underline{\mathbf{a}} \in Q_A, \underline{\mathbf{b}} \in Q_B)$, is a subset of the following set of points:

$$\partial\big(g(s, Q_A, Q_B)\big) \subset \left\{ \frac{K_n^i(s)}{\lambda K_d^j(s) + (1 - \lambda)K_d^k(s)}, \ \frac{\lambda K_n^j(s) + (1 - \lambda)K_n^k(s)}{K_d^i(s)} \right\} \quad (7.16)$$

where $\lambda \in [0, 1], i = 1, \ldots, 4, (j, k) \in (1, 2), (1, 3), (2, 3), (3, 4)$ and K_n^i, K_d^i correspond to the ith Kharitonov polynomials of the numerator and denominator of $g(s, \mathbf{a}, \mathbf{b})$, respectively. However, if a pole-zero cancellation occurs, then the set of points defined in eqn. 7.16 are no longer valid. The parameter space points where such a cancellation occurs have to be identified and studied separately. An analytical solution to the problem of identifying the parameter space points where common roots in the numerator and the denominator polynomial families exist is based on the use of the Bezout matrix [24]. Moreover, setting the determinant of this matrix equal to zero in general results in a nonlinear equation with respect to the uncertain parameters entering the polynomials symbolically, and its solutions (if any within the parameter space) define the points where such cancellations can occur. This latter problem can be treated in exactly the same way for any type of uncertainty structure of the transfer function concerned, and so it will be omitted in the sequel.

As far as affine linear systems are concerned, the boundary of their frequency response is obtained from the image of the edges of the parameter space Q [25], denoted by $E(Q)$. Furthermore, it is known that, for each frequency, the image of each edge on the complex plane is either an arc or a straight line. Hence, the computation of the boundary of the frequency response of an affine linear system involves a λ-scan ($\lambda \in [0, 1]$) of the $\ell 2^{\ell-1}$ edges of the parameter box Q.

7.5 Robust direct Nyquist array (RDNA)

Consider the modified return difference matrix $\boldsymbol{R}(s)$ defined earlier in eqn. 7.3, assuming initially that the plant $\boldsymbol{G}(s)$ in Figure 7.1 is an interval $n \times n$ MIMO system. A necessary assumption of the DNA stability theorem is the diagonal dominance of the modified return difference matrix. However, since the elements of the transfer-function matrix are uncertain, the diagonal dominance measure function, defined in eqn. 7.1, is also uncertain; i.e. $\phi_c(\omega, \underline{\mathbf{q}})$, $c = 1, \ldots, n$, and so the associated row or column Gershgorin discs are not unique at each frequency. By selecting the parameter vector $\mathbf{q}$ corresponding to the worst-case scenario for the dominance measure function, at each frequency ω, a unique-robustness measure of the diagonal dominance of the system can be defined. In DNA terms, at each frequency, the robust (worst-case) column diagonal dominance, $\phi_c^W(\omega)$, considered here corresponds to the parameter vector $\underline{\mathbf{q}}^*$, $\phi_c(\omega, \underline{\mathbf{q}}^*)$, for which the corresponding Gershgorin disc encloses, or is closest to inclusion of, the critical point the most. This means that the minimum of the diagonal dominance measure function with respect to the uncertainty is considered as the robust solution, i.e.

$$\phi_c^W(\omega) = \min_{\underline{\mathbf{q}}} \left\{ \phi_c\left(\omega, \underline{\mathbf{q}}\right) \right\}$$

Regarding this minimisation problem, first note that there is no dependency between the uncertainty of the elements, and hence it can be decomposed into n different subminimisation procedures, i.e.

$$\phi_c^W(\omega) = \min_{\underline{\mathbf{q}}} \left\{ \phi_c\left(\omega, \underline{\mathbf{q}}\right) \right\}$$

$$= \min_{\underline{\mathbf{q}} \in Q_{1c} \times Q_{2c} \times \cdots \times Q_{nc}} \left\{ \left| r_{cc}\left(j\omega, \underline{\mathbf{q}}_{cc}\right) \right| - \sum_{\substack{r=1 \\ r \neq c}}^{n} \left| r_{rc}\left(j\omega, \underline{\mathbf{q}}_{rc}\right) \right| \right\}$$

$$= \min_{\underline{\mathbf{q}}_{cc}} \left\{ \left| r_{cc}\left(j\omega, \underline{\mathbf{q}}_{cc}\right) \right| \right\} - \sum_{\substack{r=1 \\ r \neq c}}^{n} \max_{\underline{\mathbf{q}}_{rc}} \left\{ \left| r_{rc}\left(j\omega, \underline{\mathbf{q}}_{rc}\right) \right| \right\}$$

where $\underline{\mathbf{q}}_{rc} \in Q_{rc}$, $r, c = 1, \ldots, n$, corresponds to the uncertainty vector of the rc-element of the interval transfer-function matrix $\boldsymbol{G}(s)$. From the definition of the modified return difference matrix as $\boldsymbol{R} = \boldsymbol{G} + \boldsymbol{F}^{-1}$ and the fact that $\boldsymbol{F}$ is diagonal, it follows that the frequency response of $r_{rc}(s)$ is either the same as the corresponding element of $\boldsymbol{G}(s)$ ($r \neq c$), or is translated by the constant f_c^{-1} ($r = c$). Therefore, the interval transfer-function matrices $\boldsymbol{R}(s)$ and $\boldsymbol{G}(s)$ share the same boundary properties.

Using the aforementioned and the results of the preceding Section, the minimum magnitude of the frequency response of $r_{rc}(j\omega, Q_{rc})$ with respect to the uncertainty {denoted by $m_{rc}(\omega)$}, and the corresponding maximum magnitude $(M_{rc}(\omega))$, can then be defined as

$$
m_{rc}(\omega) = \begin{cases} \min\left\{|z|: z \in \partial\big(g_{rc}(j\omega, Q_{rc})\big)\right\} & r \neq c \\ \min\left\{\left|f_r^{-1} + z\right|: z \in \partial\big(g_{cc}(j\omega, Q_{cc})\big)\right\} & r = c \end{cases}
$$

$$
M_{rc}(\omega) = \begin{cases} \max\{|z| : z \in \partial(g_{rc}(j\omega, Q_{rc}))\} & r \neq c \\ \max\{|f_r^{-1} + z| : z \in \partial)g_{cc}(j\omega, Q_{cc}))\} & r = c \end{cases}
$$

(7.17)

where the notation used follows that of eqn. 7.16. Hence, it is only necessary to consider the systems corresponding to the Kharitonov segments of each of the elements of the interval transfer-function matrix $G(s)$ to establish a sufficient condition for the robust diagonal dominance of the whole family. Following this analysis, the modified return difference matrix $\mathbf{R}$ is robust column diagonal dominant if

$$
\phi_c^W(\omega) > 0 \qquad \forall \omega \tag{7.18}
$$

where $\phi_c^W(\omega)$, the worst column diagonal dominance of the cth column of $\mathbf{R}$ at a frequency ω, is given by

$$
\phi_c^W(\omega) = m_{cc}(\omega) - \sum_{\substack{r=1 \\ r \neq c}}^{n} M_{rc}(\omega) \tag{7.19}
$$

The robust column diagonal dominance measure defined in eqn. 7.19 is accurate and very fast to compute. The latter can be justified by considering the fact that the diagonal dominance is used in the DNA context, where the frequency response evaluation of the elements of the transfer-function matrix is a requirement. Hence, it follows that the calculation of $\phi_c^W(\omega)$ is equivalent to solving n times a problem of the type: 'find the minimum/maximum magnitude of a set of precalculated frequency response data'. Also, note that the maximum number of Kharitonov segments (which correspond to the 32 edges defined in eqn. 7.16) does not increase as the number of uncertain parameters increases. In other words, dimensionality is not an issue in this analysis. The robust row diagonal dominance measure can be defined in a similar fashion.

Based on the above, the DNA stability theorem can be extended to interval MIMO systems without any significant computational cost, as described in the following theorem [24].

Theorem 3 (RDNA: the interval case) *Consider the feedback configuration shown in Figure 7.1, and assume that G(s) is an interval MIMO system. Suppose that the modified return difference matrix R(s) is robustly column (row) diagonally dominant according to eqns. 7.17–7.19, $\forall s \in D$, the Nyquist contour. Then, provided that the number of open-loop unstable poles is invariant, the closed-loop system is robustly asymptotically stable if eqn. 7.4 holds.*

Proof Assuming that eqn. 7.4 holds for every member of the interval MIMO system G(s), and provided that the diagonal dominance of R(s) is robustly satisfied, the DNA applied to every single member of the transfer-function matrix family G(s) will ensure the asymptotic stability of the entire family. Hence, it is sufficient to prove that, under the assumptions made, the quantities appearing in eqn. 7.4 are invariant with respect to the uncertainty; i.e. the total number of anticlockwise encirclements of the critical points in all loops, as well as the number of unstable open-loop poles (p_0), are invariant with respect to the uncertainty. While the latter is an assumption of Theorem 3, the invariance of the total number of anticlockwise encirclements of the critical points in all loops is guaranteed from the fact that R(s) is assumed to be robustly diagonal dominant.

Consider now the affine linear case. Assume G(s, $\underline{q}$) to be an affine linear MIMO system, and define the modified return difference matrix R(s, $\underline{q}$). Define the vertices and the edges of the parameter space Q as $V(Q)$, $E(Q)$, respectively, and note that both the minimum (m_{rc}) and the maximum (M_{rc}) magnitude of the *rc*-element of the matrix R(s, $\underline{q}$ (r_{rc}(s, $\underline{q}$)) are attained at points on $E(Q)$; i.e.

$$m_{rc}(\omega) = \begin{cases} \min\limits_{\underline{q} \in E(Q)} \left\{ \left| g_{rc}(j\omega, \underline{q}) \right| \right\} & r \neq c \\[2ex] \min\limits_{\underline{q} \in E(Q)} \left\{ \left| f_c^{-1} + g_{cc}(j\omega, \underline{q}) \right| \right\} & r = c \end{cases}$$

$$\tag{7.20}$$

$$M_{rc}(\omega) = \begin{cases} \max\limits_{\underline{q} \in E(Q)} \left\{ \left| g_{rc}(j\omega, \underline{q}) \right| \right\} & r \neq c \\[2ex] \max\limits_{\underline{q} \in E(Q)} \left\{ \left| f_c^{-1} + g_{cc}(j\omega, \underline{q}) \right| \right\} & r = c \end{cases}$$

The column dominance of the *c*th column of R(s, $\underline{q}$), denoted as $\phi(\omega, \underline{q})$, is defined as

$$\phi_c(\omega, \underline{q}) = \left| f_c^{-1} + g_{cc}(j\omega, \underline{q}) \right| - \sum_{\substack{r=1 \\ r \neq c}}^{n} \left| g_{rc}(j\omega, \underline{q}) \right| \tag{7.21}$$

Following the same philosophy as in the interval case, it is then obvious that the worst-case column diagonal dominance is defined as

$$\phi_c^W(\omega) = \min_{\underline{q} \in Q} \left\{ \phi_c(j\omega, \underline{q}) \right\} \tag{7.22}$$

For the minimisation problem defined in eqn. 7.22, which is a multimodal problem in general, no accurate analytical or extreme point solution, i.e. one based only on the vertices or the edges of Q, exists to date. In the following, two solutions to this problem are given: a conservative extreme point solution, and an accurate numerical solution based on global optimisation methods [24].

Definition 4   The affine linear MIMO transfer-function matrix $\boldsymbol{R}(s, \underline{q})$ is robustly column diagonally dominant if

$$\phi_c^E(\omega) = \min_{\underline{q} \in E(Q)} \left\{ \left| \left| f_c^{-1} + g_{cc}(j\omega, \underline{q}) \right| \right| \right\} - \sum_{\substack{r=1 \\ r \neq c}}^{n} \max_{\underline{q} \in E(Q)} \left\{ \left| \left| g_{rc}(j\omega, \underline{q}) \right| \right| \right\} \tag{7.23}$$

is positive $\forall s = j\omega \in D$, the Nyquist contour.

Note that any dependencies on common uncertain parameters in the elements of the cth column of the transfer-function matrix $\boldsymbol{G}(s, \underline{q})$ are ignored, which implies that overbounding is performed. It is obvious that $\phi_c^E(\omega) \leq \phi_c(\omega, q) \forall \omega$ and $\forall q \in Q$, thus producing a lower bound on the dominance measure function. Note, however, that this is an extreme point solution to the diagonal dominance problem, and hence is easy to compute. However, in general, it is conservative, and therefore alternative solutions were sought.

An accurate numerical solution to the minimisation problem, defined in eqn. 7.23, was sought based on global optimisation techniques, since the function to be minimised is in general multimodal. Various optimisation methods were tested, among which the most successful ones were as follows:

1.   a hybrid scheme consisting of a genetic algorithm (GA) [26] to find the neighbourhood of the global minimum, and then a hill-climbing method to accurately calculate it
2.   a branch-and-bound (B&B) [27] optimisation method.

The main difference between these two methods is that the first is stochastic, whereas the second is deterministic. This reflects in both the accuracy and speed of the methods. In other words, the solution obtained using a GA is accurate if the search of the parameter space was exhaustive enough, which depends on many variables. However, the tuning of these variables to improve the accuracy reduces the speed of the algorithm. On the other hand, the B&B approach guarantees the optimal solution within ε-tolerance. This latter procedure works as follows: calculate a convex underestimator of the actual cost function and its minimum branch, calculate new convex underestimators for the two branches and compare the minima found with the previous one, and so forth. This procedure was found to be very successful for the problems tested, and an illustration of its use is given in the examples.

Having solved the diagonal dominance problem, the 'robustification' of the DNA for the affine linear case follows immediately by substituting, in Theorem 3, the diagonal dominance of $\boldsymbol{R}(s)$ by either of the criteria given above.

7.5.1 Robust generalised dominance (RGD) stability theorem

First consider the interval modified return-difference matrix $\boldsymbol{R}(s)$, and define the elements of the comparison matrix $\boldsymbol{T}(\omega)$, introduced in Definition 2 as

$$
\boldsymbol{T}(\omega, \underline{\mathbf{q}}_{rc}, \underline{\mathbf{q}}_{cc}) =
\begin{cases}
1 & r = c \\[2mm]
\dfrac{\left| g_{rc}(s, \underline{\mathbf{q}}_{rc} \in Q_{rc}) \right|}{\left| g_{cc}(s, \underline{\mathbf{q}}_{cc} \in Q_{cc}) + f_c^{-1} \right|} & r \neq c
\end{cases}
\tag{7.24}
$$

Recall that the generalised diagonal dominance condition for the modified return-difference matrix $\boldsymbol{R}(j\omega)$ is that the PF eigenvalue of $\boldsymbol{T}(\omega)$ should not exceed the value of 2. Hence, the objective is to calculate the maximum PF eigenvalue of the family of transfer-function matrices $\boldsymbol{T}(\omega)$ at each frequency, which corresponds to the worst-case scenario with respect to the uncertainty, and compare it with the critical value of 2.

It is known from the theory of non-negative matrices [13] that

$$
\mathbf{T}_1 > \mathbf{T}_2 > 0 \Rightarrow r[\mathbf{T}_1] > r[\mathbf{T}_2]
\tag{7.25}
$$

Hence, it is obvious that the maximum PF eigenvalue at each frequency corresponds to the matrix $\mathbf{T}^{M}(\omega) = [\mathrm{T}^{M}_{rc}(\omega)]_{r,c=1,\ldots,n}$, where

$$
\mathrm{T}^{M}_{rc}(\omega) = \max_{\underline{\mathbf{q}}_{rc}, \underline{\mathbf{q}}_{cc}} \left\{ \mathrm{T}_{rc}(\omega, \underline{\mathbf{q}}_{rc}, \underline{\mathbf{q}}_{cc}) \right\}
$$

$$
= \frac{\max\limits_{\underline{\mathbf{q}}_{rc}} \left\{ \left| \mathbf{G}_{rc}(j\omega, \underline{\mathbf{q}}_{rc}) \right| \right\}}{\min\limits_{\underline{\mathbf{q}}_{cc}} \left\{ \left| \mathbf{G}_{cc}(j\omega, \underline{\mathbf{q}}_{cc}) + f_c^{-1} \right| \right\}}
\tag{7.26}
$$

Note that the uncertain vectors in the numerator and denominator of the elements of the comparison matrix $\boldsymbol{T}(\omega)$ are independent, which justifies eqn. 7.26. Moreover, since the frequency response boundaries of the elements of the interval transfer-function matrix $\boldsymbol{G}(j\omega)$ can be determined accurately, both the minimum and maximum in eqn. 7.26 can also be determined rapidly and accurately. Having formed the matrix $\mathbf{T}^{M}(\omega)$, it then follows that the modified return-difference matrix $\boldsymbol{R}(s)$ is robustly generalised diagonally dominant if the following condition is satisfied for all frequencies ω:

$$
r[\mathbf{T}^{M}(\omega)] < 2
\tag{7.27}
$$

Hence, the robust generalised dominance (RGD) stability theorem can now be stated as follows.

Theorem 4   (RGD: the interval case)   *Consider the feedback cofiguration shown in Figure 7.1, and assume that the* $n \times n$ *transfer-function matrix* **G**(s) *is an interval MIMO system. Let the interval transfer function corresponding to the* cth *diagonal element of* **G** *map the Nyquist D-contour into* Γ_c, *which encircles the critical point* $(-f_c^{-1}, 0)$ N_c *times anticlockwise, where* c = 1, ..., n. *Suppose now that the modified return-difference matrix* **R**(s), *defined in eqn. 7.3, is robustly generalised diagonally dominant. Provided that the number of open-loop unstable poles (*p_0*) of the interval MIMO family is invariant, the closed-loop system is robustly asymptotically stable, if eqn. 7.4 holds.*

Consider now the affine linear case. Assume **G**(s, **q**) to be an affine linear MIMO system and define the modified return-difference matrix **R**, as well as the matrix **T** of eqn. 7.24, as before. Similarly to the interval case, the worst comparison matrix **T** is sought, which results in the worst PF eigenvalue for all possible perturbations. The solution given in eqn. 7.26 can be used in this case as well, but note that it becomes an inequality due to the dependencies of the numerator and denominator on common uncertain elements, indicating the overbounding performed. However, since the transfer-function matrix **G** is assumed to have an affine linear perturbation structure, both minima and maxima can be calculated rapidly and accurately, and the RGD stability theorem can be stated similarly.

The generalised dominance concept also serves as a way to improve or even achieve row/column diagonal dominance, as mentioned in the preliminary Section. Moreover, if robust diagonal dominance is required over a range of frequencies, a dynamic diagonal compensator **K**(s) can be designed, such that each diagonal element of **K**(s) approximates the variations of the corresponding element of the left PF eigenvector of the matrix $\mathbf{T}^M(\omega)$ defined in eqn. 7.28. Note that the uncertainty structure of **G** is preserved after its right multiplication by the matrix **K**(s), since **K**(s), is diagonal.

Note also that the least conservative definition for the matrix $\mathbf{T}^M(\omega)$ would be

$$\mathbf{T}^M(\omega) = \mathbf{T}(\omega, \underline{\mathbf{q}}_w) \quad \text{such that} \quad \underline{\mathbf{q}}_w \in Q : \lambda\left[\mathbf{T}\left(j\omega, \underline{\mathbf{q}}_w\right)\right] = \max_{\underline{\mathbf{q}} \in Q}\left\{\lambda\left[\mathbf{T}\left(j\omega, \underline{\mathbf{q}}\right)\right]\right\}$$

For the interval systems, the latter definition and eqn. 7.26 are equivalent. However, for affine linear MIMO systems this is not the case, since the vector $\underline{\mathbf{q}}_w$ that maximises the PF eigenvalue at a given frequency does not necessarily maximise the magnitude of all of the elements of the comparison matrix as well. Nevertheless, it is the latter condition that guarantees that the PF scaling will achieve robust diagonal dominance. Hence, even if eqn. 7.26 is analytically more conservative, from a design point of view it is more useful.

7.5.2 Robust fundamental dominance (RFD) stability theorem

Let the modified return difference matrix **R** have a regular splitting as defined in eqn. 7.7, where **D**, **B** consist of the diagonal and off-diagonal elements of **R**, respectively, assuming that **D** is nonsingular. Define the matrix **T**, where

$$T = B \ D^{-1} \tag{7.28}$$

Define next the fundamental dominance polynomial (FDP) as

$$p(\omega, \underline{q}, \lambda) = \text{num}\{\det[\lambda \mathbf{I}_n - T]\} \tag{7.29}$$

where $\underline{q}$ is the vector of uncertain parameters (intervals), and num$\{\cdot\}$, det$[\cdot]$ denote the numerator and the determinant, respectively. The FDP is an uncertain polynomial in λ with complex coefficients, and its uncertainty structure is any up to multilinear or polynomic, if R is an interval or affine linear transfer-function matrix, respectively. In the case of polynomic uncertainty, the FDP is overbounded using the Sideris transformation [28] to a multilinear polynomial.[4] The rest of the procedure applies to both interval and affine linear uncertainty structures and is stated below.

Recall the fundamental dominance condition

$$\rho(T) < 1 \tag{7.30}$$

where T is defined in eqn. 7.28, and note that the roots of the FDP are the eigenvalues of the matrix T. Eqn. 7.30 imposes a restriction on the roots of the FDP to lie inside the unit circle. The latter can be interpreted using the zero exclusion condition as follows: if the value sets of the FDP at a frequency ω, approximated using the mapping theorem, for all points $\lambda(\theta)$ on the unit circle, i.e.

$$\lambda(\theta) = \cos \theta + j \sin \theta, \qquad \theta = 0, \ldots, 2\pi \tag{7.31}$$

exclude the origin, and there is a member of the multilinear family having its roots inside the unit circle (e.g. the member corresponding to the nominal point in the parameter space), then $R(j\omega)$ is robustly fundamentally dominant at that frequency ω. As mentioned before, the FDP has complex coefficients and hence both positive and negative frequencies should be considered.

Any of the stability testing algorithms mentioned in the preceding Sections can be used to analyse the stability of the FDP with exact or approximated results, depending on its uncertainty structure. Consequently, the robust fundamental dominance (RFD) stability theorem can be stated as follows [29].

---

[4] The Sideris transformation is actually exact. However, the resulting parameter space is a polytope, and conflicts with the mapping theorem that is used later, which applies only to multilinear uncertainty within a box.

Theorem 5   (RDF: interval and affine linear case)   *Let the interval or affine linear transfer-function matrix* $\mathbf{R}(s)$ *have a regular splitting as defined in eqn. 7.7, and assume* $\mathbf{R}(s)$ *to be robustly fundamental diagonally dominant at all frequencies* ω *around the Nyquist contour, i.e. the condition described in eqn. 7.30 is satisfied for all possible perturbations and for all frequencies. The feedback system shown in Figure 7.1 is then closed-loop stable if eqn. 7.4 holds, provided that the number of open-loop unstable poles is invariant.*

The procedure described earlier for checking the fundamental dominance will give the information on whether the transfer-function matrix under investigation is fundamentally diagonally dominant or not. If an exact measure of dominance is required, a search for the circle of radius $k < 1$, for which the value sets of the polynomial family just exclude the origin, can be initialised, i.e. eqn. 7.31 becomes

$$\lambda(\theta) = k(\cos\theta + j\sin\theta), \qquad \theta = 0, \ldots, 2\pi \tag{7.32}$$

The radius k resulting from such a search defines the robust fundamental dominance measure at that frequency.

As suggested by Yeung [12], the splitting of the transfer-function matrix does not have to be diagonal. Actually, any splitting is valid, provided that it is regular, i.e. either the $\mathbf{D}$, or $\mathbf{B}$ matrix of eqn. 7.7 is nonsingular. For instance, $\mathbf{D}$ could correspond to the nominal system, and $\mathbf{B}$ to an additive perturbation matrix. Following this idea, assume an affine linear transfer-function matrix $\boldsymbol{G}(s, \underline{\mathbf{q}}) = [g_{rc}(s, \underline{\mathbf{q}})]_{r,c=1,\ldots,n}$ and define a regular splitting as follows:

$$\boldsymbol{G}(s, \underline{\mathbf{q}}) = \boldsymbol{G}^0(s, \underline{\mathbf{q}}_0) + \boldsymbol{G}^u(s, \underline{\mathbf{q}}) \tag{7.33}$$

where $\boldsymbol{G}^0(s, \underline{\mathbf{q}}_0)$ and $\boldsymbol{G}^u(s, \underline{\mathbf{q}})$ correspond to the nominal and uncertain parts of $\boldsymbol{G}(s, \underline{\mathbf{q}})$, respectively, and are as defined in the Appendix (Section 7.10.1). Note that $\boldsymbol{G}^0(s, \underline{\mathbf{q}}_0)$ is a certain transfer-function matrix, whereas $\boldsymbol{G}^u(s, \underline{\mathbf{q}})$ is an affine linear transfer-function matrix. Then,

$$\boldsymbol{T}(s, \underline{\mathbf{q}}) \;=\; \boldsymbol{G}^u(s, \underline{\mathbf{q}})(\boldsymbol{G}^0(s, \underline{\mathbf{q}}_0))^{-1} \tag{7.34}$$

and the fundamental dominance of $\boldsymbol{T}(s, \underline{\mathbf{q}})$ defines a measure of the accuracy of the approximation of an uncertain MIMO system by the nominal system. Furthermore, if $\boldsymbol{T}(s, \underline{\mathbf{q}})$ is robustly fundamentally dominant, then the stabilisation of the nominal system will also ensure the stability of the uncertain system as well. Hence, the design of an uncertain system $\boldsymbol{G}(s, \underline{\mathbf{q}})$, at least as far as closed-loop stability is concerned, can be simplified to the design of a controller $\mathbf{K}(s)$ for the nominal system $\boldsymbol{G}^0(s, \underline{\mathbf{q}}_0)$, ensuring that the matrix $\boldsymbol{G}(s, \underline{\mathbf{q}})\mathbf{K}(s) + \mathbf{F}^{-1}$ is robustly fundamentally dominant.

Yet another approximation of the original system by its diagonal-nominal part can be defined similarly, considering the corresponding regular splitting. In terms of carrying out the fundamental dominance tests, the procedure described earlier in this Section applies. The only difference now is the need to calculate the corresponding fundamental dominance polynomial, which can be performed using any symbolic computation package, such as Mathematica [30], or the Symbolic Toolbox of Matlab [31].

7.6 Illustrative examples

Example 1

Consider the following interval MIMO system:

$$G(s, \underline{q}) = \begin{bmatrix} \dfrac{s+4+q_1}{s^2+(6+q_2)s+5} & \dfrac{1}{5s+1+q_5} \\ \dfrac{s+1}{s^2+(10+q_3)s+100+10\,q_4} & \dfrac{2+q_6}{2s^2-s-1} \end{bmatrix} \tag{7.35}$$

where

$$|q_1| \le 1, \qquad |q_2| \le 1, \qquad |q_3| \le 3, |q_4| \le 1, \qquad |q_5| \le 0.5, \qquad |q_6| \le 0.3 \tag{7.36}$$

Assuming that the loops are closed with unity feedback, the stability of the closed-loop system will be assessed using the RDNA, the RGD and the RFD stability theorems, in turn.

Figure 7.2 shows the robust DNA of $G(s, \underline{q})$ at 100 frequencies in the range [0.01, 1000] rad/s, as well as the nominal system response, denoted by ' × 's. To use the RDNA stability theorem, the modified return difference matrix, defined as $R(s, \underline{q}) = I_2 + G(s, \underline{q})$, should be diagonally dominant. To check this, the robust column Gershgorin discs are superimposed on the DNA of $G(s, \underline{q})$ in Figure 7.3.

It is clear from Figure 7.3 that the critical point $(-1, 0)$ in the 2nd loop is included, and hence $R(s, \underline{q})$ is not robust column diagonally dominant. Hence, no conclusion can be drawn for the robust stability of the closed-loop system using the RDNA stability theorem.

Similarly, the generalised diagonal dominance discs of $R(s, \underline{q})$ are superimposed on the robust DNA plot of $G(s, \underline{q})$ in Figure 7.4, from which it is clear that the critical points are excluded in both loops, and hence $R(s, \underline{q})$ is robust generalised diagonally dominant. Consequently, the RGD stability theorem can be used to assess the stability of the closed-loop system.

A simple examination of the transfer-function matrix $G(s, \underline{q})$ reveals that the system is open-loop unstable, since element $(2,2)$ has an unstable pole at $s = 1$. Using the Smith-McMillan form of the nominal system, it can be found that the nominal multivariable system has also one unstable pole at $s = 1$. Now, since $R(s, \underline{q})$ is generalised diagonally dominant, and no pole-zero cancellations can occur in the elements of the transfer-function matrix $G(s, \underline{q})$, it is clear that the number of open-loop unstable poles is invariant, i.e. equal to one, as in the nominal case. On the other hand, it is evident in Figure 7.4 that the sum of the

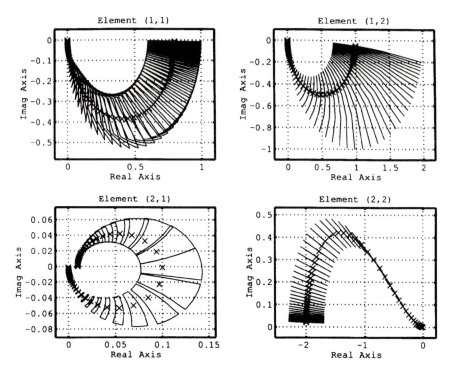

Figure 7.2   Robust and nominal DNA of **G**(s, **q**)

anticlockwise encirclements of the critical points in both loops is zero. Hence, the closed-loop system will be unstable with unity feedback.

The same conclusion can be drawn using the RFD stability theorem. The only difference in this case is the testing of the robust diagonal dominance of **R**(s, **q**), which is performed using the FD condition of eqn. 7.32. For demonstration purposes, the frequency of 1 rad/s will be used. To test the fundamental diagonal dominance, the fundamental dominance polynomial (FDP) is calculated first, in symbolic form, and is given in the Appendix (Section 7.10.2). Then, following the procedure outlined in the relevant Section, the value sets of the FDP for the frequency of 1 rad/s are drawn on the complex plane for all points $(\lambda(\theta))$ around the unit circle, as shown in Figure 7.5.

Since the origin is excluded, and there is a member of the polynomial family which satisfies the fundamental dominance condition (for instance, the system corresponding to the nominal parameter vector ($\mathbf{q}_{nom}$ = [0 0 0 0 0 0]) for which the FD was found to be 0.065), the whole family is fundamentally diagonally dominant at the frequency of 1 rad/s. Doing the same for the rest of the frequencies, it can be shown that **R**(s, **q**) is robust fundamentally diagonally dominant. Then, using the RFD stability theorem, it can be concluded that the closed-loop system will be unstable with unity feedback, since the sum of the

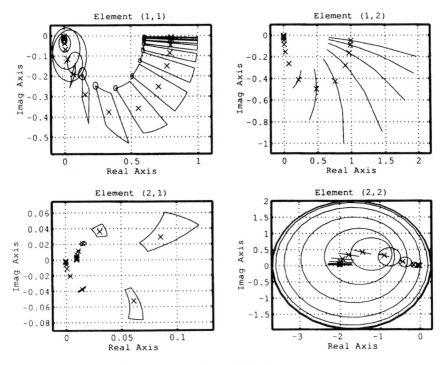

Figure 7.3 Robust column Gershgorin discs of $\mathbf{R}$(s, $\underline{\mathbf{q}}$)

anticlockwise encirclements of the critical points is zero and the number of open-loop unstable poles is one.

If, instead of a unit circle, a circle of radius 0.079 were scanned, the exact measure of dominance at the frequency of 1 rad/s would be found, as shown in Figure 7.6, since the value sets now just exclude the origin.

Since $\mathbf{R}$(s, $\underline{\mathbf{q}}$) is not robust column diagonally dominant, the PF scaling approach will now be used in an attempt to improve it.

Figure 7.7 shows the worst PF eigenvalue of $\mathbf{R}$(s, $\underline{\mathbf{q}}$) for 20 frequencies in the range [0.02, 1000] rad/s. The value of 2 is not exceeded, and hence $\mathbf{R}$(s, $\underline{\mathbf{q}}$) can be made robustly column diagonally dominant using the PF scaling technique.

To do so, a dynamic postcompensator is designed, which approximates the variations of the PF left-eigenvectors over the frequency range of interest. The design of such a compensator can be carried out using the following steps:

● Normalise the elements of the PF eigenvectors (Figure 7.8).
● Fit a simple transfer function which approximates the variation of the PF eigenvector, as shown in Figure 7.9. The order of the fitting function chosen here was 1. A higher-order function would result in a tighter approximation, but also in a more complicated controller.

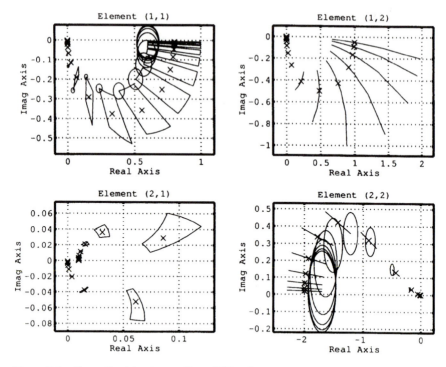

Figure 7.4   Generalised dominance discs of $\boldsymbol{R}(s, \underline{\mathbf{q}})$

Following this procedure, the post-compensator designed to achieve column diagonal dominance was found to be

$$\mathbf{K}_{\text{post}}(s) = \begin{bmatrix} 1 & 0 \\ 0 & \dfrac{0.4431\,s + 5.0421}{s + 0.2512} \end{bmatrix}$$

and the resulting system

$$\boldsymbol{G}_1(s, \underline{\mathbf{q}}) = \mathbf{K}_{\text{post}}(s)\boldsymbol{G}(s, \underline{\mathbf{q}})$$

became column diagonal dominant as shown in Figure 7.10.

In terms of stability of the closed-loop system, assuming that both loops are closed with unity feedback, the same conclusion can be drawn as before, namely that the system is closed-loop unstable, since the only encirclement of the critical point in the 2nd loop is clockwise. In an attempt to stabilise the plant, the pre-compensator $\mathbf{K}_{\text{pre}}(s)$ was determined, where

$$\mathbf{K}_{\text{pre}}(s) = \begin{bmatrix} 1 & 0 \\ 0 & \dfrac{0.5\,(s + 1)^2}{(s + 10)^2} \end{bmatrix}$$

which robustly stabilises the plant, as shown in Figure 7.11.

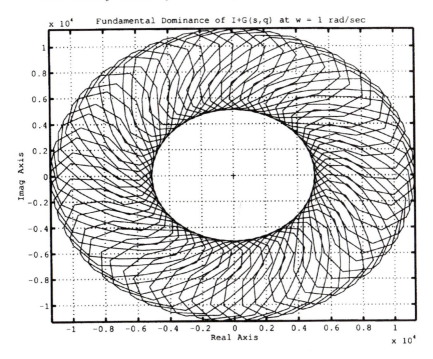

Figure 7.5   Fundamental dominance of $\mathbf{I} + \mathbf{G}(j1, \underline{\mathbf{q}})$

Since it is not clear enough whether the critical point $(-1, 0)$ in the 2nd loop is excluded, a zoom closer to that point in the $(2,2)$ element is performed in Figure 7.12.

Suppose now that the design carried out above was based on the nominal system, and the engineer wanted to check the robustness of his solution. To do so, the compensated system, $\mathbf{I} + \mathbf{K}_{\text{post}}(s)\mathbf{G}(s, \underline{\mathbf{q}})\mathbf{K}_{\text{pre}}(s)$, is split into the nominal and the uncertain part, and the roots of the fundamental dominance polynomial are checked, to see if they are all located inside the unit circle using the zero exclusion condition (Figure 7.13).

For a randomly selected member of the parameter space, e.g. for the parameter vector

$$\mathbf{q}_1 = [-1, -1, -3, -1, -0.5, -0.3]$$

the spectral radius of the corresponding matrix $\mathbf{T}$, defined in eqn. 7.34, was found to be $0.2184 < 1$. Considering the above, and the fact that the origin is excluded, the fundamental dominance condition is robustly satisfied for the given splitting, and hence the nominal system approximation, used during the design, was valid at the frequency of 1 rad/s. Doing the same for the rest of the frequencies, the robustness of such a design can be determined.

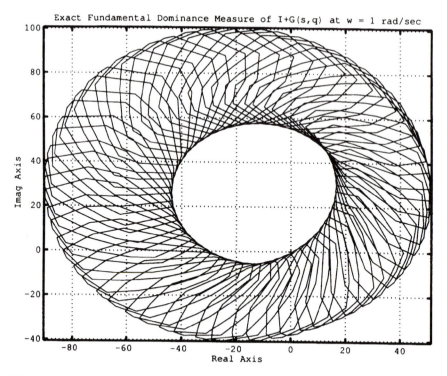

Figure 7.6   Exact fundamental dominance of $\mathbf{I} + \mathbf{G}(j1, \underline{\mathbf{q}})$

All of the Figures given in this example were created using the Parametric Systems Toolbox (PST), a collection of files written in Matlab, for the analysis and design of parametric systems [32].

Example 2

This example demonstrates the use of the spatial branch-and-bound global optimisation approach to the column diagonal dominance problem. Consider a 3×3 system whose 1st column is defined as follows:

$$\mathbf{G}_{1:3,1}(s, q) = \begin{bmatrix} \dfrac{0.987(0.12s + q_1)}{0.066s^2 + 0.366s + 1} \\ \dfrac{0.205q_2}{0.419s + 1} \\ \dfrac{0.4988(0.498s + 1)}{0.124q_3s^2 + 0.403s + q_1} \end{bmatrix} \tag{7.37}$$

The optimisation was carried out for the frequencies of 1 and 3 rad/s, yielding column dominance of 0.0391 and -0.0648, respectively. This problem was

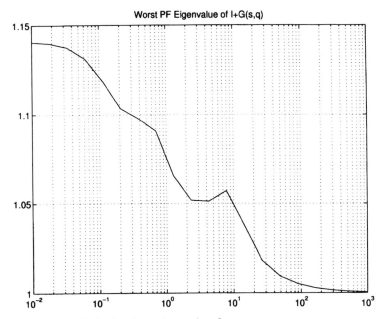

Figure 7.7   Perron-Frobenius eigenvalue against frequency

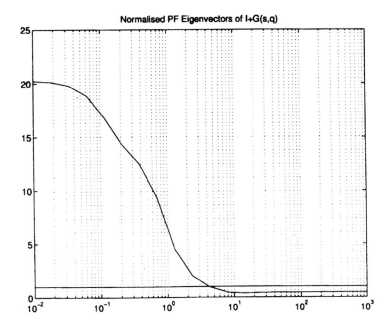

Figure 7.8   Normalised Perron-Frobenius eigenvectors against frequency

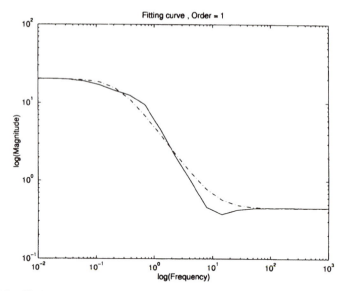

Figure 7.9   Fitting a transfer function to the magnitude data

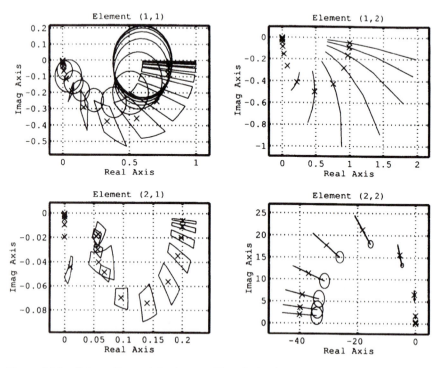

Figure 7.10   Column diagonal dominance of $I + G_1(s, \underline{q})$

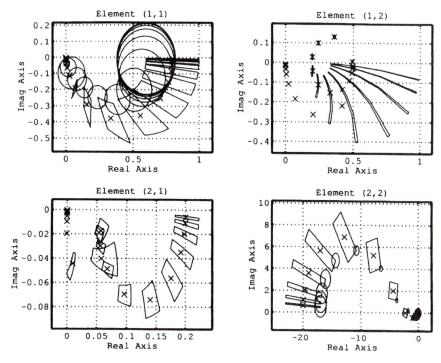

Figure 7.11 *Robust DNA of* $\mathbf{I} + \mathrm{K}_{post}(s)\mathbf{G}(s, \underline{\mathbf{q}})\mathbf{K}_{pre}(s)$

solved on a SUN SPARC workstation 10/41, and the CPU time required was 0.34 and 0.47 s, respectively. Since the column dominance for 3 rad/s is negative, it can be concluded that this column is not diagonally dominant.

7.7 Conclusions

In this paper, the robustification of well-known stability theorems for MIMO systems having interval or affine linear perturbation structure has been developed. The robust stability theorems presented are based on the diagonal dominance of the modified return difference matrix **R**, a condition which can be checked without significant computational cost for all definitions given before. The accuracy of the results presented depends on the uncertainty structure of the transfer-function matrix under consideration. In the interval case, the results are accurate, extreme point solutions of their respective problems, and hence fast to compute. For the affine linear case, the results are either conservative, extreme point solutions, or accurate, based on global optimisation techniques, which were found to be both successful and fast.

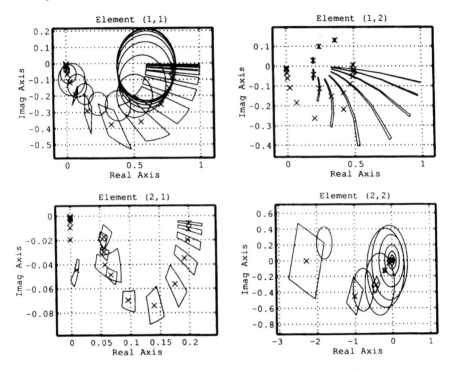

Figure 7.12   Zoom on element (2,2) reveals that critical point is excluded

The robust generalised dominance (RGD) and the robust fundamental dominance (RFD) conditions are more accurate measures of the diagonal dominance of a system. However, the robust row/column diagonal dominance gives an independent solution for each of the loops, which makes this criterion very attractive for design purposes, i.e. each loop can be designed independently. On the other hand, the compensator needs to be determined fully before any test using the RGD, or the RFD dominance can be carried out. Therefore, the latter diagonal dominance criteria are considered to be more analysis tools rather than design tools.

The RDNA stability theorem, even though proposed as an analytical, classical control-based tool for looking at the robust stability problem of MIMO systems, could well be used as a design tool, as one would normally use Nyquist or Bode plots for SISO design. In an RDNA design procedure, one has to ensure the robust diagonal dominance of the original plant. Once this has been established, a diagonal input-compensator can be designed, independently for each loop, such that the design specifications are robustly met, but ignoring any interaction caused by the off-diagonal elements. The RDNA stability theorem then ensures that closed-loop stability of the diagonal system carries over to the full

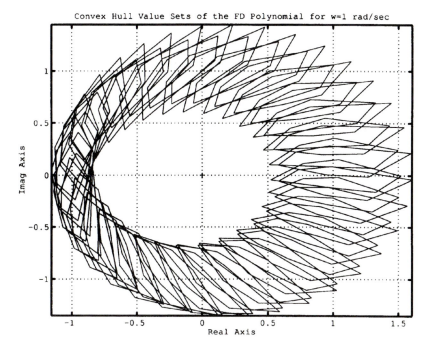

Figure 7.13 Fundamental dominance condition of compensated system for $\omega = 1$ rad/s

closed-loop system as well. In terms of performance, it is obvious that the higher the degree of diagonal dominance, the less significant the degradation of the robust performance will be, when interaction is taken into account. Hence, diagonal dominance benefits the design procedure a great deal, since the design of a controller can then be decomposed into SISO subloops.

Furthermore, as shown before, the robust fundamental dominance provides a measure of the robustness of a nominal point based design, or equivalently any other point in the parameter space.

7.8 Acknowledgments

The authors would like to thank the EPSRC for its financial support of this work.

7.9 References

1 FRANCIS, B. A.: 'A course in H_∞ control theory', *Lect. Notes Control Inf. Sci.*, 1987

2 Skogestad, S., and Postlethwaite, I.: 'Multivariable feedback control' (Wiley, 1996)

3 ZHOU, K., DOYLE, C. J., and GLOVER, K.: 'Robust and optimal control' (Prentice Hall, NJ, 1996)

4 BARMISH, B. R.: 'New tools for robustness of linear systems' (MacMillan Pub. Co., NY, 1994)

5 BHATTACHARYA, S. P., CHAPELLAT, H., and KEEL, L. H.: 'Robust control, The Parametric Approach' (Prentice Hall, 1995)

6 ACKERMANN, J.: 'Robust control, systems with uncertain physical parameters' (Springer-Verlag, 1993)

7 DE SANTIS, E., and VICINO, A.: 'Diagonal dominance for robust stability of MIMO interval systems', *IEEE Trans. Autom. Control*, 1996, **41**,(6), pp. 871–875

8 ROSENBROCK, H. H.: 'Computer-aided control systems design' (Academic Press, London, 1974)

9 MACIEJOWSKI, J. M.: 'Multivariable feedback design' (Addison-Wesley Publishing Co., 1989)

10 LIMEBEER, D. J. N.: 'The application of generalised diagonal dominance to linear systems stability theory', 1982, *Int. J. Control*, **36**,(2), pp. 185–212

11 ARAKI, M., and NWOKAH, O. I.: 'Bounds for closed-loop transfer functions', *IEEE Trans. Aut. Control*, 1975, **20**, pp. 666–670

12 YEUNG, LAM-FAT: 'Dominant and direct methods in multivariable designs'. PhD Thesis, (Imperial College, London, 1990)

13 SENETA, E.: 'Non-negative matrices' (George Allen & Unwin, London, 1973)

14 MEES, A. I.: 'Achieving diagonal dominance', *Syst. Control Lett.*, 1981, **1**,(3), pp. 155–159

15 MUNRO, N.: 'Computer-aided design I: The inverse Nyquist array design method', in O'REILLY (Ed.), *'Multivariable control for industrial applications'* (Peter Peregrinus, Stevenage, 1987), pp. 211–228

16 BRYANT, G. F., and YEUNG, L. F.: 'Multivariable control system design techniques: dominance and direct methods' (John Wiley & Sons, 1996)

17 KHARITONOV, V. L.: 'On a generalisation of a stability criterion' *Izv. Akad. Nauk Kaz. SSR Ser. Fis. Math.*, 1978, (1), pp. 53–57

18 DASGUPTA, S.: 'Kharitonov's theorem revisited', *Syst. Control Lett.*, 1988, **11**, (5), pp. 381–384

19 ZADEH, L. A., and DESOER, C. A.: 'Linear systems theory–a state-space approach' (McGraw Hill, New York, 1963)

20 DE GASTON, R. R. E.: 'Exact calculation of the multiloop stability margin', *IEEE Trans. Autom. Control*, 1988, **33**,(2), pp. 156–171

21 BARTLETT, A. C., HOLLOT, C. V., and HUANG, L.: 'Root locations of an entire polytope of polynomials: it suffices to check the edges', *Math. Control, Signals Syst.*, 1988, **1**, pp. 61–71

22 KRAUS, F. J., ANDERSON B. D. O., and MANSOUR, M.: 'Robust stability of polynomials with multilinear parameter dependence', *Int. J. Control*, 1989, **50**,(5), pp. 1745–1762

23 POLYAK, B. T., and KOGAN, J.: 'Necessary and sufficient conditions for robust stability of linear systems with multiaffine uncertainty structure', *IEEE Trans. Autom. Control*, 19995, **40**,(7), pp. 1255–1260

24 KONTOGIANNIS, E., and MUNRO, N.: 'Extreme point solutions to the diagonal dominance problem and stability analysis of uncertain systems'. Proc. ACC '97, Albuquerque, NM, 1997, pp. 3936–3940

25 FU, M.: 'Computing the frequency response of linear systems with parametric uncertainty, *Syst. Control Lett.*, 1990, **15**, (1), pp. 45–52

26 GOLDBERG, D. E.: 'Genetic algorithms in search, optimisation & machine learning' (Addison-Wesley, 1989)

27 SMITH, E. M. B., and PANTELIDIS, C. C.: 'Global optimisation of general process models, in GROSSMAN, I. E. (Ed.), 'Global optimisation in engineering design' (Kluwer Academic Publishers, 1996, Dordrecht, pp. 355–386)

28 SIDERIS, A., and SANCHEZ PENA, R. S.: 'Fast computation of the multivariable stability margin for real interrelated uncertain parameters', *IEEE Trans. Autom. Control*, 1989, **34**, (12), pp. 1272–1276

29 KONTOGIANNIS, E., and MUNRO, N.: 'The fundamental dominance condition for MIMO systems with parametric uncertainty'. Proc. IEE/IFAC Control '96, Exeter, UK, 1996, pp. 1202–1207

30 WOLFRAM RESEARCH: 'Mathematica Version 3', 1996

31 THE MATHWORKS INC.: 'Matlab Version 4.2', 1992

32 KONTOGIANNIS, E., and MUNRO, N.: 'The parametric systems toolbox'. Proceedings of the 7th IFAC Symposium on CACSD, Gent, Belgium, 1997, pp. 299–303

7.10 Appendixes

7.10.1 Appendix A

Consider the affine-linear TFM $\mathbf{G}(s, q)$, where $\mathbf{q} = [q_1, q_2, \ldots, q_\ell]$, $\mathbf{q}_0$ denote the uncertainty vector and its nominal value, respectively, and assume the regular splitting defined in eqn. 7.33. Let $n_{rc}^0(s)$, $d_{rc}^0(s)$ denote the nominal numerator and denominator polynomials, respectively, corresponding to the rc-element $\mathbf{g}_{rc}(s, \underline{\mathbf{q}})$, and $n_{rc}^i(s)$, $d_{rc}^i(s)$, $i = 1, \ldots, \ell$, the corresponding uncertain polynomials. Following this notation, the nominal system $\mathbf{G}^0(s, \underline{\mathbf{q}}_0)$ is given by

$$\mathbf{G}^0(s, \underline{\mathbf{q}}_0) = \left[\frac{n_{rc}^0(s)}{d_{rc}^0(s)} \right]_{r,c=1,\ldots,n}$$

and $\mathbf{G}^u(s, q) = \mathbf{G}(s, q) - \mathbf{G}^0(s, q_0)$. After some easy algebraic manipulations, it can be found that the rc-element of the additive perturbation matrix $\mathbf{G}^u(s, \underline{\mathbf{q}})$ is given by

$$\mathbf{G}_{rc}^u(s, \underline{\mathbf{q}}) = \frac{\displaystyle\sum_{i=1}^{\ell} \left(n_{rc}^i(s) \, d_{rc}^0(s) - n_{rc}^0(s) \, d_{rc}^i(s) \right) q_i}{d_{rc}^0(s) \, d_{rc}^0(s) + \displaystyle\sum_{i=1}^{\ell} \left(d_{rc}^0(s) \, d_{rc}^i(s) \right) q_i}$$

which is also an affine linear transfer function.

7.10.2 *Appendix B*

The fundamental dominance polynomial (FDP) for the system $\mathbf{I} + \mathbf{G}(s, q)$ was found to be:

$$\begin{aligned}
\boldsymbol{P}(s, \underline{\mathbf{q}}, \lambda) = {} & 9q_5sq_3 + 10q_4sq_2 + 9q_6sq_3 + 4s^2q_3q_1 + 790q_5q_6s + 179q_5q_6s^2 \\
& + 2q_5s^4q_1 + 5s^3q_6q_1 + 90q_5q_6q_4 + 100q_5q_6q_1 + 100q_1 + 12q_5s^3q_3 \\
& - 3s^3q_3q_1 + 40sq_4q_1 - 30s^2q_4q_1 + 51s^2q_6q_1 + 191q_5s^2q_1 + 10q_6q_4q_1 \\
& + 10s^4q_3q_1 - 2q_5s^2q_3 + 520sq_6q_4 + 52s^2q_6q_3 + 100s^3q_4q_1 + 10q_5q_4q_1 \\
& + 120q_5s^2q_4 + 19q_5s^3q_1 + 510sq_6q_1 - 20q_5sq_4 + 100q_6sq_2 + 100q_5sq_2 \\
& + 5s^2q_6q_3q_1 + 10q_5q_6q_4q_1 + 4390s + 639s^2 + 900q_6 + sq_3q_1 - 90q_5sq_1 \\
& + 90q_4 + 900q_5 + 100sq_2 + 36q_6s^3q_3 + 17q_5q_6s^3 + q_5q_6s^4 + 2q_5s^5q_3 \\
& + 13q_5s^4q_3 + 50s^3q_6q_4 + 360s^2q_6q_4 + 5s^4q_6q_3 + 20q_5s^4q_4 + 130q_5s^3q_4 \\
& + s^2q_3q_2 + 510s^2q_6q_2 + 191q_5s^3q_2 - 90q_5s^2q_2 + 10s^5q_3q_2 + 100s^4q_4q_2 \\
& - 3s^4q_3q_2 - 30s^3q_4q_2 + 4s^3q_3q_2 + 40s^2q_4q_2 + 5s^4q_6q_2 + 51s^3q_6q_2 \\
& + 2q_5s^5q_2 + 19q_5s^4q_2 + 9sq_3 + 7363s^3 + 1743s^5 + 7432s^4 + 5290sq_6 \\
& + 1189q_5s^2 - 110q_5s + 900q_5q_6 + 73s^4q_3 + 730s^3q_4 + 2s^3q_3 + 20s^2q_4 \\
& + 43s^2q_3 + 430sq_4 + 912s^3q_6 + 4129s^2q_6 + 90q_6q_4 + 342q_5s^4 \\
& + 1418q_5s^3 + 90q_5q_4 + 410s^2q_2 + 410sq_1 - 259s^3q_2 - 259s^2q_1 \\
& + 100q_6q_1 + 10q_4q_1 + 100q_5q_1 + 974s^4q_2 + 974s^3q_1 + 10s^6q_2 \\
& + 10s^5q_1 + 97s^5q_2 + 97s^4q_1 + 10s^6q_3 + 67s^5q_3 + 100s^5q_4 + 670s^4q_4 \\
& + 5s^5q_6 + 86s^4q_6 + 2q_5s^6 + 33q_5s^5 + 900 + q_5s^2q_3q_2 + q_6s^2q_3q_2 \\
& + 9q_5q_6sq_3 + 10q_5q_6s^2q_2 + q_5q_6s^3q_2 + 10q_5q_6q_4s^2 + 10q_5q_6q_4sq_2 \\
& + 2q_5s^4q_3q_2 - q_5s^3q_3q_2 + 50s^2q_6q_4q_2 + 10s^7 + 167s^6 + 5s^3q_6q_3q_2 \\
& + 20q_5s^3q_4q_2 - 10q_5s^2q_4q_2 + 100q_5q_6sq_2 + 10q_6q_4sq_2 + 10q_5q_4sq_2 \\
& + q_5q_6s^3q_3 + 7q_5q_6s^2q_3 + q_5q_6s^2q_3q_2 + q_5sq_3q_1 + q_6sq_3q_1 + 10q_5q_6sq_1 \\
& + q_5q_6s^2q_1 + 70q_5q_6q_4s + 2q_5s^3q_3q_1 - q_5s^2q_3q_1 + 50sq_6q_4q_1 \\
& + 20q_5s^2q_4q_1 - 10q_5sq_4q_1 + q_5q_6sq_3q_1)\lambda^2 + 5 + 16s + 8s^2 + sq_2 \\
& - 14s^3 - 2s^5 - 13s^4 + 2s^2q_2 - s^3q_2 - 2s^4q_2
\end{aligned}$$

which has multilinear uncertainty structure.

Substituting $s = j\omega$ in the above expression, the corresponding polynomial for the different frequencies ω can be found, which is then analysed as described before.

Part III

Design and synthesis methods

Chapter 8

Robust control under parametric uncertainty. Part II: design

L. H. Keel and S. P. Bhattacharyya

8.1 Introduction

In chapter 4 we presented some fundamental results that are aids in analysing the behaviour of control systems subject to real parameter uncertainty. In this chapter we illustrate the application of these results to design by: (a) showing how classical design techniques can be robustified, by using these results and (b) developing a new linear programming approach to controller design that exploits these results. These examples should suggest to the reader how to formulate other design questions that take advantage of the fundamental results in real parametric robust control theory (RPRCT).

The design of control systems is an art that should combine practical experience, analytical results and computational capabilities blended with imagination and control expertise to produce efficient systems. In this process the role of the analytical computations is to supply ready answers to various questions that the designer may pose. A typical design procedure may involve iterative loops within which the analytical calculations are embedded. In the following sections we show how the theory described in Chapter 4 can be employed in this fashion.

8.2 Robust classical controller design using RPRCT

In this section we give some methods that may be used in conjunction with the previous theory to design robust classical control systems. We describe how classical design techniques may be robustified using the frequency-domain template generating properties developed in Chapter 4. We consider an interval plant $G(s)$ described by a transfer function along with coefficient uncertainty lying in intervals. The requirement of robust design is that the design specifications must be satisfied over the entire parameter set. Thus the worst case values must be acceptable. As shown in Chapter 4, these worst case values occur over the extremal set $\mathbf{G}_E(s)$. Thus it suffices to verify that the specifications are met over this set. The design procedure is best illustrated with an example.

Example 1

Consider the interval plant

$$\mathbf{G}(s) = \frac{n_0}{d_3 s^3 + d_2 s^2 + d_1 s + d_0}$$

with $n_0 \in [10, 20]$, $d_3 \in [0.06, 0.09]$, $d_2 \in [0.2, 0.8]$, $d_1 \in [0.5, 1.5]$, $d_0 = 0$. The objective is to design a controller so that the closed-loop system

(a) is robustly stable under all parameter perturbations
(b) possesses a guaranteed phase margin of at least 45°. In other words, the worst-case phase margin over the set of uncertain parameters must be better than 45°
(c) possesses a bandwidth ≥ 0.1 rad/s with a reasonable value of resonant peak M_p.

We adapt the standard classical control design techniques to the present situation by first constructing the Bode envelopes of the open-loop interval plant $\mathbf{G}(s)$. From the magnitude and phase envelopes, we observe that the worst-case phase margin of 50° is achieved if the gain crossover frequency ω_c' is at 0.5 rad/s. To bring the magnitude envelope down to 0 dB at the new crossover frequency ω_c', the phase lag compensator of the form

$$C(s) = \frac{1 + aTs}{1 + Ts}, \qquad a < 1$$

must provide the amount of attenuation equal to the minimum value of the magnitude envelope at ω_c':

$$\max_{G(j\omega_c') \in \mathbf{G}(j\omega_c')} |G(j\omega_c')| = -20 \log_{10} a \text{ dB}$$

Thus, we have $a = 0.01$. Now choose the corner frequency $1/aT$ to be one decade below ω_c':

$$\frac{1}{aT} = \frac{\omega_c'}{10}\bigg|_{\omega_c' = 0.5}$$

and we have $T = 2000$. The resulting phase lag compensator is

$$C(s) = \frac{1 + 20s}{1 + 2000s}$$

Here we give some analysis to check whether the controller $C(s)$ satisfies the given design requirements. The robust stability of the closed-loop system with the controller $C(s)$ may be easily determined by applying the GKT. Moreover, by GKT, since we use a first-order controller it is only necessary to check the stability of the following vertex polynomials:

$$(1 + 2000s)K_d^i(s) + (1 + 20s)K_n^j(s), \qquad i = 1, 2, 3, 4; \; j = 1, 2$$

where

$$K_n^1(s) := 10 \qquad\qquad K_n^2(s) := 20$$
$$K_d^1(s) := 0.5s + 0.8s^2 + 0.09s^3 \qquad K_d^2(s) := 1.5s + 0.8s^2 + 0.06s^3$$
$$K_d^3(s) := 0.5s + 0.2s^2 + 0.09s^3 \qquad K_d^4(s) := 1.5s + 0.2s^2 + 0.06s^3$$

The six polynomials above are stable, and this shows that the closed-loop system remains stable under all parameter perturbations within the given ranges. Figure 8.1 shows the frequency response (Bode envelopes) of $\mathbf{G}(s)$ (uncompensated system) and $C(s)\mathbf{G}(s)$ (compensated system).

Clearly, the guaranteed phase margin requirement of $45°$ is satisfied. The guaranteed gain margin of the system is $12\,\text{dB}$. The closed-loop response $|\mathbf{M}(j\omega)|$, called $\mathbf{T}^y(j\omega)$ in Theorem 6 of Chapter 4, is shown in Figure 8.2, where

$$\mathbf{M}(j\omega) := \mathbf{M}(s)|_{s=j\omega} := \left\{ M(s) : \frac{C(s)G(s)}{1 + C(s)G(s)}, \; G(s) \in \mathbf{G}(s) \right\} \qquad (8.1)$$

Note that the $|\mathbf{M}(j\omega)|$ envelope shown in this figure is calculated from the result of the theorem. Figure 8.2 shows that the M_p of every system in the family lies between 1.08 and 1.3886. This also shows that the bandwidth of every system in the family lies in between 0.12 and 0.67 rad/s. Thus, the design objective is achieved. Figure 8.3 shows the Nyquist plot of $C(s)\mathbf{G}(s)$.

The centre of the M-circle in Figure 8.3 is

$$\left(\frac{M_p^2}{1 - M_p^2}, \; 0 \right) = \left(\frac{1.3886^2}{1 - 1.3886^2}, \; 0 \right)$$

$$= (-2.772, \; 0)$$

and the radius of the circle is

$$r = \left| \frac{M - p}{1 - M_p^2} \right| = 1.496$$

Example 2 *(Lead-lag compensation)*

We give an example of lead-lag compensation design utilising the developments described above. Let us consider the interval plant

$$G(s) = \frac{a_0}{b_3 s^3 + b_2 s^2 + b_1 s + b_0}$$

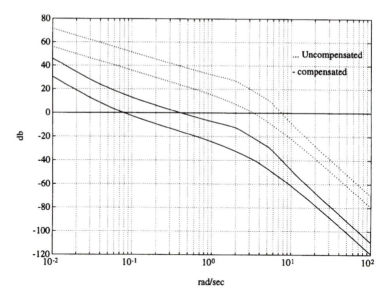

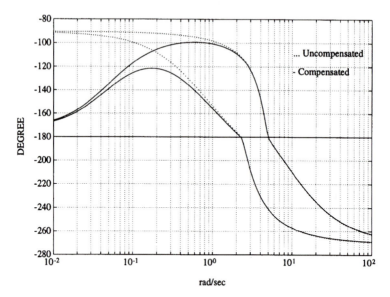

Figure 8.1   Bode envelopes of **G**(*s*) *and* C(*s*)**G**(*s*) *(Example 1)*

with coefficients bounded as follows:

$$a_0 \in [5, 7], \quad b_3 \in [0.09, 0.11], \quad b_2 \in [0.9, 1.2], \quad b_1 \in [0.8, 1.5], \quad b_0 \in [0.1, 0.3]$$

The objective of the design is to guarantee that the entire family of systems has a phase margin of at least 60° and a gain margin of at least 30 dB. From Figure

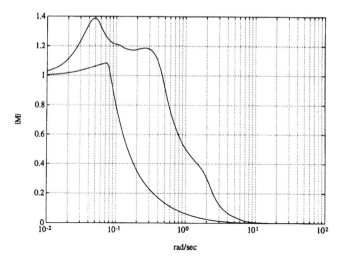

Figure 8.2   Closed-loop frequency response (Example 1)

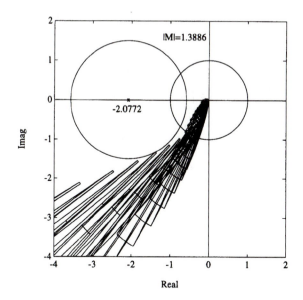

Figure 8.3   Nyquist envelope of $C(s)\mathbf{G}(s)$ with M-circle (Example 1)

8.4 we observe that the phase margin 70° which is equal to the desired phase margin of 60° plus some safety factor can be obtained if the new gain-crossover frequency ω'_c is at 0.35 rad/s. This means that the phase-lag compensator must reduce the maximum magnitude of $\mathbf{G}(j\omega'_c)$ over the interval family to 0 dB.

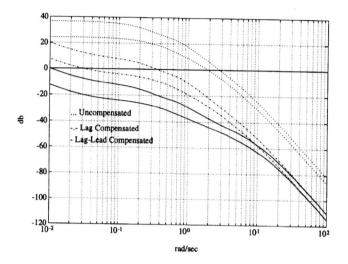

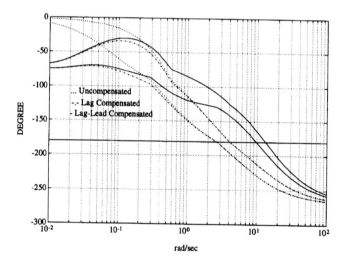

Figure 8.4   Bode envelopes (Example 2)

Thus we solve

$$-20 \log_{10} a = \max_{G(j\omega'_c) \in \mathbf{G}(j\omega'_c)} |G(j\omega'_c)| = 28 \text{ dB}$$

and we have

$$a = 0.0398$$

In order that the phase lag of the compensator does not affect the phase at the new gain-crossover frequency ω'_c, we choose the value of $1/aT$ to be one decade below ω'_c. Thus,

$$T = \frac{10}{a\omega'_c} = \frac{10}{(0.0398)(0.35)} = 717.875$$

Therefore, the lag compensator obtained is

$$C_1(s) = \frac{1 + aTs}{1 + Ts} = \frac{28.5714s + 1}{717.875s + 1}$$

We have now achieved $\approx 70°$ of guaranteed phase margin and 24 dB of guaranteed gain margin.

To achieve the desired gain margin, we now wish to move the phase-crossover frequency w''_c to 4.7 rad/s. If the magnitude plot does not move, we can achieve the gain margin of 35 dB at this frequency. Thus, we solve

$$-10\log_{10} a = -(35 - 25) = -10 \text{ dB}$$

and we have

$$a = 10$$

Then,

$$\frac{1}{T} = \sqrt{a}\omega''_c = \sqrt{10}(4.7) = 14.86$$

and $T = 0.0673$. Therefore, the cascaded lead compensator is

$$C_2(s) = \frac{1}{a}\frac{1 + aTs}{1 + Ts}$$

$$= \frac{1}{10}\frac{(10)(0.0673)s + 1}{0.0673s + 1} = \frac{s + 1.485}{s + 14.859}$$

From Figure 8.4 we verify that the compensated system provides $\approx 105°$ of guaranteed phase margin and 50 dB of guaranteed gain margin. Therefore, the controller

$$C(s) = C_2(s)C_1(s) = \frac{s + 1.5}{s + 15}\frac{28.5714s + 1}{717.682s + 1}$$

attains the design specifications robustly. Figures 8.5, 8.4 and 8.6 show the Nyquist, Bode and Nichols envelopes of the uncompensated and compensated systems.

In the following Section we describe a somewhat different approach to the design problem.

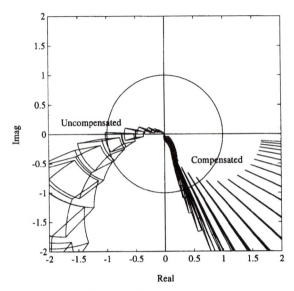

Figure 8.5   Nyquist envelope (Example 2)

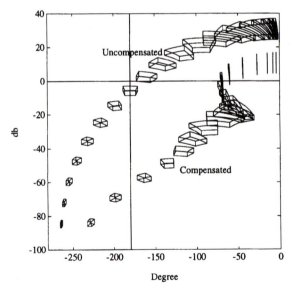

Figure 8.6   Nichols templates (Example 2)

8.3 Linear programming approach to design

Consider the feedback control system shown below (Figure 8.7).

Let $\mathbf{p} := [p_1, p_2, \ldots, p_s]$ denote the vector of uncertain real parameters representing the plant and $\mathbf{x}$ be the vector of real parameters representing the

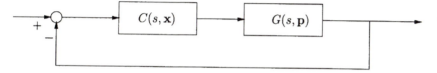

Figure 8.7 A standard feedback system

controller. In general $\mathbf{p} = \mathbf{p}^0$ represents the nominal value of the plant parameters and we deal with uncertainty in the plant parameters by letting $\mathbf{p}$ lie in a
box $\mathbf{\Pi}$ defined as

$$\mathbf{\Pi} := \{\mathbf{p} \colon p_i^- \leq p_i \leq p_i^+, i = 1, 2, \ldots, s\}$$

We assume that the controller in question is of fixed dynamic order and in
such a case $\mathbf{x} := [x_1, x_2, \ldots, x_r]$ is a vector of fixed dimension.

The performance of the control system can often be characterised by a
vector $\mathbf{a} = [a_1, \ldots, a_l]$ of real numbers which are functions of the plant and controller parameters:

$$a_i = a_i(\mathbf{x}, \mathbf{p}), \qquad i = 1, 2, \ldots, l$$

The a_i may represent closed-loop characteristic polynomial coefficients or the
ordered list of numerator and denominator coefficients of a closed-loop transfer
function such as the sensitivity function or complementary sensitivity function in
the model matching problem. In this Chapter we specifically take the a_i to
represent the coefficients of the closed-loop characteristic polynomial, although
for convenience we refer to the vector $\mathbf{a}(\mathbf{x}, \mathbf{p})$ as the *closed-loop performance vector*.

Suppose that the desired or target value of the closed-loop performance
vector is

$$\Delta_T := [\delta_1^T, \delta_2^T, \delta_3^T, \ldots, \delta_l^T]$$

which is a point in $\mathbb{R}^l$. Then, ideally, the goal of control system design or the controller synthesis problem could be stated as the problem of choosing, if possible,
the controller parameter vector $\mathbf{x}$, such that the set of equalities

$$a_i(\mathbf{p}^0, \mathbf{x}) = \delta_i^T, \qquad i = 1, 2, \ldots, l \tag{8.2}$$

is attained exactly. Note that in eqn. 8.2 we have used the nominal parameter
$\mathbf{p}^0$ to indicate that the performance vector is evaluated at the nominal value of
the plant parameters. In general the performance specification represented by
the target vector must explicitly or implicitly include the stability of the closed-
loop system. If the above equations can be satisfied the desired performance is
achieved under the ideal condition of zero uncertainty in the plant. Under plant
perturbations, such ideal performance will deteriorate and the robust performance problem is to ensure that the controller design vector $\mathbf{x}$ is chosen so that
the deterioration that could occur as $\mathbf{p}$ ranges over the uncertainty set $\mathbf{\Pi}$
remains within acceptable limits.

However, even under the assumption of zero plant uncertainty, the performance goal represented by eqn. 8.2 is in general not achievable due to the restrictions imposed on the controller such as structure and dynamic order. For example, in the pole assignment problem, Δ_T represents the coefficients of the closed-loop characteristic polynomial which is selected as a polynomial whose roots are *arbitrarily* prespecified. It is known in the SISO case that in general a solution to the problem given in eqn. 8.2 does not exist unless the order of the controller is $n - 1$ or greater when the plant is of order n. On the other hand, the likelihood of existence of a solution, when the controller order is restricted and is less than $n - 1$, will be greatly increased if we relax the desired target performance Δ_T from a point to a box-like set such as the interval family

$$\mathbf{\Delta}_T := \left\{ \Delta_T : \delta_i^{T-} \leq \delta_i^T \leq \delta_i^{T+}, i = 1, 2, , \ldots, l \right\} \tag{8.3}$$

In other words the set of equalities in eqn. 8.2 is replaced by the set of *inequalities*

$$\delta_i^{T+} \leq a_i(\mathbf{p}, \mathbf{x}) \leq \delta_i^{T-}, \qquad i = 1, 2, , \ldots, l$$

The robust design problem then is to choose $\mathbf{x}$ if possible so that the above set of inequalities are satisfied for all $\mathbf{p} \in \mathbf{\Pi}$.

Of course the size of the target box would have to be adjusted to ensure that the system performance remains satisfactory whenever the actual performance vector or set of vectors lie inside the target box $\mathbf{\Delta}_T$. Thus, for instance, nominal as well as robust stability must be guaranteed as long as the performance vector lies in the target performance set $\mathbf{\Delta}_T$. As we show, this can be done quite systematically using several extremal results from robust parametric control theory [1].

Based on these ideas, we formulate the nominal fixed-order pole assignment and its robust version, namely the case when the plant parameter $\mathbf{p}$ belongs to the uncertainty set given as the parameter space box $\mathbf{\Pi}$. We assume that the parameter $\mathbf{p}$ appears linearly or multilinearly in the closed-loop performance vector $\mathbf{a}(\mathbf{x}, \mathbf{p})$, whereas $\mathbf{x}$ appears linearly or affinely. In Theorem 1 below, we show by using the mapping theorem [1, 2] that, in these cases, the nominal as well as the robust versions of the controller design problem reduce to linear programming problems, obtained by simply replacing the uncertain parameter $\mathbf{p}$ by its vertex values in the design inequalities in eqn. 8.3. Let $\mathbf{v}_j, j = 1, 2, \ldots, n$, denote the vertices of the parameter box $\mathbf{\Pi}$.

Theorem 1   *Assuming that $a_i(\mathbf{p}, \mathbf{x})$ are multilinear functions of $\mathbf{p}$ then*

$$\delta_i^{T-} \leq a_i(\mathbf{p}, \mathbf{x}) \leq \delta_i^{T+}, \qquad i = 1, 2, , \ldots, l$$

holds for all $\mathbf{p} \in \mathbf{\Pi}$ if and only if

$$\delta_i^{T-} \leq a_i(\mathbf{v}_j, \mathbf{x}) \leq \delta_i^{T+}, \qquad i = 1, 2, \ldots, l, \quad j = 1, 2, \ldots, n. \tag{8.4}$$

Proof   The proof of this theorem follows immediately from the mapping theorem [1, 2], which states that the set

$$\{\mathbf{a}(\mathbf{p}, \mathbf{x}): \mathbf{p} \in \Pi\}$$

is contained in the convex hull of the points $\mathbf{a}(\mathbf{v}_j, \mathbf{x})$, $j = 1, 2, \ldots, n$.   □

Assuming that the performance vector is linear or affine in the vector $\mathbf{x}$, the above theorem shows that the robust design problem reduces to a linear programming problem. Sometimes it is necessary to test whether a box of controllers achieves the desired specification. If $\mathbf{x}$ appears multilinearly in $\mathbf{a}(\mathbf{p}, \mathbf{x})$ the conditions given in eqn. 8.4 need be verified only at the vertices of the controller box. These issues are discussed in greater detail in the following Sections.

8.3.1 Fixed-order pole assignment and robust stabilisation

Consider the SISO linear time-invariant system shown in Figure 8.8.

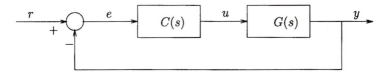

Figure 8.8   A SISO feedback system

Let the plant transfer function $G(s)$ of dynamic order n and the controller transfer function $C(s)$ of dynamic order r be given, respectively, by

$$G(s) := \frac{n(s)}{d(s)} = \frac{n_n s^n + n_{n-1} s^{n-1} + \cdots + n_1 s + n_0}{d_n s^n + d_{n-1} s^{n-1} + \cdots + d_1 s + d_0} \tag{8.5}$$

$$C(s) := \frac{n_c(s)}{d_c(s)} = \frac{a_r s^r + a_{r-1} s^{r-1} + \cdots + a_1 s + a_0}{b_r s^r + b_{r-1} s^{r-1} + \cdots + b_1 s + b_0} \tag{8.6}$$

The closed-loop characteristic polynomial is

$$\delta(s) := n(s)n_c(s) + d(s)d_c(s)$$

$$= (n_n a_r + d_n b_r)s^{n+r} + (n_n a_{r-1} + n_{n-1} a_r + d_n b_{r-1} + d_{n-1} b_r)s^{n+r-1}$$

$$+ \cdots + (n_0 a_0 + d_0 b_0) \tag{8.7}$$

Let us now introduce a desired (or target) characteristic polynomial $\delta_T^0(s)$ of degree $n + r$ which is Hurwitz and which has the desired set of closed-loop characteristic roots:

$$\delta_T^0(s) = \delta_{n+r}^0 s^{n+r} + \delta_{n+r-1}^0 s^{n+r-1} + \cdots + \delta_1^0 s + \delta_0^0 \tag{8.8}$$

To attain the desired or target characteristic polynomial by choosing the controller it is necessary and sufficient to solve the following set of linear equations:

$$\delta_{n+r}^0 = n_n a_r + d_n b_r$$

$$\delta_{n+r-1}^0 = n_n a_{r-1} + n_{n-1} a_r + d_n b_{r-1} + d_{n-1} b_r \tag{8.9}$$

$$\vdots$$

$$\delta_0^0 = n_0 a_0 + d_0 b_0$$

It is known in the literature that when $r = n - 1$ the above equations admit a solution for the controller coefficients for arbitrary $\delta_T^0(s)$ whenever the plant has no common pole-zero pairs. In general for $r < n - 1$ exact attainment of the target polynomial is impossible, and here we suggest the following procedure.

Let us relax the requirement of attaining the target polynomial exactly and enlarge the target to a box in coefficient space containing the point representing the original desired characteristic polynomial. This corresponds to choosing an interval polynomial family as the target rather than a single point. Such a procedure facilitates a solution, as we shall see, and also makes sense from a root space point of view.

Suppose that the desired region of closed-loop poles coincides with or is over-bounded by the interval polynomial family, given as

$$\Delta_T(s) := \left\{ \delta_T(s) = \delta_{n+r}^T s^{n+r} + \cdots + \delta_1^T s + \delta_0^T : \delta_i^{T-} \leq \delta_i^T \leq \delta_i^{T+}, \quad \text{for all } i \right\} \tag{8.10}$$

The root space corresponding to this family can be effectively computed by the edge theorem [3]. Then one can easily formulate the following set of linear inequalities that constrains controller coefficients so that the closed-loop system shown in Figure 8.8 has its closed-loop poles inside the root space of the interval polynomial $\delta_T(s)$:

$$\delta_{n+r}^{T-} \leq n_n a_r + d_n b_r \leq \delta_{n+r}^{T+}$$

$$\delta_{n+r-1}^{T-} \leq n_n a_{r-1} + n_{n-1} a_r + d_n b_{r-1} + d_{n-1} b_r \leq \delta_{n+r-1}^{T+} \tag{8.11}$$

$$\vdots$$

$$\delta_0^{T-} \leq n_0 a_0 + d_0 b_0 \leq \delta_0^{T+}$$

Equivalently,

$$
\underbrace{\begin{bmatrix} \delta^{T-}_{n+r} \\ \delta^{T-}_{n+r-1} \\ \delta^{T-}_{n+r-2} \\ \vdots \\ \vdots \\ \vdots \\ \delta^{T-}_{1} \\ \delta^{T-}_{0} \end{bmatrix}}_{\mathbf{b}_l} \leq
\underbrace{\begin{bmatrix}
n_n & & & & & d_n & & & \\
n_{n-1} & n_n & & & & d_{n-1} & d_n & & \\
\vdots & n_{n-1} & \ddots & & & \vdots & d_{n-1} & \ddots & \\
\vdots & \vdots & & n_n & & \vdots & \vdots & & d_n \\
n_0 & \vdots & & n_{n-1} & d_0 & \vdots & & & d_{n-1} \\
& n_0 & & \vdots & & d_0 & & & \vdots \\
& & \ddots & \vdots & & & \ddots & & \vdots \\
& & & n_0 & & & & & d_0
\end{bmatrix}}_{A}
\underbrace{\begin{bmatrix} a_r \\ a_{r-1} \\ \vdots \\ a_0 \\ b_r \\ b_{r-1} \\ \vdots \\ b_0 \end{bmatrix}}_{x}
$$

$$
\leq \underbrace{\begin{bmatrix} \delta^{T+}_{n+r} \\ \delta^{T+}_{n+r-1} \\ \delta^{T+}_{n+r-2} \\ \vdots \\ \vdots \\ \vdots \\ \delta^{T+}_{1} \\ \delta^{T+}_{0} \end{bmatrix}}_{\mathbf{b}_h}
\tag{8.12}
$$

Clearly, solving the following LP problem:

$$
\min_{A_c \mathbf{x} \leq \mathbf{b}} f(\mathbf{x}) \tag{8.13}
$$

where

$$
A_c = \begin{bmatrix} A \\ -A \end{bmatrix}, \qquad \mathbf{b} = \begin{bmatrix} \mathbf{b}_h \\ -\mathbf{b}_l \end{bmatrix} \tag{8.14}
$$

and $f(\mathbf{x})$ is an arbitrary linear function in $\mathbf{x}$, for example $f(\mathbf{x}) = \sum_i x_i$, gives a desired controller whenever one exists. Consequently,

$$
\mathcal{R}(\delta(s)) \subset \mathcal{R}(\mathbf{\Delta}_T(s)) \tag{8.15}
$$

where $\mathcal{R}(\delta(s))$ is a set of roots of $\delta(s)$ and $\mathcal{R}(\mathbf{\Delta}_T(s))$ is the root space of the interval polynomial family $\mathbf{\Delta}_T(s)$.

Note that when $r = n - 1$, there, of course, always exists a pole assignment controller [4]. That means that any characteristic polynomial can be attained exactly. However, the method described above will in general give much lower-order controllers r that satisfy the requirement given in eqn. 8.15.

8.3.2 Robust pole assignment

Let us now consider the case where the plant is subject to parameter uncertainty. We represent this by supposing that the given plant transfer function is an interval transfer function as follows:

$$\mathbf{G}(s) := \frac{\mathbf{n}(s)}{\mathbf{d}(s)}$$

$$:= \left\{ \frac{n(s)}{d(s)} : n(s) \in \mathbf{n}(s), \ d(s) \in \mathbf{d}(s) \right\}$$

$$:= \left\{ G(s) : \frac{n_n s^n + n_{n-1} s^{s-1} + \cdots + n_1 s + n_0}{d_n s^n + d_{n-1} s^{n-1} + \cdots + d_1 s + d_0}, \right.$$

$$\left. n_i^- \leq n_i \leq n_i^+, \ d_i^- \leq d_i \leq d_i^+, \ \forall i \right\} \tag{8.16}$$

The characteristic polynomial family of the closed-loop system is

$$\mathbf{\Delta}(s) = \{ n(s) n_c(s) + d(s) d_c(s) : n(s) \in \mathbf{n}(s), \ d(s) \in \mathbf{d}(s) \} \tag{8.17}$$

We require that the root space of the actual system lies in a prescribed target root space specified in terms of the root space of an interval polynomial. If we consider eqn. 8.10, the set of linear inequalities to be satisfied to achieve

$$\mathcal{R}(\mathbf{\Delta}(s)) \subseteq \mathcal{R}(\mathbf{\Delta}_T(s)) \tag{8.18}$$

is as follows:

$$\delta_{n+r}^{T-} \leq n_n a_r + d_n b_r \leq \delta_{n+r}^{T+}$$

$$\delta_{n+r-1}^{T-} \leq n_n a_{r-1} + n_{n-1} a_r + d_n b_{r-1} + d_{n-1} b_r \leq \delta_{n+r-1}^{T+}$$

$$\vdots$$

$$\delta_0^{T-} \leq n_0 a_0 + d_0 b_0 \leq \delta_0^{T+} \tag{8.19}$$

for all $n_i^- \leq n_i \leq n_i^+$ and $d_i^- \leq d_i \leq d_i^+$. Since this set of inequalities is linear it will be satisfied if and only if the vertices of the parameters satisfy them. Accordingly they can be replaced by the equivalent set obtained by substituting vertices as follows:

$$\delta_{n+r}^{T-} \leq n_n^- a_r + d_n^- b_r \leq \delta_{n+r}^{T+}$$

$$\delta_{N+r}^{T-} \leq n_n^+ a_r + d_n^- b_r \leq \delta_{n+r}^{T+}$$

$$\delta_{n+r}^{T-} \leq n_n^- a_r + d_n^+ b_r \leq \delta_{n+r}^{T+}$$

$$\delta_{n+r}^{T-} \leq n_n^+ a_r + d_n^+ b_r \leq \delta_{n+r}^{T+}$$

$$\vdots$$

$$\delta_{n+r-1}^{T-} \leq n_n^- a_{r-1} + n_{n-1}^- a_r + d_n^- b_{r-1} + d_{n-1}^- b_r \leq \delta_{n+r-1}^{T+}$$

$$\delta_{n+r-1}^{T-} \leq n_n^+ a_{r-1} + n_{n-1}^- a_r + d_n^- b_{r-1} + d_{n-1}^- b_r \leq \delta_{n+r-1}^{T+}$$

$$\delta_{n+r-1}^{T-} \leq n_n^- a_{r-1} + n_{n-1}^+ a_r + d_n^- b_{r-1} + d_{n-1}^- b_r \leq \delta_{n+r-1}^{T+}$$

$$\delta_{n+r-1}^{T-} \leq n_n^+ a_{r-1} + n_{n-1}^+ a_r + d_n^- b_{r-1} + d_{n-1}^- b_r \leq \delta_{n+r-1}^{T+}$$

$$\delta_{n+r-1}^{T-} \leq n_n^- a_{r-1} + n_{n-1}^- a_r + d_n^+ b_{r-1} + d_{n-1}^- b_r \leq \delta_{n+r-1}^{T+}$$

$$\delta_{n+r-1}^{T-} \leq n_n^+ a_{r-1} + n_{n-1}^- a_r + d_n^+ b_{r-1} + d_{n-1}^- b_r \leq \delta_{n+r-1}^{T+}$$

$$\delta_{n+r-1}^{T-} \leq n_n^- a_{r-1} + n_{n-1}^+ a_r + d_n^+ b_{r-1} + d_{n-1}^- b_r \leq \delta_{n+r-1}^{T+}$$

$$\delta_{n+r-1}^{T-} \leq n_n^+ a_{r-1} + n_{n-1}^+ a_r + d_n^+ b_{r-1} + d_{n-1}^- b_r \leq \delta_{n+r-1}^{T+}$$

$$\delta_{n+r-1}^{T-} \leq n_n^- a_{r-1} + n_{n-1}^- a_r + d_n^- b_{r-1} + d_{n-1}^+ b_r \leq \delta_{n+r-1}^{T+}$$

$$\delta_{n+r-1}^{T-} \leq n_n^+ a_{r-1} + n_{n-1}^- a_r + d_n^- b_{r-1} + d_{n-1}^+ b_r \leq \delta_{n+r-1}^{T+}$$

$$\delta_{n+r-1}^{T-} \leq n_n^- a_{r-1} + n_{n-1}^+ a_r + d_n^- b_{r-1}^+ + d_{n-1}^+ b_r \leq \delta_{n+r-1}^{T+}$$

$$\delta_{n+r-1}^{T-} \leq n_n^+ a_{r-1} + n_{n-1}^+ a_r + d_n^- b_{r-1} + d_{n-1}^+ b_r \leq \delta_{n+r-1}^{T+}$$

$$\delta_{n+r-1}^{T-} \leq n_n^- a_{r-1} + n_{n-1}^- a_r + d_n^+ b_{r-1} + d_{n-1}^+ b_r \leq \delta_{n+r-1}^{T+}$$

$$\delta_{n+r-1}^{T-} \leq n_n^+ a_{r-1} + n_{n-1}^- a_r + d_n^+ b_{r-1} + d_{n-1}^+ b_r \leq \delta_{n+r-1}^{T+}$$

$$\delta_{n+r-1}^{T-} \leq n_n^- a_{r-1} + n_{n-1}^+ a_r + d_n^+ b_{r-1} + d_{n-1}^+ b_r \leq \delta_{n+r-1}^{T+}$$

$$\delta_{n+r-1}^{T-} \leq n_n^+ a_{r-1} + n_{n-1}^+ a_r + d_n^+ b_{r-1} + d_{n-1}^+ b_r \leq \delta_{n+r-1}^{T+}$$

(8.20)

$$\vdots$$

$$\delta_0^{T-} \leq n_0^- a_0 + d_0^- b_0 \leq \delta_0^{T+}$$

$$\delta_0^{T-} \leq n_0^+ a_0 + d_0^- b_0 \leq \delta_0^{T+}$$

$$\delta_0^{T-} \leq n_0^- a_0 + d_n^+ b_0 \leq \delta_0^{T+}$$

$$\delta_0^{T-} \leq n_0^+ a_0 + d_0^+ b_0 \leq \delta_0^{T+}$$

The number of inequalities is in general much less than listed above since there will be a lot of redundant inequalities. However, the maximum number of inequalities will be the same as the number of vertices.

We illustrate the above approach in the following example.

Example 3

Consider the interval transfer function:

$$\mathbf{G}(s) = \frac{\beta_1 s + \beta_0}{s^2 + \alpha_1 s + \alpha_0} \tag{8.21}$$

where

$$\alpha_0 \in [\alpha_0^-, \alpha_0^+] = [-1, 1], \qquad \alpha_1 \in [\alpha_1^-, \alpha_1^+] = [0.5, 1]$$
$$\beta_0 \in [\beta_0^-, \beta_0^+] = [1, 1.5], \qquad \beta_1 \in [\beta_1^-, \beta_1^+] = [0.5, 1]$$

In this example, we want to obtain a constant feedback controller, if it exists, so that the closed-loop system satisfies the given specification. Let the controller be

$$C(s) = a_0 \tag{8.22}$$

and the desired characteristic interval polynomial be chosen as

$$\mathbf{\Delta}_{\mathrm{T}}(s) = s^2 + [1, 3]s + [1, 3] \tag{8.23}$$

so that the closed poles remain in a neighbourhood of $-0.75 \pm j$.

The root space $\mathcal{R}(\mathbf{\Delta}_{\mathrm{T}}(s))$ of the desired characteristic interval polynomial $\mathbf{\Delta}_{\mathrm{T}}(s)$ is shown in Figure 8.9.

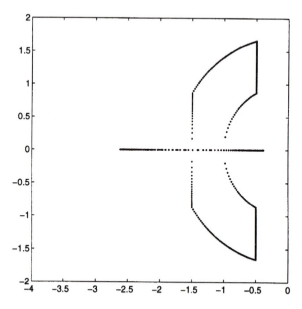

Figure 8.9   Root clusters of $\mathbf{\Delta}_{\mathrm{T}}(s)$ (Example 3)

Now let us consider the characteristic polynomial family of the plant $\mathbf{G}(s)$ in eqn. 8.21 and $C(s)$ in eqn. 8.22:

$$\mathbf{\Delta}(s) = s^2 + (\alpha_1 + a_0 \beta_1)s + (\alpha_0 + a_0 \beta_0) \tag{8.24}$$

Note that our objective is to design a controller for the given interval plant in eqn. 8.21 so that the root clusters of the family of the closed-loop characteristic polynomials are contained in those of the family $\boldsymbol{\Delta}_T(s)$. In other words, we want to select α_0 such that

$$\mathcal{R}(\boldsymbol{\Delta}(s)) \subseteq \mathcal{R}(\boldsymbol{\Delta}_T(s)) \tag{8.25}$$

where $\mathcal{R}(\cdot)$ denotes the root cluster space of $(\cdot)$. This objective will be accomplished if

$$\mathcal{F}(\boldsymbol{\Delta}(s)) \subseteq \mathcal{F}(\boldsymbol{\Delta}_T(s)) \tag{8.26}$$

where $\mathcal{F}(\cdot)$ denotes the family of polynomials.

From this consideration, we construct the following set of linear inequalities that the controller parameter a_0 should satisfy:

$$\begin{bmatrix} 1 \\ 1 \end{bmatrix} \leq \begin{bmatrix} \beta_1 \\ \beta_0 \end{bmatrix} a_0 + \begin{bmatrix} \alpha_1 \\ \alpha_0 \end{bmatrix} \leq \begin{bmatrix} 3 \\ 3 \end{bmatrix} \tag{8.27}$$

Replacing interval parameters as before by their vertices, we have

$$\begin{bmatrix} 1 - \alpha_1^- \\ 1 - \alpha_1^+ \\ 1 - \alpha_1^- \\ 1 - \alpha_1^+ \\ 1 - \alpha_0^- \\ 1 - \alpha_0^+ \\ 1 - \alpha_0^- \\ 1 - \alpha_0^+ \end{bmatrix} \leq \begin{bmatrix} \beta_1^- \\ \beta_1^- \\ \beta_1^+ \\ \beta_1^+ \\ \beta_0^- \\ \beta_0^- \\ \beta_0^+ \\ \beta_0^+ \end{bmatrix} a_0 \leq \begin{bmatrix} 3 - \alpha_1^- \\ 3 - \alpha_1^+ \\ 3 - \alpha_1^- \\ 3 - \alpha_1^+ \\ 3 - \alpha_0^- \\ 3 - \alpha_0^+ \\ 3 - \alpha_0^- \\ 3 - \alpha_0^+ \end{bmatrix} \tag{8.28}$$

After simplification by eliminating redundant inequalities, we have

$$\begin{bmatrix} 0.5 \\ 0.5 \end{bmatrix} \leq \begin{bmatrix} 0.5 \\ 1 \end{bmatrix} a_0 \leq \begin{bmatrix} 2 \\ 2 \end{bmatrix}$$

$$\begin{bmatrix} 1 \\ 1.5 \end{bmatrix} a_0 = \begin{bmatrix} 2 \\ 2 \end{bmatrix}$$

which are impossible to satisfy for any choice of α_0. Therefore, we conclude that no constant controller for $\mathbf{G}(s)$ exists to meet the objective stated in eqn. 8.25.

We now choose to use a first-order controller

$$C(s) = \frac{b_1 s + b_0}{s + a_0} \tag{8.29}$$

and we pick the desired locations of the closed-loop poles to be

$$-2, \qquad -1 \pm j$$

which are not, of course, attainable exactly. Thus we expand the coefficients of the target polynomial to

$$\Delta_{\mathrm{T}}(s) = s^3 + [2.5, 5.5]s^2 + [3.5, 8.5]s + [1, 7] \qquad (8.30)$$

so that the constraining root space $\mathcal{R}(\Delta_d(s))$ is acceptable. This root space is depicted in Figure 8.10.

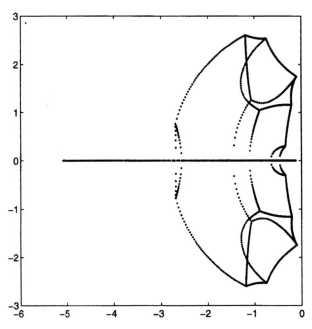

Figure 8.10   Root clusters of $\Delta_{\mathrm{T}}(s)$ (Example 3)

The family of actual characteristic polynomials with a first-order $C(s)$ is

$$\Delta(s) = s^3 + (a_0 + b_1\beta_1)s^2 + (a_0\alpha_1 + \alpha_0 + b_0\beta_1 + b_1\beta_0)s + (a_0\alpha_0 + b_0\beta_0) \qquad (8.31)$$

From these, we have the following linear inequalities to solve for (a_0, b_0, b_1):

$$\begin{bmatrix} 2.5 \\ 3.5 - \alpha_0 \\ 1 \end{bmatrix} \leq \begin{bmatrix} 1 & 0 & \beta_1 \\ \alpha_1 & \beta_1 & \beta_0 \\ \alpha_0 & \beta_0 & 0 \end{bmatrix} \begin{bmatrix} a_0 \\ b_0 \\ b_1 \end{bmatrix} \leq \begin{bmatrix} 5.5 \\ 8.5 - \alpha_0 \\ 7 \end{bmatrix} \qquad (8.32)$$

We now construct a linear programming problem. The linear constraints given in eqn. 8.32 are rewritten as

$$
\underbrace{\begin{bmatrix} 2.5 \\ 2.5 \\ 4.5 \\ 4.5 \\ 4.5 \\ 4.5 \\ 4.5 \\ 4.5 \\ 4.5 \\ 4.5 \\ 1 \\ 1 \\ 1 \\ 1 \end{bmatrix}}_{\mathbf{b}_l} \leq \underbrace{\begin{bmatrix} 1 & 0 & 0.5 \\ 1 & 0 & 1 \\ 0.5 & 0.5 & 1 \\ 1 & 0.5 & 1 \\ 0.5 & 1 & 1 \\ 1 & 1 & 1 \\ 0.5 & 0.5 & 1.5 \\ 1 & 0.5 & 1.5 \\ 0.5 & 1 & 1.5 \\ 1 & 1 & 1.5 \\ -1 & 1 & 0 \\ 1 & 1.0 & 0 \\ -1 & 1.5 & 0 \\ 1 & 1.5 & 0 \end{bmatrix}}_{A} \underbrace{\begin{bmatrix} a_0 \\ b_0 \\ b_1 \end{bmatrix}}_{\mathbf{x}} \leq \underbrace{\begin{bmatrix} 5.5 \\ 5.5 \\ 7.5 \\ 7.5 \\ 7.5 \\ 7.5 \\ 7.5 \\ 7.5 \\ 7.5 \\ 7.5 \\ 7 \\ 7 \\ 7 \\ 7 \end{bmatrix}}_{\mathbf{b}_h}
\tag{8.33}
$$

Now we set up a standard LP problem. Since we want a controller to satisfy the requirement

$$
\mathcal{R}(\mathbf{\Delta}(s)) \subseteq \mathcal{R}(\mathbf{\Delta}_{\mathrm{T}}(s))
\tag{8.34}
$$

we arbitrarily choose the minimisation index

$$
f(\mathbf{x}) = \sum_i x_i
\tag{8.35}
$$

By solving the LP problem

$$
\min_{A_c \mathbf{x} \leq \mathbf{b}} f(\mathbf{x})
\tag{8.36}
$$

where

$$
A_c = \begin{bmatrix} A \\ -A \end{bmatrix} \qquad \mathbf{b} = \begin{bmatrix} \mathbf{b}_h \\ -\mathbf{b}_l \end{bmatrix}
\tag{8.37}
$$

we obtain the controller

$$
C(s) = \frac{3s + 2}{s + 1}
\tag{8.38}
$$

With this $C(s)$, we construct the root space of $\mathbf{\Delta}(s)$ which is shown in Figure 8.11. Figure 8.12 also shows that the root space $\mathcal{R}(\mathbf{\Delta}(s))$ is clearly contained in $\mathcal{R}(\mathbf{\Delta}_{\mathrm{T}}(s))$.

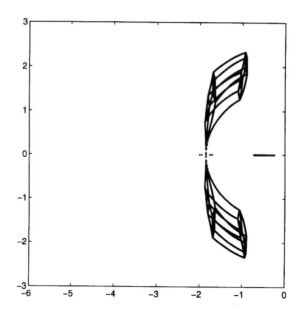

Figure 8.11 Root clusters of $\Delta(s)$ with first-order $C(s)$ (Example 3)

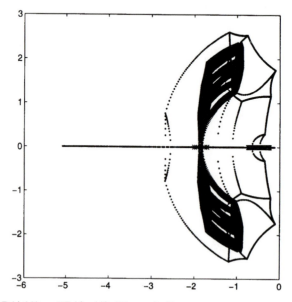

Figure 8.12 $\mathcal{R}(\Delta(s))$ and $\mathcal{R}(\Delta_T(s))$ (Example 3)

Example 4

Consider the plant family described by the interval transfer function

$$\mathbf{G}(s) = \frac{\beta_2 s^2 + \beta_1 s + \beta_0}{s^3 + \alpha_2 s^2 + \alpha_1 s + \alpha_0}$$

(8.39)

where

$$\alpha_0 \in [-1.1, -0.9] \qquad \alpha_1 \in [-5.1, -4.9] \qquad \alpha_2 \in [-3.1, -2.9]$$
$$\beta_0 \in [1.9, 2.1] \qquad \beta_1 \in [-3.1, -2.9] \qquad \beta_2 \in [1.9, 2.1]$$

In this example, we want to obtain a first-order feedback controller, that robustly D-stabilises the closed-loop system. To achieve this we first pick a desired set of closed-loop poles and construct a nominal polynomial. Intervals are built around these nominal coefficients in such a way that the roots do not escape a desired region in the complex plane. The target interval polynomial constructed in this way for D-stability is taken in this example as follows:

$$\boldsymbol{\Delta}_T(s) := \left\{ \delta_T(s) = s^4 + \delta_3^T s^3 + \delta_2^T s^2 + \delta_1^T s + \delta_0^T \right\}$$

where

$$\delta_0^T \in \left[\delta_0^{T-}, \delta_0^{T+}\right] = [38.25, 54.25] \qquad \delta_1^T \in \left[\delta_1^{T-}, \delta_1^{T+}\right] = [57, 77]$$
$$\delta_2^T \in \left[\delta_2^{T-}, \delta_2^{T+}\right] = [31.25, 45.25] \qquad \delta_3^T \in \left[\delta_0^{T-}, \delta_0^{T+}\right] = [6, 14]$$

The controller is represented as

$$C(s) = \frac{b_1 s + b_0}{s + a_0}$$

(8.40)

and we define the parameter vectors

$$\mathbf{p} := \begin{bmatrix} \alpha_0 & \alpha_1 & \alpha_2 & \beta_0 & \beta_1 & \beta_2 \end{bmatrix}$$
$$\mathbf{x} := \begin{bmatrix} a_0 & b_0 & b_1 \end{bmatrix}$$

(8.41)

We also define respective parameter space boxes $\mathbf{\Pi}$ and $\boldsymbol{\Delta}$. If we assume that the characteristic polynomial family of the closed-loop system is

$$\boldsymbol{\Delta}(\mathbf{p}, \mathbf{x}) := \left\{ \delta(s, \mathbf{p}, \mathbf{x}) = s^4 + \delta_3(\mathbf{p}, \mathbf{x})s^3 + \delta_2(\mathbf{p}, \mathbf{x})s^2 + \delta_1(\mathbf{p}, \mathbf{x})s + \delta_0(\mathbf{p}, \mathbf{x}) \right\}$$

(8.42)

Then using the vertices of $\mathbf{\Pi}$ and $\boldsymbol{\Delta}$, one can form the set of linear inequality constraints,

$$\underbrace{\begin{bmatrix} \delta_3^{T-} \\ \delta_2^{T-} \\ \delta_1^{T-} \\ \delta_0^{T-} \end{bmatrix}}_{\mathbf{b}_l} \leq \underbrace{\begin{bmatrix} \delta_3(\mathbf{p}, \mathbf{x}) \\ \delta_2(\mathbf{p}, \mathbf{x}) \\ \delta_3(\mathbf{p}, \mathbf{x}) \\ \delta_0(\mathbf{p}, \mathbf{x}) \end{bmatrix}}_{A(\mathbf{p})\mathbf{x}+\mathbf{c}(\mathbf{p})} \leq \underbrace{\begin{bmatrix} \delta_3^{T+} \\ \delta_2^{T+} \\ \delta_1^{T+} \\ \delta_0^{T+} \end{bmatrix}}_{\mathbf{b}_h}$$

(8.43)

where $\mathbf{p} \in \mathbf{\Pi}_v$.

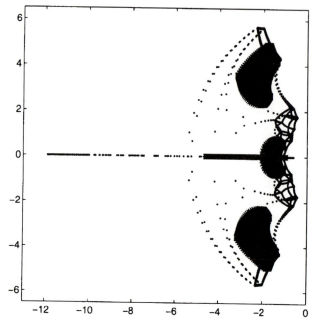

Figure 8.13 $\mathcal{R}(\mathbf{\Delta}(s, \mathbf{p}))$ *and* $\mathcal{R}(\mathbf{\Delta}_T(s))$ *(Example 4)*

From this, we formulate a linear programming problem. The solution obtained is

$$C(s) = \frac{14.2271s + 14.9751}{s - 17.9315} \tag{8.44}$$

With this $C(s)$, we construct the root space of $\mathbf{\Delta}(s, \mathbf{p})$ for $\mathbf{p} \in \mathbf{\Pi}$. Figure 8.13 clearly verifies that the root space of $\mathbf{\Delta}(s, \mathbf{p})$ lies inside that of $\mathbf{\Delta}_T(s)$.

Example 5

We take the plant dealt with in Example 4. In this example, we impose the additional design requirement that each controller parameter should be allowed to perturb by the amount $\Delta x = \pm 0.01$. Therefore, the goal of this problem is to obtain a first-order controller of the form in eqn. 8.40 that robustly D-stabilises the closed-loop system under given plant parameter as well as controller parameter perturbations. As we described earlier, the mapping theorem allows us to construct a similar set of linear inequalities by letting

$$\hat{\mathbf{p}} = [\mathbf{p} \quad \Delta x] \tag{8.45}$$

Then

$$
\underbrace{\begin{bmatrix} \delta_3^{T-} \\ \delta_2^{T-} \\ \delta_1^{T-} \\ \delta_0^{T-} \end{bmatrix}}_{\mathbf{b}_l} \leq \underbrace{\begin{bmatrix} \delta_3(\hat{\mathbf{p}}, \mathbf{x}) \\ \delta_2(\hat{\mathbf{p}}, \mathbf{x}) \\ \delta_3(\hat{\mathbf{p}}, \mathbf{x}) \\ \delta_0(\hat{\mathbf{p}}, \mathbf{x}) \end{bmatrix}}_{A(\hat{\mathbf{p}})\mathbf{x}+\mathbf{c}(\mathbf{p})} \leq \underbrace{\begin{bmatrix} \delta_3^{T+} \\ \delta_2^{T+} \\ \delta_1^{T+} \\ \delta_0^{T+} \end{bmatrix}}_{\mathbf{b}_h} \tag{8.46}
$$

where $\hat{\mathbf{p}}$ ranges over the vertex set of $\mathbf{\Pi}$.

From this, we formulate a linear programming problem. The solution we obtained is

$$
C(s) = \frac{14.0628s + 16.5103}{s - 17.1803} \tag{8.47}
$$

With this $C(s)$, we construct the root space of $\mathbf{\Delta}(s, \hat{\mathbf{p}})$ for $\hat{\mathbf{p}} \in \hat{\mathbf{\Pi}}$. Figure 8.14 clearly verifies that the root space of $\mathbf{\Delta}(s, \hat{\mathbf{p}})$ lies inside that of the target root space of the family $\mathbf{\Delta}_T(s)$.

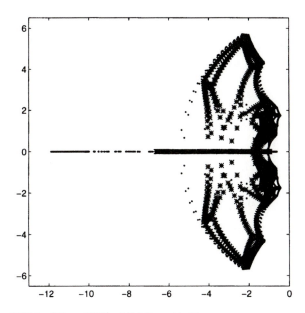

Figure 8.14 $\mathcal{R}(\mathbf{\Delta}(s, \hat{\mathbf{p}}))$ *and* $\mathcal{R}(\mathbf{\Delta}_T(s))$ *(Example 5)*

8.4 Conclusions and future directions

We have presented (a) some fundamental results in the area of robust control under parametric uncertainty in Chapter 4 and (b) some approaches to design illustrated through examples in this chapter. We have proposed some procedures

dures to deal with the fixed-order design problem. This is a long-standing open problem, and in our opinion the conspicuous lack of results on this problem has prevented modern design methods from being applicable to real-world problems which operate under the constraint of fixed order or structure. Our methods lead to linear programming problems, and thus standard and efficient software can be used to solve the computational part. The design task can be formulated simply and clearly: it is to select the controller to position the performance vector or set of vectors in the case of robust performance within a prescribed box within which the performance deterioration is acceptable. This formulation has a chance to work because it draws heavily on many fundamental bounding results from robust parametric control. Such bounds can be used to position the closed-loop root space within prescribed sets and also to achieve guaranteed and robust performance in the H_∞ sense.

We believe that the techniques given here are among a handful of results available at this time which address the fixed-order design problem, and we hope that this chapter will stimulate vigorous research in this important applied area of control.

8.5 Acknowledgments

This research is supported in part by NASA Grant NCC-5228 and NSF Grants ECS-9417004 and HRD-9706268.

8.6 References

1 BHATTACHARYYA, S. P., CHAPELLAT, H., and KEEL, L. H.: 'Robust control: the parametric approach' (Prentice Hall, Upper Saddle River, NJ, 1995)
2 ZADEH, L. A., and DESOER, C. A.: 'Linear systems theory' (McGraw Hill Book Co., New York, NY, 1963)
3 BARTLETT, A. C., HOLLOT, C. V., and LIN, H.: 'Root location of an entire polytope of polynomials: it suffices to check the edges', *Math. Control Signals Syst.*, 1988, **1**, pp. 61–71
4 PEARSON, J. B.: 'Compensator design for dynamic optimization', *Int. J. Control*, 1968, **9**, pp. 473

Chapter 9
Dynamic sliding mode control design using symbolic algebra tools

S. K. Spurgeon and X. Y. Lu

9.1 Introduction

Sliding mode control has long been recognised as providing an appropriate solution to the robust control problem. Up to this point the majority of design methodologies have been based around linear uncertain systems [1] or specific types of nonlinear systems. The latter may involve particular application areas, such as robotics [2], or require that relatively stringent conditions are met by members of the system class; for example, the system class may be required to be feedback linearisable [3]. It is obviously desirable to have a sliding mode control methodology which will be applicable to a fairly broad class of nonlinear system representations, exhibit robustness while yielding appropriate performance and lend itself to the development of appropriate toolboxes for controller design. It has been shown that dynamic sliding mode policies based around differential input-output (IO) system representations are sufficiently general to meet this remit [4−6].

This Chapter gives a brief description of the dynamic sliding mode control methodology and shows how it can be implemented in a straightforward manner using symbolic algebra. A particular Mathematica implementation is described and some examples of design sessions presented.

The following notation is used throughout:

$$\mathcal{N}_\delta(0) = \{\mathbf{x} | \mathbf{x} \in \mathbf{R}^n, \|\mathbf{x}\| < \delta\}$$

where $\|\cdot\|$ is the Euclidian norm.

9.2 Dynamic sliding mode control

For sliding mode controller design using static feedback, it is necessary that the system assumes a *regular form* and that the control variables appear linearly in the system to recover the control parameters from the chosen sliding condition [7]. In general, this is not practically implementable for general nonlinear systems with nonlinear control. To develop the sliding mode control method to include dynamic policies, and hence to ensure it becomes applicable to an extended class of nonlinear systems, differential IO system representations are employed.

For a given system in state-space form which is locally observable,

$$\dot{\mathbf{x}} = f(\mathbf{x}, \mathbf{u})$$
$$\mathbf{y} = h(\mathbf{x}, \mathbf{u}) \tag{9.1}$$

where $\mathbf{x} \in \mathbf{R}^n, \mathbf{u} \in \mathbf{R}^m$ and $f(\mathbf{x}, \mathbf{u})$ and $h(\mathbf{x}, \mathbf{u})$ are smooth vector functions, the following locally equivalent differential IO system exists [8]:

$$y_1^{(n_1)} = \varphi_1(\widehat{\mathbf{y}}, \hat{\mathbf{u}}, t)$$
$$\ldots\ldots \tag{9.2}$$
$$y_p^{(n_p)} = \varphi_p(\widehat{\mathbf{y}}, \hat{\mathbf{u}}, t)$$

where

$$\hat{\mathbf{u}} = (u_1, \ldots, u_1^{(\beta_1)}, \ldots, u_m, \ldots, u_m^{(\beta_m)}), \text{ and } \widehat{\mathbf{y}} = (y_1, \ldots, y_1^{(n_1-1)}, \ldots, y_p, \ldots, y_p^{(n_p-1)})$$
with $n_1 + \cdots + n_p = n$.

Definition 1   A differential IO system (eqn. 9.2) is called *proper* if

(i) $p = m$
(ii)   all $\varphi_i(., ., .)$, $i = 1, \ldots, m$, are C^3-functions
(iii)   (regularity condition)

$$\det\left[\frac{\partial(\varphi_1, \ldots, \varphi_m)}{\partial(u_1^{(\beta_1)}, \ldots, u_m^{(\beta_m)})}\right] \neq 0 \tag{9.3}$$

is satisfied with $\widehat{\mathbf{y}} \in \mathcal{N}_\delta(0)$ for all $t \geq 0$, some $\delta > 0$ and generically for $\hat{\mathbf{u}}$.

Whether or not the resulting closed-loop system is minimum-phase will be pertinent to the design methodology.

Definition 2   The *zero dynamics* corresponding to eqn. 9.2 is defined as

$$\varphi_1(0, \hat{\mathbf{u}}, t) = 0$$
$$\ldots\ldots \tag{9.4}$$
$$\varphi_p(0, \hat{\mathbf{u}}, t) = 0$$

Eqn. 9.2 is called *minimum-phase* if there exist $\delta > 0$ and $\widetilde{\mathbf{u}}_0 \in \mathbf{R}^\beta$, $\beta = \beta_1 + \cdots + \beta_m$, such that eqn. 9.4 is uniformly asymptotically (exponentially) stable for initial condition $\widetilde{\mathbf{u}}(0) \in \mathcal{N}_\delta(\widetilde{\mathbf{u}}_0)$, where $\widetilde{\mathbf{u}} = (u_1, \ldots, u_1^{(\beta_1 - 1)}, \ldots, u_m, \ldots, u_m^{(\beta_m - 1)})$. Otherwise, it is *non-minimum-phase*. Note that, in this case, minimum-phaseness is a property of the chosen control signal.

To address robustness, uncertain systems of the following form may be considered:

$$y_1^{(n_1)} = \varphi_1(\widehat{\mathbf{y}}, \widehat{\mathbf{u}}, t) + \Delta_1(\widehat{\mathbf{y}}, t)$$

$$\cdots\cdots$$

$$y_p^{(n_p)} = \varphi_p(\widehat{\mathbf{y}}, \widehat{\mathbf{u}}, t) + \Delta_p(\widehat{\mathbf{y}}, t)$$

$$\text{(9.5)}$$

The uncertainties are Lebesgue measurable and satisfy

$$\|\Delta_i(\widehat{\mathbf{y}}, t)\| \leq \rho_i \|\widehat{\mathbf{y}}\| + l_i, \qquad \rho_i \geq 0, \qquad l_i \geq 0, \qquad i = 1, \ldots, p \qquad \text{(9.6)}$$

The uncertainty may be due to external uncertainties, internal parameter uncertainties, measurement noise, system identification error, or indeed the elimination procedure used to generate a differential input-output model from a state space model as in [8]. □

It is often convenient to consider the generalised controller canonical form (GCCF) representation of (eqn. 9.2). Without loss of generality, suppose that $n_1, \ldots, n_{m_1} > 1$ and $n_{m_1} = \cdots = n_{m_1 + m_2} = 1$, $m_1 + m_2 = m$. The system 9.2 may be expressed in the following GCCF [9]:

$$\dot{\zeta}_1^{(1)} = \zeta_2^{(1)}$$

$$\cdots\cdots$$

$$\dot{\zeta}_{n_1 - 1}^{(1)} = \zeta_{n_1}^{(1)}$$

$$\dot{\zeta}_{n_1}^{(1)} = \varphi_1(\zeta, \widehat{u}, t)$$

$$\text{(9.7)}$$

$$\cdots\cdots$$

$$\dot{\zeta}_{n_m - 1}^{(m)} = \zeta_{n_m}^{(m)}$$

$$\dot{\zeta}_{n_m}^{(m)} = \varphi_m(\zeta, \widehat{u}, t)$$

where $\zeta^{(i)} = (\zeta_1^{(i)}, \ldots, \zeta_{n_i}^{(i)}) = (y_i, \ldots, y_i^{(n_i - 1)})$, $i = 1, \ldots, m$ and $\zeta = (\zeta^{(1)}, \ldots, \zeta^{(m)})^T$ represent the system outputs and their derivatives.

Any sliding mode control design strategy may be divided into two independent procedures; the choice of sliding surfaces according to the system structure and desired performance and the choice of sliding reachability conditions to ensure the sliding surfaces are reached in finite time or asymptotically. Two popular choices of sliding surface are:

(i)   direct sliding surface [5, 6, 10, 11],

$$s_i = \sum_{j=1}^{n_i} a_j^{(i)} \zeta_j^{(i)}, \qquad i = 1, \ldots, m, \tag{9.8}$$

where $\sum_{j=1}^{n_i} a_j^{(i)} \lambda^{j-1}$ are Hurwitz polynomials with $a_{n_i}^{(i)} = 1, i = 1, \ldots, m$.

(ii)   indirect sliding surface [4, 12],

$$s_i = \sum_{j=1}^{n_i+1} a_j^{(i)} \zeta_j^{(i)} + \varphi_i(\zeta, \hat{\mathbf{u}}, t), \qquad i = 1, \ldots, m. \tag{9.9}$$

where $\sum_{j=1}^{n_i+1} a_j^{(i)} \lambda^{j-1}$ are Hurwitz polynomials with $a_{n_i+1}^{(i)} = 1, i = 1, \ldots, m$.

It is interesting to note that, with the indirect choice, the system becomes equivalent to an nth-order linear system, with dynamics prescribed by the choice of Hurwitz polynomial, in the sliding mode.

Having chosen the required sliding surface, it is then necessary to choose a reachability condition to ensure that the sliding surface is reached. One choice of sliding reachability condition is $\mathbf{s}\dot{\mathbf{s}} \leqslant 0$, if $\mathbf{s} \neq 0$, where the derivative is understood to be taken along the trajectories of the system. To guarantee the closed-loop stability of the dynamic sliding mode control method, it is necessary to employ a particular sliding reachability condition as follows.

Definition 3   A *general sliding reachability condition* is defined as

$$\dot{\mathbf{s}} = -\gamma(\kappa, \mathbf{s}) \tag{9.10}$$

where $\mathbf{s} = [s_1, \ldots, s_m]$, and $\kappa = [\kappa_1, \ldots, \kappa_l]$ is a set of design parameters such that for fixed κ, $\gamma(\kappa, \mathbf{s}) = [\gamma_1(\kappa, \mathbf{s}), \ldots, \gamma_m(\kappa, \mathbf{s})]$ satisfies the following conditions:

(a) $\gamma(\kappa, 0) = 0$
(b) $\gamma_i(\kappa, \mathbf{s})$ is a bounded C^1-function of $\mathbf{s}$ if $s_i \neq 0, i = 1, \ldots, m$
(c)   eqn. 9.10 is globally uniformly asymptotically stable, and is called a *properly coupled sliding reachability condition* if it satisfies (a), (b) and
(d) $s_i \gamma_i(\kappa, \mathbf{s}) > 0$ if $s_i \neq 0$
(e)   there exists a potential function $\Gamma(\kappa, \mathbf{s})$ such that for fixed κ

$$\frac{\partial \Gamma}{\partial s_i} = \gamma_i(\kappa, \mathbf{s}), \qquad i = 1, \ldots, m$$

The *decoupled sliding reachability condition* is a special case of the properly coupled sliding reachability condition for which $s_i = \gamma_i(\kappa, s_i), i = 1, \ldots, m$.   □

Examples of properly coupled reachability conditions, which will be necessary for the MIMO dynamic sliding mode controller design method considered later, are given below.

Example 1

If switching is adopted,

$$\gamma(\kappa, \mathbf{s}) = (\kappa_{ij})\,\mathbf{s} + \begin{bmatrix} \kappa_{1\,m+1}\,\mathrm{sgn}(s_1) \\ \vdots \\ \kappa_{m\,m+1}\,\mathrm{sgn}(s_m) \end{bmatrix}$$

where $(\kappa_{ij}) \in \mathbf{R}^{m \times m}$ is a positive definite real matrix and $\kappa_{i\,m+1} \geq 0$, $i = 1, \ldots, m$, which may be discontinuous at $s_i = 0$. Its potential function is $\Gamma(\kappa, \mathbf{s}) = \frac{1}{2}\mathbf{s}^T(\kappa_{ij})\,\mathbf{s} + \sum_{i=1}^{m} \kappa_{i\,m+1}s_i\,\mathrm{sgn}(s_i)$, which is positive definite. Note that $\Gamma(\kappa, \mathbf{s})$ is radially unbounded. □

A continuous roundoff may be obtained by replacing $\mathrm{sgn}(\cdot)$ with $\mathrm{sat}_\varepsilon(\cdot)$, where

$$\mathrm{sat}_\varepsilon(\mathbf{x}) = \varepsilon\,\mathrm{sat}\!\left(\frac{\mathbf{x}}{\varepsilon}\right) = \begin{cases} 1, & \mathbf{x} > \varepsilon \\ \mathbf{x}, & |\mathbf{x}| \leq \varepsilon \\ -1, & \mathbf{x} < -\varepsilon \end{cases}$$

Example 2

For general m, let

$$\Gamma(\kappa, \mathbf{s}) = \frac{1}{2p}\left(\sum_{i=1}^{m} \kappa_i s_i^2\right)^p + \sum_{i=1}^{m}|s_i|, \qquad p > 0$$

Then

$$\gamma_j = \frac{\partial \Gamma}{\partial s_j} = \kappa_j\,s_j\left(\sum_{i=1}^{m} \kappa_i s_i^2\right)^{p-1} + \mathrm{sgn}(s_j), \qquad j = 1, \ldots, m$$

To satisfy Definition 3 it is required that $\kappa_i > 0$, $i = 1, \ldots, m$. Obviously, the γ_j's satisfy the conditions in Definition 3, and therefore define a properly coupled sliding reachability condition. When $p > 0$, γ_i is a C^1-function if $s_i \neq 0$; when $p > \frac{1}{2}$, a globally continuous reachability condition results; when $p > 1\frac{1}{2}$, γ_j is a global C^1-function. Note that $\Gamma(\kappa, \mathbf{s})$ is positive definite and radially unbounded for $p > 0$. Similarly, a continuous roundoff is obtained by replacing $\mathrm{sgn}(\cdot)$ with $\mathrm{sat}_\varepsilon(\cdot)$. □

9.3 Design method

To implement the design method in a symbolic package it is first necessary to input details of the nonlinear differential IO system: the order of the system output, the number of inputs and the right-hand side of eqn. 9.2. The only free design parameters are the roots of the characteristic equations relating to the choice of sliding surface. These must be selected *a priori*. There are two fundamental computational stages which must be carried out to determine the control. The first involves the numerical solution of a Lyapunov equation. This is used to parametrise some constants in the controller. The second stage involves symbolically recovering the control from eqn. 9.10.

Algorithmic descriptions for the choices of both *direct* and *indirect* sliding surface for the nominal system description (eqn. 9.2) will first be presented. A description of the robust design module, based on the system description (eqn. 9.5) then follows. This latter method involves an additional stage concerned with appropriately parametrising the control design problem to ensure robustness.

9.3.1 Direct sliding mode control

Step 1:   Choose the poles which are desired roots of the characteristic equation relating to the Hurwitz polynomial in eqn. 9.8.

Step 2:   Choose a properly coupled sliding reachability condition.

Step 3:   Choose the *sliding gain matrix* $\mathbf{K}$ such that

$$\mathbf{K} - [\mathbf{BD}]^T[\mathbf{BD}] > 0 \qquad (9.11)$$

where $\mathbf{D} = \mathrm{diag}[\mathbf{D}_1, \ldots, \mathbf{D}_{m_1}, \mathbf{D}_{m_1+1}, \ldots, \mathbf{D}_m]^T$ with $\mathbf{D}_i = [0, \ldots, 0, 1]^T$ of dimension $n_i - 1$ for $i = 1, \ldots, m_1$ and $\mathbf{D}_j = [0]$ of dimension   1   for $j = m_1 + 1, \ldots, m$; $\mathbf{A} = \mathrm{diag}[\mathbf{A}_1, \ldots, \mathbf{A}_{m_1}]$ with $\mathbf{A}_i$ the companion matrix of the Hurwitz polynomial $\sum_{j=1}^{n_i} a_j^{(i)} \lambda^{j-1}$ with $a_{n_i}^{(i)} = 1$; $\mathbf{A}$ and $\mathbf{B}$ satisfy the Lyapunov equation

$$\mathbf{A}^T\mathbf{B} + \mathbf{B}\mathbf{A} = -I_{n-m} \qquad (9.12)$$

Step 4:   Differentiating eqn. 9.8 with respect to time t along the trajectories of eqn. 9.7 leads to

$$\dot{s}_i\big|_{(1.2.7)} = \sum_{j=1}^{n_i-1} a_j^{(i)} \zeta_{j+1}^{(i)} + \varphi_i(\zeta, \hat{u}, t), \qquad i = 1, \ldots, m \qquad (9.13)$$

Now set

$$\sum_{j=1}^{n_i-1} a_j^{(i)} \zeta_{j+1}^{(i)} + \varphi_i(\zeta, \hat{\mathbf{u}}, t) = -\gamma_i(\kappa, \mathbf{s}), \qquad i = 1, \ldots, m \qquad (9.14)$$

where $\mathbf{s}$ is as defined in eqn. 9.8, to determine the feedback control. Eqn. 9.13 becomes

$$\dot{\mathbf{s}} = -\gamma(\kappa, \mathbf{s})$$

Step 5:   From eqn. 9.14 the highest-order derivatives of the control $[u_1^{(\beta_1)}, \ldots, u_m^{(\beta_m)}]$, can be solved out by the implicit function theorem as

$$u_i^{(\beta_i)} = p_i(\zeta, \hat{\mathbf{u}}, t), \qquad i = 1, \ldots, m$$

if the regularity condition is satisfied. Note that $p_i(\zeta, \hat{\mathbf{u}}, t)$ is a continuous function if $s_i \neq 0$ because φ_i is C^1 and γ_i is C^1 if $s_i \neq 0$. This dynamic feedback can be realised in canonical form by introducing the pseudo-state variables as

$$\dot{z}_1^{(1)} = z_2^{(1)}$$

$$\cdots\cdots$$

$$\dot{z}_{\beta_1 - 1}^{(1)} = z_{\beta_1}^{(1)}$$

$$\dot{z}_{\beta_1}^{(1)} = p_1(\zeta, \mathbf{z}, t)$$

$$\cdots\cdots \tag{9.15}$$

$$\dot{z}_1^{(m)} = z_2^{(m)}$$

$$\cdots\cdots$$

$$\dot{z}_{\beta_m - 1}^{(m)} = z_{\beta_m}^{(m)}$$

$$\dot{z}_{\beta_m}^{(m)} = p_m(\zeta, \mathbf{z}, t)$$

where

$$z^{(i)} = (z_1^{(i)}, \ldots, z_{\beta_i}^{(i)}) = (u_i, \dot{u}_i, \ldots, u_i^{(\beta_i - 1)}), \quad i = 1, \ldots, m$$

$$\mathbf{z} = (z^{(1)}, \ldots, z^{(m)})^T \tag{9.16}$$

Eqns. 9.15, together with eqn. 9.7, yields a closed-loop system of dimension $\sum_{i=1}^{m} n_i + \sum_{i=1}^{m} \beta_i$, where β_i is the highest-order derivative of u_i. With abuse of notation, the closed-loop system can be written as:

$$\dot{\zeta} = \Phi(\zeta, \mathbf{z}, t)$$

$$\dot{\mathbf{z}} = P(\zeta, \mathbf{z}, t)$$

$$\Phi(\zeta, \mathbf{z}, t) = [\zeta_2^{(1)}, \ldots, \zeta_{n_1}^{(1)}, \varphi_1; \ldots; \zeta_2^{(m)}, \ldots, \zeta_{n_m}^{(m)}, \varphi_m]^T \tag{9.17}$$

$$P(\zeta, \mathbf{z}, t) = [z_2^{(1)}, \ldots, z_{\beta_1}^{(1)}, p_1; \ldots; z_2^{(m)}, \ldots, z_{\beta_m}^{(m)}, p_m]^T$$

Step 6: Choose $\tilde{u}_0 \in \mathbf{R}^\beta$ and a $\delta > 0$ such that, for initial condition $\tilde{u}(0) \in \mathcal{N}_\delta(\tilde{u}_0)$:

(i)   the regularity condition is satisfied
(ii) the zero dynamics are uniformly asymptotically stable
(iii) all the initial conditions for eqn. 9.17 are compatible.

The details of the formal stability analysis relating to this method can be found in [11].

9.3.2 Indirect sliding mode control

Step 1:   Choose the Hurwitz polynomial specifying the sliding surface (eqn. 9.9).
Step 2:   Choose a properly coupled sliding reachability condition (eqn. 9.10).
Step 3:   Determine the sliding gain matrix $\mathbf{K} \in \mathbf{R}^{m \times m}$ as follows:

$$\mathbf{K} - [\mathbf{BD}]^T C^{-1} [\mathbf{BD}] > 0 \qquad (9.18)$$

where $\mathbf{A} = \mathrm{diag}[\mathbf{A}_1, \ldots, \mathbf{A}_m]$, $\mathbf{A}_i$ is the companion matrix of the Hurwitz polynomial $\sum_{j=1}^{n_i+1} a_j^{(i)} \lambda^{j-1}$ with $a_{n_i+1}^{(i)} = 1$, $i = 1, \ldots, m$; C is chosen positive definite; $\mathbf{A}, \mathbf{B}$ and $\mathbf{C}$ satisfy the Lyapunov equation

$$\mathbf{A}^T \mathbf{B} + \mathbf{B}\mathbf{A} = -\mathbf{C} \qquad (9.19)$$

and $\mathbf{D} = \mathrm{diag}[d_1, \ldots, d_m]$, $d_i = [0, \ldots, 1]^T$, $\dim(d_i) = n_i, i = 1, \ldots, m$.

Step 4:   From eqn. 9.9 and $\dot{s} + \gamma(\kappa, \mathbf{K}s) = 0$, the highest-order derivatives of the control $[u_1^{(\beta_1+1)}, \ldots, u_m^{(\beta_m+1)}]$, which appear linearly, can be solved out as

$$u_i^{(\beta_i+1)} = p_i(\zeta, \hat{\mathbf{u}}, t), \qquad i = 1, \ldots, m$$

This dynamic feedback can be realised in canonical form by introducing the pseudo-state variables as $\hat{\mathbf{z}}^{(i)} = (z_1^{(i)}, \ldots, z_{\beta_i+1}^{(i)}) = (u_i, \dot{u}_i, \ldots, u_i^{(\beta_i)})$, $i = 1, \ldots, m$, $\hat{z} = (\hat{z}^{(1)}, \ldots, \hat{z}^{(m)})^T$ which, together with eqn. 9.7, yields a closed-loop system of dimension $m + \sum_{i=1}^{m} n_i + \sum_{i=1}^{m} \beta_i$:

$$\dot{\zeta} = \mathbf{A}\zeta + \mathbf{D}s$$
$$\dot{\hat{z}} = \mathbf{P}(\zeta, \hat{z}, t) \qquad (9.20)$$
$$\mathbf{P} = [z_2^{(1)}, \ldots, z_{\beta_1+1}^{(1)}, p_1(\zeta, \hat{z}, t), \ldots, z_2^{(m)}, \ldots, z_{\beta_m+1}^{(m)}, p_m(\zeta, \hat{z}, t)]^T$$

Step 5:   Choose $\hat{\mathbf{u}}_0 \in \mathbf{R}^{\beta+m}$ and a $\delta > 0$ such that, for initial condition $\hat{\mathbf{u}}(0) \in \mathcal{N}_\delta(\hat{\mathbf{u}}_0)$:

(a)   the regularity condition is satisfied
(b)   the zero dynamics (eqn. 9.4) are uniformly asymptotically stable
(c)   all the initial conditions for eqn. 9.20 are compatible.

The stability analysis of this method appears in [4, 12].

9.3.3 Robust design method

The robust design method is described below. The stability analysis is presented in [5]. The system 9.5 may be expressed in the following generalised controller canonical form:

$$\dot{\zeta}_1^{(1)} = \zeta_2^{(1)}$$

$$\cdots\cdots$$

$$\dot{\zeta}_{n_1-1}^{(1)} = \zeta_{n_1}^{(1)}$$

$$\dot{\zeta}_{n_1}^{(1)} = \varphi_1(\zeta, \hat{\mathbf{u}}, t) + \Delta_1(\zeta, t)$$

$$\cdots\cdots \tag{9.21}$$

$$\dot{\zeta}_1^{(m)} = \zeta_2^{(m)}$$

$$\cdots\cdots$$

$$\dot{\zeta}_{n_m-1}^{(m)} = \zeta_{n_m}^{(m)}$$

$$\dot{\zeta}_{n_m}^{(m)} = \varphi_m(\zeta, \hat{\mathbf{u}}, t) + \Delta_m(\zeta, t)$$

where $\zeta^{(i)} = (\zeta_1^{(i)}, \ldots, \zeta_{n_i}^{(i)}) = (y_i, \ldots, y_i^{(n_i-1)})$, $i = 1, \ldots, m$, and $\zeta = (\zeta^{(1)}, \ldots, \zeta^{(m)})^T$.

Step 1:   Choose design parameters to define the sliding surface eqn. 9.8. For $i = 1, \ldots, m$, if $n_i > 1$, choose both $(a_1^{(i)}, \ldots, a_{n_i-1}^{(i)}, 1)$ and $(a_1^{(i)}, \ldots, a_{n_i-1}^{(i)})$ to be Hurwitz. This is always possible according to the result in [13].

Without loss of generality, suppose that $n_1, \ldots, n_{m_1} > 1$ and $n_{m_1+1} = \cdots = n_{m_1+m_2} = 1$, $m_1 + m_2 = m$.

Step 2:   Estimate the uncertainty bound as in eqn. 9.6 when the system is in the GCCF. Choose θ_0 and θ, where $0 < \theta < 1$, $\theta_0 + \theta = 1$ and define

$$\rho^{(0)} = \sqrt{\sum_{i=1}^{m} \rho_i^2}, \quad l_0 = \sqrt{\sum_{i=1}^{m} l_i^2}$$

$$\rho^{(1)} = \rho_0 \left(1 + \max_{1 \le i \le m_1, 1 \le j \le n_i} \left\{ \left| a_j^{(i)} \right| \right\} \max_{1 \le i \le m_1} \left\{ \sqrt{n_i - 1} \right\} \right) \tag{9.22}$$

$$\rho = \rho_0 + \frac{\left(\rho^{(1)} \right)^2}{4\theta}$$

Step 3:   Choose $\gamma(\kappa, \mathbf{s}) = \gamma_0(\kappa, \mathbf{s}) + \mathbf{K}_0 \mathrm{sat}_\varepsilon(s)$ in the sliding reachability condition such that $s^T \gamma_0(\kappa, \mathbf{s}) \ge s^T \mathbf{K}s$, where $\mathbf{K}$ satisfies

$$\lambda_{\min}(K) - \left[\frac{1}{\theta_0} [\mathbf{BD}]^T [\mathbf{BD}] + \rho I_{m_1} \right] > 0 \tag{9.23}$$

where $\mathbf{K}_0 = \text{diag}[k_{01}, \ldots, k_{0m}]$, $\text{sat}_\varepsilon(\mathbf{s}) = [\text{sat}_\varepsilon(s_1), \ldots, \text{sat}_\varepsilon(s_m)]^T$; $\mathbf{D} = \text{diag}[\mathbf{D}_1,$
$\ldots, \mathbf{D}_{m_1}]^T$ with $\mathbf{D}_i = [0, \ldots, 0, 1]^T$ of dimension $n_i - 1$ for $i = 1, \ldots, m_1$;
$\mathbf{A} = \text{diag}[\mathbf{A}_1, \ldots, \mathbf{A}_m]$ with $\mathbf{A}_i$ the companion matrix of the Hurwitz polynomial
$\sum_{j=1}^{n_i-1} a_j^{(i)} \lambda^{j-1}$ and $\mathbf{A}$ and $\mathbf{B}$ satisfy the Lyapunov equation (eqn. 9.12).

Step 4:   Differentiating eqn. 9.8 with respect to time t along the trajectories of
eqn. 9.21 leads to

$$\dot{s}_i\big|_{(1.3.11)} = \sum_{j=1}^{n_i-1} a_j^{(i)} \zeta_{j+1}^{(i)} + \varphi_i(\zeta, \hat{\mathbf{u}}, t) + \Delta_i(\zeta, t), \qquad i = 1, \ldots, m \qquad (9.24)$$

Now set

$$\sum_{j=1}^{n_i-1} a_j^{(i)} \zeta_{j+1}^{(i)} + \varphi_i(\zeta, \hat{u}, t) = -\gamma_{0i}(\kappa, \mathbf{s}) - k_{0i} \, \text{sat}_\varepsilon(s_i), \qquad i = 1, \ldots, m \qquad (9.25)$$

where $k_{0i} > l_0$ and $\mathbf{s}$ is as defined in eqn. 9.8, to determine the feedback control.
Eqn. 9.13 becomes

$$\dot{\mathbf{s}} = -\gamma_0(\kappa, \mathbf{s}) - \mathbf{K}_0 \, \text{sat}_\varepsilon(s) + \Delta(\zeta, t)$$

Step 5:   From eqn. 9.25 the highest-order derivatives of the control, $[u_1^{(\beta_1)}, \ldots,$
$u_m^{(\beta_m)}]$, can be solved out by the implicit function theorem as

$$u^{(\beta_i)} = p_i(\zeta, \hat{\mathbf{u}}, t), \qquad i = 1, \ldots, m$$

if the regularity condition is satisfied. Note that $p_i(\zeta, \hat{\mathbf{u}}, t)$ is a continuous
function if $s_i \neq 0$ because φ_i is C^1 and γ_i is C^0 if $s_i \neq 0$. This dynamic feedback can
be realised in canonical form by introducing the pseudo-state variables as

$$\dot{z}_1^{(1)} = z_2^{(1)}$$

$$\cdots\cdots$$

$$\dot{z}_{\beta_1-1}^{(1)} = z_{\beta_1}^{(1)}$$

$$\dot{z}_{\beta_1}^{(1)} = p_1(\zeta, \mathbf{z}, t)$$

$$\cdots\cdots \qquad\qquad\qquad (9.26)$$

$$\dot{z}_1^{(m)} = z_2^{(m)}$$

$$\cdots\cdots$$

$$\dot{z}_{\beta_m-1}^{(m)} = z_{\beta_m}^{(m)}$$

$$\dot{z}_{\beta_m}^{(m)} = p_m(\zeta, \mathbf{z}, t)$$

where

$$z^{(i)} = (z_1^{(i)}, \ldots, z_{\beta_i}^{(i)}) = (u_i, \dot{u}_i, \ldots, u_i^{(\beta_i-1)}), \qquad i = 1, \ldots, m$$
$$z = (z^{(1)}, \ldots, z^{(m)})^T \qquad\qquad (9.27)$$

Eqn. 9.26 together with eqn. 9.21 yields a closed-loop system of dimension
$\sum_{i=1}^{m} n_i + \sum_{i=1}^{m} \beta_i$, where β_i is the highest-order derivative of u_i.

Step 6: Choose $\hat{\mathbf{u}}_0 \in \mathbf{R}^\beta$ and a $\delta > 0$ such that, for initial condition $\hat{\mathbf{u}}(0) \in \mathcal{N}_\delta(\hat{\mathbf{u}}_0)$:

(i)   the regularity condition is satisfied
(ii) the zero dynamics (eqn. 9.4), (or (eqn. 9.26) when $\zeta = 0$) are uniformly asymptotically stable
(iii) all the initial conditions for eqns. 9.21 and 9.26 are compatible.

9.4 Mathematica implementation

The three algorithms for dynamic sliding mode controller design described in the preceding Section have been implemented in Mathematica. The solutions to Lyapunov equations are computed in Matlab, and the Mathematica and Matlab packages are interfaced via the Mathlink package MaMa.[1] It has been necessary to add a number of functions and packages to the available built-in functions and packages. At the top level, the toolbox comprises three packages:

dslid1::usage=" for *dynamic direct sliding mode control* design of general nonlinear proper IO systems;" [11]
dslid2::usage=" for *dynamic indirect sliding mode control* design of nonlinear proper IO systems;" [4]
rdslid1::usage=" for *robust dynamic direct sliding mode control* design of general nonlinear proper IO systems with uncertainty;" [5]

An important concern in using the packages is the choice of design parameters relating to the coefficients of those Hurwitz polynomials which define the sliding surfaces, the *sliding gain matrix* **K** and *switching gain vector* **K**$_0$. One can either choose **K** and **K**$_0$ at the design stage, or choose them at the simulation stage to improve the performance. However, the choice of zeros of the Hurwitz polynomials must be made at the design stage because the related Lyapunov equation cannot be solved symbolically in general. If **K** and **K**$_0$ are chosen at the simulation stage, the appropriate lower bound will be saved together with the sliding surface and resulting controller, in a specified file in order to assist the design engineer in selection of **K** and **K**$_0$.

These packages for dynamic sliding mode control design are supported by various functions which are collected in two libraries. These functions, described briefly below, may also be used independently or for the purpose of building further symbolic packages.

funclib contains the following functions:

Sat: the saturation function of real x
DSat: the derivative of $Sat[x]$

[1] Download from *http://www.mathsource/cgi-bin/ MathSource/Enhancements/Interfacing/ Matlab/0206-143*

LieD: to calculate the Lie derivative of the function v along the vectorfield w. This may be used to calculate the time derivative of a Lyapunov function candidate along the trajectories of the closed-loop system

ad: lie bracket operation of two vector fields

jmatrix: form the Jacobean matrix of a vector field of dimension n

pdef: to check the definitiveness of a matrix

zerod: to determine the zero dynamics of a proper differential IO system

hord: to calculate the highest-order derivative of a given variable with respect to t in an expression

pcg: to generate the coefficient list of a polynomial from a given list of zeros

cmg: to generate the companion matrix of a polynomial from a given list of zeros

gdiag: to generate a block diagonal matrix from some composite matrices.

srclib is a library of sliding mode reachability conditions:

(i) *src[1]:* properly coupled sliding reachability condition with discontinuous switching

(ii) *src[2]:* properly coupled sliding reachability condition with discontinuous switching replaced by continuous approximation

(iii) *src[3]:* properly coupled sliding reachability condition generated from a potential function.

9.5 Design examples

Example 3

Consider the SISO nonlinear model:

$$y^{(4)}(t) = (y^{(3)}(t))^5 \sin(y(t)) + (\cos t + 1.5)(u^{(2)}(t) + 1)^3 + 4$$

(i) Its zero dynamics are

$$(\cos t + 1.5)(u^{(2)}(t) + 1)^3 + 4 = 0$$

which is Lagrange-stable.

(ii) The regularity condition is satisfied because

$$\det\left[\frac{\partial(\varphi)}{\partial(u^{(2)})}\right] = 3(\cos t + 1.5)(u^{(2)}(t) + 1)^2 \neq 0 \tag{9.28}$$

(iii)   It has a GCCF representation as

$$\dot{\zeta}_1 = \zeta_2$$

$$\dot{\zeta}_2 = \zeta_3$$

$$\dot{\zeta}_3 = \zeta_4$$

$$\dot{\zeta}_4 = \varphi = \zeta_4^5 \sin \zeta_1 + (\cos t + 1.5)\,(u^{(2)}(t) + 1)^3 + 4$$

(iv)   Choose a sliding surface

$$s = a_1\zeta_1 + a_2\zeta_2 + a_3\zeta_3 + a_4\zeta_4 + \varphi$$

(v)   The controller can be solved from the sliding reachability condition

$$\dot{s} = -ks - k_0 \operatorname{sgn} s.$$

Now **dslid2** is used for *indirect sliding mode* controller design. **K** and **K**$_0$ are left as parameters to be chosen at the simulation stage. The print-out of the design run including user-defined input information is presented in the Appendix (Section 9.9.1).

Example 4

The control of a planar vertical take-off and landing (PVTOL) aircraft model is now used to demonstrate direct sliding mode control of MIMO systems. To use the *direct sliding mode* module, it is essential that the m outputs should be separated into two groups: the first group consists of those m_1 outputs with derivative order strictly greater than 1; the second group contains those $m - m_1$ outputs with derivative order 1. Without external uncertainty, the model of the PVTOL system may be expressed as [14]:

$$\ddot{x} = -\sin\theta\, u_1 + \tau \cos\theta\, u_2$$
$$\ddot{y} = \cos\theta\, u_1 + \tau \sin\theta\, u_2 - 1 \qquad (9.29)$$
$$\ddot{\theta} = u_2$$

where '-1' is the gravitational acceleration and τ is the (small)coefficient representing the coupling between the rolling moment and the lateral acceleration of the aircraft. $\tau > 0$ means that applying a (positive) moment to roll left produces an acceleration to the right (positive x). In practice, τ is chosen to satisfy $\tau \geq 0$.

This is a non-minimum-phase system since if we let $u = u_1$ and $v = \theta$, its zero dynamics are:

$$-u \sin v + \tau \ddot{v} \cos v = 0$$

$$u \cos v + \tau \ddot{v} \sin v - 1 = 0$$

which is unstable. Its GCCF representation is

$$
\begin{aligned}
\dot{\zeta}_1 &= \zeta_2 \\
\dot{\zeta}_2 &= -\sin v\, u + \tau \cos v\, \ddot{v} - \ddot{X}_d \\
\dot{\zeta}_3 &= \zeta_4 \\
\dot{\zeta}_4 &= \cos v\, u + \tau \sin v\, \ddot{v} - \ddot{Y}_d - 1
\end{aligned}
\tag{9.30}
$$

(i)   Choose a sliding surface

$$
\begin{aligned}
s_1 &= \zeta_1 + 2\zeta_2 \\
s_2 &= \zeta_3 + 2.3\zeta_4
\end{aligned}
$$

(ii)   Choose a sliding reachability condition as $\dot{\mathbf{s}} = -\mathbf{K}\mathbf{s} - \mathbf{K}_0\,\mathrm{sat}_\varepsilon(\mathbf{s})$.

A design run using the direct module *dslid1* appears in the Appendix (Section 9.9.2).

Example 5

rdslid1 is used to design a robust direct sliding mode controller for the PVTOL aircraft system. The nominal system is the same as for Example 4. Stability analysis is carried out in [10].

Its zero dynamics are the same as in Example 4. Other design parameters are chosen on-line according to the uncertainty bounds.

The reference trajectories are chosen as

$$(X_d, Y_d) = (\sin t,\, e^{-t}\cos t)$$

The additive uncertainties are chosen as

$$
\begin{aligned}
\Delta_1 &= \zeta_1 \sin(5t) + \zeta_2 + \zeta_3 + \mathrm{rand}(1) \\
\Delta_2 &= \zeta_1 + \zeta_3 \cos(3t) + \zeta_4 + \mathrm{rand}(1)
\end{aligned}
$$

where $\mathrm{rand}(1)$ is a random function of one variable in MATLAB with a value in $[0, 1]$. Although it is not Lebesgue measurable, it is practically integrable. The uncertainty bound is given as

$$\|\Delta\| < \sqrt{2}\|\zeta\| + \sqrt{2}$$

i.e. $\rho^{(0)} = \sqrt{2}$, $l_0 = \sqrt{2}$. Choose $a_1 = a_2 = 1$;

$$\rho^{(1)} = \sqrt{2}(1 + \max\{a_1, a_2\}) = 2\sqrt{2}$$

$$\rho = \sqrt{2} + \frac{2}{\theta}$$

Let $\theta = \frac{1}{2}$. Then $\rho = 4 + \sqrt{2}$ and $\theta_0 = \frac{1}{2}$.

Choose $k_{10} = k_{20} = 1.5 > l_0 = \sqrt{2}$. Now determine the gain matrix $\mathbf{K}$ from Section 9.3.1. $\mathbf{B} = \mathrm{diag}[0.5, 0.5]$ is obtained from the Lyapunov equation with $\mathbf{A} = \mathrm{diag}[-1, -1]$

$$\frac{1}{\theta_0}[\mathbf{BD}]^T[\mathbf{BD}] + \rho\mathbf{I}_2 = 2\mathbf{B}^T\mathbf{B} + (4 + \sqrt{2})\mathbf{I}_2 = \begin{bmatrix} 4.5 + \sqrt{2} & 0 \\ 0 & 4.5 + \sqrt{2} \end{bmatrix}$$

Thus choose

$$\mathbf{K} = \begin{bmatrix} 6.5 & 0.5 \\ 0.5 & 6.5 \end{bmatrix}$$

Then

$$\lambda_{\min}(\mathbf{K})\mathbf{I}_2 - \left(\frac{1}{\theta_0}[\mathbf{BD}]^T[\mathbf{BD}] + \rho\mathbf{I}_2\right) = \begin{bmatrix} 1.5 - \sqrt{2} & 0 \\ 0 & 2.5 - \sqrt{2} \end{bmatrix} > 0$$

In simulation define $e = |\zeta_1| + |\zeta_2| + |\zeta_3| + |\zeta_4|$. A layer of thickness $e_0 = 0.6$ is introduced around the sliding surface.

If $e \geq e_0$ the dynamic sliding mode controller calculated above is used.

If $e < e_0$ use the following controller with $k_3 = 10$ and $(b_1, b_2, 1) = (6, 5, 1)$ which is Hurwitz.

If $|z_1| > \sigma$, let

$$\dot{z}_1 = -k_3 \, \mathrm{sgn} \, z_1 \tag{9.31}$$
$$\dot{z}_2 = 0$$

such that $k_3 > 0$ is big enough.

Otherwise, let

$$\dot{z}_1 = z_2 \tag{9.32}$$
$$\dot{z}_2 = -b_1 z_1 - b_2 z_2$$

The implementation using this variable structure control strategy is necessary to overcome the non-minimum-phase behaviour of the PVTOL.

Initial conditions are chosen as $x(0) = (5.42, -1.9, 4.21, -7.5)$; $(v(0), \dot{v}(0)) = (0.4, 0.51)$. All the figures are in the original state variables $(x, \dot{x}, y, \dot{y})$; $\tau = 1.5, \sigma = 1$.

A design run is presented in the Appendix (Section 9.9.3). Simulation results are shown in Figure 9.1.

9.6 Conclusion

This Chapter has presented a number of different algorithms for dynamic sliding mode control design. The procedures are applicable to a fairly general class of nonlinear systems and have the added advantage that robustness issues can be addressed directly. It has been demonstrated that the methods lend themselves to straightforward implementation using symbolic algebra tools.

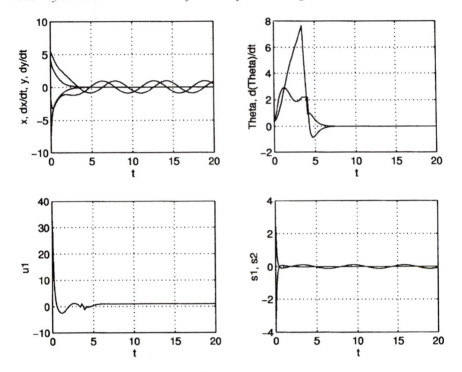

Figure 9.1   PVTOL aircraft control, θ is well regulated

9.7 Acknowledgment

Financial support by UK EPSRC (grant reference GR/J08362) is gratefully acknowledged.

9.8 References

1 SPURGEON, S. K., and DAVIES, R.: 'A nonlinear control strategy for robust sliding mode performance in the presence of unmatched uncertainties', *Int. J. Control*, 1993, **57**, pp. 1107–1123
2 SLOTINE, J. J., and SASTRY, S. S.: 'Tracking control of nonlinear systems using sliding surfaces, with application to robot manipulators', *Int. J. Control*, 1983, **38**, pp. 465–492
3 SPURGEON, S. K., and DAVIES, R.: 'Nonlinear control via sliding modes for uncertain nonlinear systems', *CTAT Special Issue on Sliding Mode Control*, 1994, **10**, (4) pp. 713–736
4 LU, X. Y., and SPURGEON, S. K.: 'Asymptotic feedback linearisation of multiple input systems via sliding modes.' *Proc. of 13th IFAC World Congress*, 1–5 July 1996, San Francisco, USA, Vol. F, pp. 211–216

5 LU, X. Y., and SPURGEON, S. K.: 'Robust sliding mode control of uncertain nonlinear systems', *Syst. Control Lett.*, 1997, **32**,(2), pp. 75–90
6 SIRA-RAMIREZ, H.: 'On the dynamical sliding mode control of nonlinear systems', *Int. J. Control*, **57**, (5), pp. 1039–1061
7 UTKIN, V. I.: 'Sliding modes in control and optimisation' (Springer-Verlag, Berlin, 1992)
8 VAN DER SCHAFT, A. J.: 'Representing a nonlinear state space system as a set of higher-order differential equations in the inputs and outputs', *Syst. Control Lett.*, 1989, **12**, pp. 151–160
9 FLIESS, M.: 'What the Kalman state variable representation is good for', *Proc. IEEE CDC*, Honolulu, Hawaii, 1990, pp. 1282–1287
10 LU, X. Y., SPURGEON, S. K., and POSTLETHWAITE, I.: 'Robust variable structure control of PVTOL aircraft', *Int. J. Syst. Sci.*, 1997, **28**,(6), pp. 547–558
11 LU, X. Y., and SPURGEON, S. K.: 'Dynamic sliding mode control for nonlinear systems: stability'. *Proc. of ECC97*, 1-4 July 1997, Brussels, Belgium
12 LU, X. Y., and SPURGEON, S. K.: 'Asymptotic stabilisation of multiple input nonlinear systems via sliding modes', *Dyn. Control*, 1998, **8**, (3), pp. 231–254
13 HAHN, W.: 'Stability of motion' (Springer-Verlag, NewYork, 1967)
14 HAUSER, J., SASTRY, S., and MEYER, G.: 'Nonlinear control design for slightly non-minimum phase systems: application to V/STOL aircraft', *Automatica*, 1992, **28**,(4), pp. 665–679

9.9 Appendixes

9.9.1 *Appendix A*

```
dudley% math
Mathematica 2.2 for SPARC
Copyright 1988-93 Wolfram Research, Inc.
 -- SunView graphics initialized --

In[1]:= <<dslid2
```

Suppose the system is in a proper differential I-O form.
The nonlinear right hand side is:

```
Pj[y,D[y,t],...D[y,{t,n}];u,D[u,t],,,,,D[u,{t,r}]],{j,1,m}
```
where m is the number of inputs.

Enter the number of inputs.
```
In:= 1
```

Enter the order of y[1]
```
In:= 4
```

Enter RHS of Proper Differential I-O system.
Enter RHS of equation 1
```
In:=(D[y[1][t],{t,3}])^5*Sin[y[1][t]]+(Cos[t]+1.5)*\
> (D[u[1][t],{t,2}]+1)^3+4
```

Choose a number to determine sliding reachability conditions:
1 Properly coupled with discontinuous switching;
2 Properly coupled with continuous round-off switching;

3 Properly coupled but generated from a potential function.

Enter a number.
In=: 1

Enter zero lists of Hurwitz polynomials a[k], k=1,2,...,m.

Complex zeros need to be in conjugate pairs to make a\
> real polynomial.

Enter zero list 1 of dimension 4 .
In:= {-1.2,-3.14,-2.71,-4.5}

Transpose[BP].BP is:

229.95

To determine Sliding Gain Matrix K which is positive /
> definite and satisfies:

K-Transpose[BP].BP>0.

One can choose Sliding Gain Matrix and Switching Gain Vector
1. in controller design stage; or
2. in simulation stage.

Please choose a number : 2

Please specify the output file name.
In:= controller1

The Controller is :
(3)
{(u[1]) [t] ->
> (-1.66667 (3.17625 + 0.5775 Cos[t] + 0.05 K[1, 1] s[1] +
> 0.05 KO[1]Sign[K[1, 1] s[1]]-0.2 Sin[t]+2.29754 (y[1])'[t]+
> 2.59875 (u[1])''[t] + 1.7325 Cos[t] (u[1])''[t] -
> 0.6 Sin[t] (u[1])''[t] + 2.59875 (u[1])''[t]^2 +
> 1.7325 Cos[t] (u[1])''[t]^2 - 0.6 Sin[t] (u[1])''[t]^2 +
> 0.86625 (u[1])''[t]^3 + 0.5775 Cos[t] (u[1])''[t]^3 -
> 0.2 Sin[t] (u[1])''[t]^3 + 4.00468 (y[1])''[t] +
> 2.36272 (y[1]) $^{(3)}$ [t] + 5.5 Sin[y[1][t]] (y[1]) $^{(3)}$ [t]^4 +
> Cos[t] Sin[y[1][t]] (y[1]) $^{(3)}$ [t]^4 +
> 4.5 Sin[y[1][t]] (u[1])''[t] (y[1]) $^{(3)}$ [t]^4 +
> 3. Cos[t] Sin[y[1][t]] (u[1])''[t] (y[1]) $^{(3)}$ [t]^4 +
> 4.5 Sin[y[1][t]] (u[1])''[t]^2 (y[1]) $^{(3)}$ [t]^4 +
> 3. Cos[t] Sin[y[1][t]] (u[1])''[t]^2 (y[1]) $^{(3)}$ [t]^4 +
> 1.5 Sin[y[1][t]] (u[1])''[t]^3 (y[1]) $^{(3)}$ [t]^4 +
> Cos[t] Sin[y[1][t]] (u[1])''[t]^3 (y[1]) $^{(3)}$ [t]^4 +
> 0.5775 Sin[y[1][t]] (y[1]) $^{(3)}$ [t]^5 +
> 0.2 Cos[y[1][t]] (y[1])'[t] (y[1]) $^{(3)}$ [t]^5 +

```
> Sin[y[1][t]]² (y[1]) ⁽³⁾[t]⁹ )) / ((1.5 + Cos[t]) \
> (1. + (u[1])''[t]) ²) }
  dudley%
```

Note that $K[i,j]$ and $KO[i]$, $(i=1,...,m)$ are design
parameters to be chosen at the simulation stage. A unique
controller is obtained because the highest order derivative
of the control appears linearly in the sliding reachability
condition.

9.9.2 *Appendix B*

```
dudley% math
Mathematica 2.2 for SPARC
Copyright 1988-93 Wolfram Research, Inc.
-- SunView graphics initialised --
In[1]:= <<dslid1
  Suppose the system is in a proper differential I-O form.
  The nonlinear right hand side is :
  Pj[y,D[y,t],...D[y,{t,n}];u,D[u,t],,,D[u,{t,r}]], {j,1,m}
  where m is the number of inputs.

  Suppose y[k], k=1,...,m1, have derivative order >1;
  and y[k], k=m1+1,...,m, have derivative order 1.

          Enter m1.
  In:= 2

          Enter the number of inputs.
  In:= 2

          Enter the order of y[1]
  In:= 2

          Enter the order of y[2]
  In:= 2

          Enter RHS of Proper Differential I-O system.
          Enter RHS of equation 1
  In:= -
  Sin[u[2][t]]*u[1][t]+Ep*Cos[u[2][t]]*D[u[2][t],{t,2}]
          Enter RHS of equation 2
  In:=
  Cos[u[2][t]]*u[1][t]+Ep*Sin[u[2][t]]*D[u[2][t],{t,2}]-1

  Choose a number to determine sliding reachability conditions.
    1 Properly coupled with discontinuous switching
    2 Properly coupled with continuous round-off switching
    3 Properly coupled but generated from a potential function

          Enter a number:
```

In=: 2
Enter a small positive number which is the thickness of the
layer around sliding surfaces.
In:= 0.1

        Enter zero lists of Hurwitz polynomials a[k],
k=1,2,...,m1.

Complex zeros need to be in conjugate pairs to make a real
  polynomial.
Enter zero list 1 of dimension 1.
In:= {-2}
Enter zero list 2 of dimension 1.
In:= {-2.3}

Transpose[BP].BP is:
    0.0625   0.
    0.   0.047259

To determine Sliding Gain Matrix K which is positive \
62; definite and satisfies:
                    K-Transpose[BP].BP>0.

One can choose Sliding Gain Matrix and Switching Gain Vector
  1. in controller design stage; or
  2. in simulation stage.

  Please choose a number : 1
  Please enter gain matrix K of dimension 2.
In:= {{1,0},{0,2}}

To select non-negative switching gain in a list form 2.
K0={ K0[[1]], ..., K0[[m]] } of dimension
In:= {1, 3.1}

    Please specify the output file name.
In:= con3

The Controller is :
{u[1][t] -> Csc[u[2][t]] (If[Abs[10. s[1]] >= 1, Sign[10.
s[1]]],
> 10. s[1]] + s[1] + 2. (y[1])'[t]) -
> (1. Ep Cot[u[2][t]] (-1. Cos[u[2][t]]
> (If[Abs[10. s[1]] >= 1, Sign[10. s[1]], 10. s[1]] + s[1] +
> 2. (y[1])'[t]) - 1. Sin[u[2][t]]
> (-1. + 3.1 If[Abs[20. s[2]] >= 1, Sign[20. s[2]], 20. s[2]] +
> 2. s[2] + 2.3 (y[2])'[t]))) /
> (-1. Ep Cos[u[2][t]]^2 - 1. Ep Sin[u[2][t]]^2),
> (u[2])''[t] ->
> (-1. (-1. Cos[u[2][t]] (If[Abs[10.s[1]] >= 1,Sign[10. s[1]],

```
> 10. s[1]] + s[1] + 2. (y[1])'[t]) -
> 1. Sin[u[2][t]] (-1. +
> 3.1 If[Abs[20. s[2]] >= 1,Sign[20. s[2]], 20. s[2]]+2.
2 2 2
> 2.3 (y[2])'[t])))/(-1. Ep Cos[u[2][t]] - 1. Ep Sin[u[2][t]] )}
  dudley%
```

Remark 1 It should be noted that, because of the use of the continuous roundoff approximation of the sgn[·] function, the controller is not evaluated until simulation.

9.9.3 Appendix C

```
dudley% math
Mathematica 2.2 for SPARC
Copyright 1988-93 Wolfram Research, Inc.
-- SunView graphics initialized --

In[1]:= <<rdslid1
```

Suppose the nominal system is in a proper differential I-O form. The nonlinear right hand side is :
```
Pj[y,D[y,t],...D[y,{t,n}];u,D[u,t],,,,,D[u,{t,r}]]+Delta
[j],
{j,1,m}, where m is the number of inputs.
```

Suppose in the nominal system y[k], k=1,...,m1, have \
```
> derivative order >1;
  and y[k], k=m1+1,...,m, have derivative order 1.
  Enter m1.
  In:= 2
        Enter the number of inputs.
  In:= 2
        Enter the order of y[1]
  In:= 2
        Enter the order of y[2]
  In:= 2
        Enter RHS of the nominal differential I-O system.
        Enter RHS of equation 1
  In:=-Sin[u[2][t]]*u[1][t] +
  Ep*Cos[u[2][t]]*D[u[2][t],{t,2}]
> +Xr[t]
        Enter RHS of equation 2
  In:= Cos[u[2][t]]*u[1][t] +
  Ep*Sin[u[2][t]]*D[u[2][t],{t,2}]
> -1+Yr[t]
```

Suppose the uncertainty satisfies the following cone bounded\
> condition.
||Delta[i](Y,t)|| <= p[i]*||Y|| + l[i], p[i], l[i] >= 0,
i=1,...,m, where Y=(y,D[y,t],...,D[y,{t,Beta}]) and Beta is\
> the highest order derivative of input in the nominal system.
        Enter parameter bound p[1]
In:= 1
        Enter parameter bound p[2]
In:= 1
        Enter parameter bound l[1]
In:= 1
        Enter parameter bound l[2]
In:= 1

Choose a number to determine sliding reachability conditions.
1. Strong sliding reachabiliy condition with discontinuous
   switching
2. Strong sliding reachability condition with continuous
   round-off switching
3. Other choice of strong sliding reachability condition
Enter a number:
In=: 1
Enter zero lists of Hurwitz polynomials a[k],
k=1,2,...,m1.
Complex zeros need to be in conjugate pairs to make
a real polynomial.
Enter zero list 1 of dimension 1.
In:= {-1}
Enter zero list 2 of dimension 1.
In:= {-1}
  The lower bound Kmat for Sliding Gain Matrix is:
  4.5 + Sqrt[2]   0.
  0.   4.5 + Sqrt[2]
To determine Sliding Gain Matrix K which is positive
definite and satisfies:
      Lmin(K)-(2*Transpose[BP].BP+p*I)>0.
and to determine switching Gain Vector K0 which satisfies:
                    K0[i]>L0, i=1,...,m.
One can choose K and K0
1. in controller design stage; or
2. in simulation stage.
Please choose a number : 1
Please enter gain matrix K of dimension 2.
In:= {{6.5,0.5},{0.5,6.5}}
The lower bound for switching gain K0 is:
Sqrt[2]

```
To select non-negative switching gain in a list form\
> KO={ KO[[1]], ..., KO[[2]] }.
  In:= {1.5,1.5}
  Please specify the output file name.
  In:= pvt
  The Controller is :
  {u[1][t] -> Csc[u[2][t]] (6.5 s[1] + 0.5 s[2] +
> 1.5 Sign[6.5 s[1] + 0.5 s[2]] + Xr[t] + (y[1])'[t]) -
> (1. Ep Cot[u[2][t]] (-1. Cos[u[2][t]]
> (6.5 s[1] + 0.5 s[2] + 1.5 Sign[6.5 s[1]+0.5 s[2]]+Xr[t] +
> (y[1])'[t]) - 1. Sin[u[2][t]]
> (-1. + 0.5 s[1] + 6.5 s[2] + 1.5 Sign[0.5 s[1] + 6.5 s[2]] +
> Yr[t] + (y[2])'[t]))) /
> (-1. Ep Cos[u[2][t]]² - 1. Ep Sin[u[2][t]]² ),
> (u[2])''[t] ->
> (-1. (-1. Cos[u[2][t]] (6.5 s[1] + 0.5 s[2] +
> 1.5 Sign[6.5 s[1] + 0.5 s[2]] + Xr[t] + (y[1])'[t]) -
> 1. Sin[u[2][t]] (-1. + 0.5 s[1] + 6.5 s[2] +
> 1.5 Sign[0.5 s[1] + 6.5 s[2]] + Yr[t] + (y[2])'[t]))) /
> (-1. Ep Cos[u[2][t]]² - 1. Ep Sin[u[2][t]]² )}
dudley%
```

## Chapter 10

Pole assignment for uncertain systems

M. T. Söylemez and N. Munro

10.1 Introduction

Many systems in the real world cannot be modelled with great accuracy. There are often uncertainties arising from unmodelled dynamics, manufacturing tolerances, time delays or approximations to nonlinearities. Therefore, a more realistic approach, in general, is to model the systems concerned with a few uncertain parameters whose values are not known exactly but for which lower and upper bounds are known. We shall call systems modelled in this way 'parametric uncertain systems'. More formally an uncertain system[1] description can be given in the following form:

$$\dot{\mathbf{x}} = A(\mathbf{q})\,\mathbf{x} + B(\mathbf{q})\,\mathbf{u}$$
$$\mathbf{y} = C(\mathbf{q})\,\mathbf{x} + D(\mathbf{q})\,\mathbf{u} \tag{10.1}$$

where $\mathbf{x} \in \mathcal{R}^n$, $\mathbf{u} \in \mathcal{R}^m$ and $\mathbf{y} \in \mathcal{R}^l$ are the state, input and output vectors, respectively, and the state-space matrices A, B, C and D are functions of the uncertainty vector $\mathbf{q}$:

$$\mathbf{q} = [\,q_1 \quad q_2 \quad \cdots \quad q_r\,] \tag{10.2}$$

whose elements are known to be interval uncertain parameters, i.e.

$$q_i^- \leq q_i \leq q_i^+ \qquad \text{for } i = 1, 2, \ldots r \tag{10.3}$$

The value of the $\mathbf{q}$ vector under nominal working conditions is denoted as

$$\mathbf{q}_n = [\,q_{1n} \quad q_{2n} \quad \cdots \quad q_{rn}\,] \tag{10.4}$$

---

[1] The shorter term 'uncertain system' is used here instead of 'parametric uncertain system'.

The r-dimensional box defined by the parametric uncertainties of the system is called *the parameter box* and denoted by Q, i.e.

$$Q \triangleq \{\mathbf{q}/\mathbf{q} \in \mathcal{R}^r, \ q_i^- \leq q_i \leq q_i^+\} \tag{10.5}$$

The parameter box of a system with three uncertain parameters is shown in Figure 10.1.

The poles of a parametric uncertain system, in general, are not a single set of complex numbers, but are spread over n regions in the complex plane. These regions can be viewed as the image of the parameter box in the root space of the characteristic polynomial $p(s, \mathbf{q}) = |sI - A(\mathbf{q})|$, and they are called *the pole spread* of the uncertain system [1].

When designing a pole assignment controller for an uncertain system such as that defined in eqn. 10.1, the classical approach is to find a compensator, F, that assigns the closed-loop nominal plant $(\mathbf{q} \rightarrow \mathbf{q}_n)$ poles to some desired locations (Γ). When this compensator is applied to the uncertain system, the closed-loop poles will be at the desired locations under nominal working conditions. As the system drifts away from the nominal working conditions, however, the closed-loop system poles move away from the desired locations, possibly resulting in undesired dynamical responses and instability of the closed-loop system. The fact that if there exists a solution to the MIMO pole assignment problem, it is in general not unique, is helpful at this point. Since there are usually more than one (generally an infinite number of) pole assign-

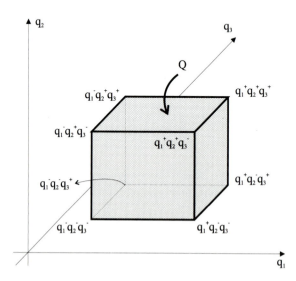

Figure 10.1   Parameter box for a system with three uncertain parameters

ment compensators, it is possible that for a certain subset of these the closed-loop system poles are not perturbed very much, and the closed-loop system's dynamical response is *acceptable* under all possible uncertainties. An *acceptable* dynamical response can generally be represented by a set of regions in the complex plane which the poles are supposed to stay inside. These regions are called D-stability regions or *D-regions* for short. If all the poles of a system lie inside the associated *D-regions* the system is called *D-stable*. The concept of 'robust pole clustering' can then be defined as follows:

> Given an uncertain system as defined in eqn. 10.1, a set of complex numbers, $\Gamma = \{\gamma_1, \gamma_2, \ldots, \gamma_{n+q}\}$, closed under complex conjugation, and a set of stability regions (called *D-regions*), find a compensator, F, of order q such that the closed-loop system poles are equal to Γ under nominal working conditions and furthermore the closed-loop system is *D-stable* for all possible perturbations being considered.

One could try to find different pole assignment compensators randomly and check for *D-stability*, when trying to find a solution to the robust pole clustering problem. A more reasonable approach, as stated in Söylemez and Munro [2] is as follows:

1. Find the general form of compensators that assign the nominal plant poles, $F(\mathbf{k})$.
2. Calculate either the closed-loop characteristic polynomial as a function of the parametric uncertainties $(\mathbf{q})$, free controller variables $(\mathbf{k})$, and the frequency variable (s), $p_c(s, \mathbf{q}, \mathbf{k})$, or the general form of the closed-loop A matrix, $A_c(\mathbf{q}, \mathbf{k})$.
3. Find the set of $\mathbf{k}$ vectors that satisfy the predefined performance criteria $(D\text{-}stability)$ for all given perturbations $\mathbf{q}$.

The first step considered above involves the generalisation of the classical pole assignment methods, and this Chapter is mainly concerned with this part of the approach. Here, instead of producing a random numeric vector required by many methods, we keep the parameters that contain the extra freedom available in symbolic form, and the general form of the pole assignment compensator is calculated with the help of a symbolic algebra language (i.e. *Mathematica* [3] in our case).

The main purpose of the second step is to reduce the computational complexity of the optimisation process involved in the third step. In many cases, the symbolic form of the closed-loop system A matrix or characteristic polynomial is in a relatively simple form. Therefore, it is faster to find the closed-loop system poles when $\mathbf{k}$ and $\mathbf{q}$ are fixed.

The third step of finding robust pole clustering compensators is basically an optimisation process. Here, a cost function, $\mathcal{J}(\mathbf{k})$, is defined to assess the robustness of the closed-loop system for different fixed values of the vector of free variables $\mathbf{k}$. Then, this cost function is minimised to find *D-stable* solutions.

A suitable cost function and some discussions on how to minimise this cost function are given in Söylemez and Munro [4] and are not dealt with in depth here.

10.2 General state-feedback pole assignment problem

The easiest case to consider is when a constant state feedback is to be applied to the uncertain system. In this case, the first step of finding a *robust* compensator is to find the general form of the state-feedback pole assignment compensator for the nominal system $[A, B]$. This problem is called 'the general state-feedback pole assignment problem' and can be formally stated as follows:

Given a system

$$\dot{x} = A\mathbf{x} + B\mathbf{u} \tag{10.6}$$

where $A \in \mathcal{R}^{n \times n}$ and $B \in \mathcal{R}^{n \times m}$, and a set of n complex numbers, $\Gamma = \{\gamma_1, \gamma_2, \ldots, \gamma_n\}$, closed under complex conjugation, find the general form of the state-feedback matrix, $K_s(\mathbf{k})$, such that

$$\Gamma = \text{Eig}\,(A - BK_s(\mathbf{k})) \qquad \text{(for almost all } \mathbf{k}) \tag{10.7}$$

where Eig denotes the eigenvalue set of a matrix.

It has long been known that a solution to the state-feedback pole assignment problem exists if, and only if, the system $[A, B]$ is completely controllable [5]. Therefore, without loss of generality, we shall assume that the nominal system is completely controllable.

The methods of solving pole assignment problems are divided into two categories in the literature as dyadic and full-rank methods [6]. We shall follow this tradition and give dyadic and full-rank solutions for the general state-feedback pole assignment problem in the following.

10.2.1 Dyadic methods

The easiest case for an arbitrary pole assignment problem is when statefeedback is considered for a single-input system. The equations for assigning the poles become linear in this case and the solution, if it exists, is unique due to the fact that the number of equations and the number of the free variables are the same.

If the system considered has multiple inputs it is always possible to recast the pole assignment problem as an equivalent pole assignment problem for a single-input system by writing the state-feedback matrix, K_s, as an outer product of two dyads (vectors), i.e.

$$K_s = \mathbf{f}\mathbf{k}' \tag{10.8}$$

where $\mathbf{f} \in \mathcal{R}^{m \times 1}$ and $\mathbf{k} \in \mathcal{R}^{n \times 1}$.

Hence, the pole assignment problem can be rewritten as follows:

Find $K_s = \mathbf{f}\,\mathbf{k}'$ such that

$$\Gamma = \mathrm{Eig}(A - B\,K_s) \tag{10.9}$$

$$= \mathrm{Eig}(A - B\,\mathbf{f}\,\mathbf{k}') \tag{10.10}$$

$$= \mathrm{Eig}(A - \mathbf{b}\,\mathbf{k}') \tag{10.11}$$

where $\mathbf{b} \triangleq B\,\mathbf{f}$.

Considering this last equation, we can see that this is the same as finding a pole assignment compensator, $\mathbf{k}'$, for the single-input system $[A, \mathbf{b}]$, provided that $[A, \mathbf{b}]$ is controllable. As a result, many of the methods in the literature first write the state-feedback matrix as in eqn. 10.8 and then find a vector $\mathbf{f}$ that satisfies the controllability condition of $[A, \mathbf{b}]$,[2] hence converting the problem into the simpler single-input case. Since these methods produce a rank one feedback matrix as an outer product of two dyads, they are conveniently called *dyadic methods* or *unity-rank methods* [6].

When searching for the general form of the dyadic state-feedback pole assignment compensator, instead of fixing the $\mathbf{f}$ vector we shall leave it in symbolic form and calculate $\mathbf{k}'$ in terms of the elements of $\mathbf{f}$. We note that, although there are m elements in the vector $\mathbf{f}$, it can be shown that at most $m - 1$ of these elements can be selected independently to yield different answers for K_s [9, 10]. Therefore, to relieve the complexity of the symbolic calculations, we usually fix one of the elements of the $\mathbf{f}$ vector.

There are many methods in the literature, all with their own advantages and disadvantages, to find the state-feedback pole assignment compensator for a single input system, $[A, \mathbf{b}]$. These seemingly different methods have recently been unified by Söylemez [11], and it has been shown that the mapping approach [12] outperforms the other methods in general, when symbolic calculations are required.[3]

Using the mapping approach, the required state-feedback matrix is given as

$$\mathbf{k} = [\Phi']^{-1} X^{-1} \delta \tag{10.12}$$

where Φ is the controllability matrix of the pair $[A, \mathbf{b}]$, i.e.

$$\Phi \triangleq [\,\mathbf{b} \quad A\,\mathbf{b} \quad A^2\,\mathbf{b} \quad \ldots \quad A^{n-1}\,\mathbf{b}\,], \tag{10.13}$$

---

[2] If $[A, B]$ is controllable and A is cyclic then almost all $\mathbf{f}$ vectors satisfy this condition (see Kailath [7, p. 504]). If A is not cyclic it can be made cyclic by application of almost any compensator in the feedback [8].

[3] We note that this method is long known to have numerical problems and there are better methods if the calculations are to be done wholly numerically. An interesting discussion on this issue can be found in Söylemez and Munro [13].

X is a lower-triangular Toeplitz matrix given as

$$X \triangleq \begin{bmatrix} 1 & 0 & 0 & \cdots & 0 \\ a_{n-1} & 1 & 0 & \cdots & 0 \\ a_{n-2} & a_{n-1} & 1 & \cdots & 0 \\ \vdots & \vdots & \vdots & \ddots & \vdots \\ a_1 & a_2 & a_3 & \cdots & 1 \end{bmatrix} \tag{10.14}$$

where the a_i are the coefficients of the open-loop system characteristic polynomial,

$$p(s) = |sI_n - A| \tag{10.15}$$

$$= s^n + a_{n-1}s^{n-1} + \cdots + a_0 \tag{10.16}$$

and δ is the difference vector given by the coefficients of the difference polynomial, $\delta(s) \triangleq p_c(s) - p(s)$, i.e.

$$\delta \triangleq [\alpha_{n-1} - a_{n-1} \quad \alpha_{n-2} - a_{n-2} \quad \cdots \quad \alpha_0 - a_0]' \tag{10.17}$$

where the α_i are the coefficients of the desired closed-loop system characteristic polynomial, $p_c(s) = \prod_{i=1}^{n}(s - \gamma_i)$.

Example 1

Let us consider the following system where the A and B matrices are given as

$$A = \begin{bmatrix} q_1 & 0 & q_2 \\ 0 & 1 & 2 \\ 3 & 1 & 3 \end{bmatrix} \qquad B = \begin{bmatrix} 1 & q_3 \\ 0 & 1 \\ 1 & 1 \end{bmatrix} \tag{10.18}$$

where the q_i are interval uncertain parameters bounded as follows:

$$0.9 \leq q_1 \leq 1.1$$

$$3.5 \leq q_2 \leq 4.5$$

$$-0.25 \leq q_3 \leq 0.25 \tag{10.19}$$

and have nominal values $q_{1n} = 1$, $q_{2n} = 4$ and $q_{3n} = 0$. The poles of the nominal system are therefore $\{-1.87, 1.00, 5.87\}$. Now, suppose the closed-loop system poles under nominal working conditions are required to be at $\{-1+j, -1-j, -7\}$. Furthermore, other design criteria state that the pole spread of the closed-loop system poles has to be interior to the *D-regions* shown in Figure 10.2.

As a first step, we find the general form of the dyadic state-feedback compensator that assigns the closed-loop system poles to the desired locations under

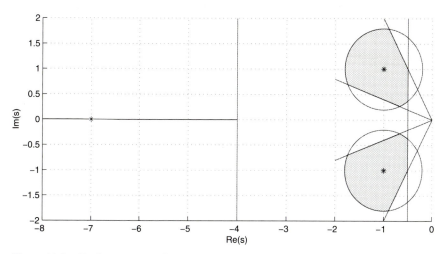

Figure 10.2 Performance specifications for Example 1

nominal working conditions. Using a symbolic algebra language, it is easy to determine this compensator through eqn. 11.12 as

$$
K_s = \begin{bmatrix}
\dfrac{13 - 77\,k + 202\,k^2}{1 - 16\,k + 17\,k^2 + 22\,k^3} & \dfrac{11 - 119\,k - 6\,k^2}{1 - 16\,k + 17\,k^2 + 22\,k^3} & \dfrac{-1 + 157\,k}{-1 + 14\,k + 11\,k^2} \\[3mm]
\dfrac{k\left(13 - 77\,k + 202\,k^2\right)}{1 - 16\,k + 17\,k^2 + 22\,k^3} & -\dfrac{k\left(-11 + 119\,k + 6\,k^2\right)}{1 - 16\,k + 17\,k^2 + 22\,k^3} & \dfrac{k\left(-1 + 157\,k\right)}{-1 + 14\,k + 11\,k^2}
\end{bmatrix}
$$

$$(10.20)$$

Hence, the closed-loop system characteristic polynomial for this compensator is found as:

$$
\begin{aligned}
p(s, \mathbf{q}, \mathbf{k}) =\ & + \left(1 - 16\,k + 17\,k^2 + 22\,k^3\right)s^3 \\
& + \left(10\left(1 - 16\,k + 17\,k^2 + 22\,k^3\right) - \left(1 - 16\,k + 17\,k^2 + 22\,k^3\right)q_1 \right. \\
& \left. + k\left(13 - 77\,k + 202\,k^2\right)q_3\right)s^2 \\
& + \left(-27 - 275\,k - 56\,k^2 + 28\,k^3 + \left(3 + 83\,k + 32\,k^2 - 220\,k^3\right)q_1\right. \\
& + 2\left(5 - 8\,k + 37\,k^2 + 68\,k^3\right)q_2 - 49\,k\,q_3 - 169\,k^2\,q_3 + 134\,k^3\,q_3\right)s \\
& + \left(-2\left(\left(11 - 53\,k - 95\,k^2 + 14\,k^3\right)q_1 + \left(5 + 2\,k - 103\,k^2 - 42\,k^3\right)q_2\right.\right. \\
& \left.\left. + \left(-38 + 157\,k + 388\,k^2\right)\left(1 + k\,q_3\right)\right)\right)
\end{aligned}
$$

$$(10.21)$$

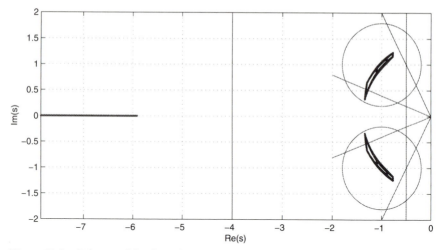

Figure 10.3   Pole spread for closed-loop system with a dyadic compensator (Example 1)

Now, we can optimise the free parameter k such that a robust solution for the given performance criteria is obtained [4]. Using a genetic algorithm [14, 15], it is found that the most robust compensator is obtained when $k = -0.15$. The pole spread of the closed-loop system as shown in Figure 10.3, however, reveals that not all of the performance criteria are met even for this value of k. Hence, we need to use more of the degrees of freedom available in the more general full-rank state-feedback pole assignment problem.

10.2.2 Full-rank methods

Since the state-feedback compensator is unique for the single-input case and the formulations given to find this compensator are relatively simple, converting the multi-input problem into a single-input problem (i.e. using a dyadic method) is very attractive. Furthermore, implementing dyadic compensators is very easy on a real system, as they require at most $n + m - 1$ gain elements and just a few connections.

However, dyadic methods have some drawbacks. First of all, it has been shown by Belletrutti [16] that dyadic compensators have bad noise rejection properties compared to full-rank compensators. Therefore, the use of such a compensator is not recommended unless the disturbance rejection property is not important. Secondly, and more importantly from our point of view, the degrees of freedom that are available in pole assignment can be severely restricted when using dyadic compensators. Since there are nm elements in the state-feedback compensator and the assignment of n poles imposes n equations to be solved, the degrees of freedom available in the general state-feedback pole assignment problem are $(n - 1)m$. When using dyadic compensators, as we mentioned above, the degrees of freedom available are reduced to $m - 1$. This is due to the assumption made on the structure of the feedback compensator, and can be quite important when searching for a robust compensator. Last, but not least, dyadic methods require the open-loop system to be cyclic. However, there are many real systems which do not satisfy this assumption, and trying to solve such a problem by introducing an additional compensator in the feedback path may increase the complexity of the uncertainties of the system.

For all of these reasons, we may want to use a full-rank compensator in the feedback path. Finding a general full-rank compensator, however, is not easy as the equations involved are of a multilinear nature. Nevertheless, there are methods which allow us to linearise these equations. In the following, we describe such a method given by Söylemez and Munro [17].

Let us write the desired full-rank state-feedback matrix K_s as a sum of m unity-rank matrices each corresponding to a feedback from the states to one of the inputs of the system, i.e.

$$K_s = \sum_{i=1}^{m} \mathbf{e}_i \mathbf{k}_i' \tag{10.22}$$

where $\mathbf{e}_i$ is the ith column of I_m and k_i' is the ith row of K_s.

Then, by selecting an arbitrary input (let us say the jth input), we can write

$$K_s = \sum_{i \neq j} \mathbf{e}_i \mathbf{k}_i' + \mathbf{e}_j \mathbf{k}_j' \tag{10.23}$$

$$= \bar{K}_j + \mathbf{e}_j \mathbf{k}_j'. \tag{10.24}$$

Note that $\bar{K}_j$ shows the state-feedback matrix with the jth row zero.

The pole assignment problem can then be rewritten as finding $\mathbf{k}_j$ such that

$$\Gamma = \mathrm{Eig}(A - B K_s) \tag{10.25}$$

$$= \mathrm{Eig}(A - B\bar{K}_j - B\mathbf{e}_j\mathbf{k}_j') \tag{10.26}$$

$$= \mathrm{Eig}(\bar{A}_j - \beta_j\mathbf{k}_j') \tag{10.27}$$

where

$$\bar{A}_j \triangleq A - B\bar{K}_j$$
$$\beta_j \triangleq B\mathbf{e}_j \tag{10.28}$$

i.e. jth column of B.

A further investigation of eqn. 10.27 reveals that this is equivalent to finding a pole assignment state-feedback compensator for the single-input system $[\bar{A}_j, \beta_j]$. This means that we can use any dyadic state-feedback pole assignment method to find the vector $\mathbf{k}_j$.

An immediate question that comes to mind is whether this formulation would preserve the generality of the solution. That is, is it always possible to find a pole assignment compensator for the system $[\bar{A}_j, \beta_j]$? The following theorem helps us to answer this question.

Theorem 1   *Given a multi-input system as defined in eqn. 10.6, with a full-rank input matrix B, the determinant of the controllability matrix Φ_j of the single input system $[\bar{A}_j, \ \beta_j]$, defined as*

$$\Phi_j \triangleq [\beta_j \quad \bar{A}_j \beta_j \quad \bar{A}_j^2 \beta_j \quad \cdots \quad \bar{A}_j^{n-1} \beta_j] \tag{10.29}$$

is identically zero if, and only if, the given system $[A, \ B]$ is uncontrollable.

The proof of this theorem can be found in Söylemez and Munro [17], and is not repeated here. A logical justification of the above theorem, however, can be given as follows. If the system $[A, B]$ is uncontrollable then there exists a mode of the system which is not controllable through the jth input and which cannot be made controllable through the application of feedback. Hence, this mode is uncontrollable for the system $[\bar{A}_j, \ \beta_j]$ as well (sufficiency part). On the other hand, if the system $[A, B]$ is controllable, then even if there exist some modes of the system that are not controllable through the jth input (i.e. there are some decoupling zeros of the system $[A, \ \beta_j]$), these can be moved away from the decoupling zeros by the application of a feedback law to the other inputs (and hence made controllable), since these modes are controllable through other inputs (necessity part).

We remark that although it is shown in the above theorem that there exists at least one $\bar{K}_j$ matrix for which a solution can be found, almost all selections of $\bar{K}_j$ are allowed, and the $\bar{K}_j$ matrices for which $[\bar{A}_j, \ \beta_j]$ is not controllable (and hence not pole assignable) form a variety in $\mathcal{R}^{(m-1) \times n}$. This means that we can keep $\bar{K}_j$ in symbolic form and solve for $\mathbf{k}_j$ when looking for a general solution to the state-feedback pole assignment problem. This way, we lose almost no freedom in the design process and reduce the general full-rank state-feedback problem to a dyadic state-feedback pole assignment problem, for which linear solutions exist.

We also note that the free parameters obtained using this method are elements of the state-feedback matrix. This can be very useful, owing to the fact that the elements of the feedback matrix cannot be infinitely large (due to the small gain theorem), and hence the search space for the robust compensator design decreases radically.

Example 2

Let us consider the problem given in the previous example, and assume that we have decided to use a full-rank state-feedback compensator, in order to provide good disturbance rejection properties.

If we choose to use the first input in the algorithm presented above ($j = 1$), then

$$\bar{K}_j = \begin{bmatrix} 0 & 0 & 0 \\ k_1 & k_2 & k_3 \end{bmatrix} \tag{10.30}$$

and hence from eqn. 10.28

$$\bar{A}_j = \begin{bmatrix} 1 & 0 & 4 \\ -k_1 & 1 - k_2 & 2 - k_3 \\ 3 - k_1 & 1 - k_2 & 3 - k_3 \end{bmatrix} \qquad \beta_j = \begin{bmatrix} 1 \\ 0 \\ 1 \end{bmatrix} \tag{10.31}$$

A general full-rank state-feedback pole assignment compensator can then be found from eqn. 10.12 by replacing A with A_j and $\mathbf{b}$ with β_j in the definitions given previously,

$$K_s = \begin{bmatrix} \dfrac{r_1(\mathbf{k})}{p(\mathbf{k})} & \dfrac{r_2(\mathbf{k})}{p(\mathbf{k})} & \dfrac{r_3(\mathbf{k})}{p(\mathbf{k})} \\ k_1 & k_2 & k_3 \end{bmatrix} \tag{10.32}$$

where

$$p(\mathbf{k}) = 3\,k_1^2 - k_1\,(6 + k_2 - 5\,k_3) - (1 + k_2 - 2\,k_3)(-2 + k_3)$$

$$r_1(\mathbf{k}) = -((13 + 3\,k_2 - 6\,k_3)(-2 + k_3)) - 3\,k_1^2\,(-10 + k_2 + k_3)$$
$$\qquad + k_1\left(k_2^2 - k_2\,(10 + k_3) - 2\left(22 - 16\,k_3 + k_3^2\right)\right)$$

$$r_2(\mathbf{k}) = 22 + k_2^3 - 35\,k_3 + 2\,k_3^2 - k_2^2\,(10 + k_3) + k_2\left(1 + 22\,k_3 - 2\,k_3^2\right) \tag{10.33}$$
$$\qquad + k_1\left(-41 - 3\,k_2^2 - 3\,k_2\,(-10 + k_3) + 3\,k_3\right)$$

$$r_3(\mathbf{k}) = 12\,k_1^2 + k_1\left(-40 + k_2\,(2 - 3\,k_3) + 44\,k_3 - 3\,k_3^2\right)$$
$$\qquad + (-2 + k_3)\left(-1 + k_2^2 + 23\,k_3 - 2\,k_3^2 - k_2\,(10 + k_3)\right)$$

As in the previous example, finding the closed-loop characteristic polynomial and optimising for the free variables $(k_1, k_2$ and $k_3)$, it is found that the feedback matrix

$$K_s = \begin{bmatrix} 2.34 & -1.45 & -0.33 \\ 0.00 & 2.60 & 9.39 \end{bmatrix} \tag{10.34}$$

is a robust compensator. The pole spread in the resulting closed-loop system, when this compensator is used, is shown in Fig. 10.4. From this Figure, it is clear that all the performance criteria are met by this compensator.

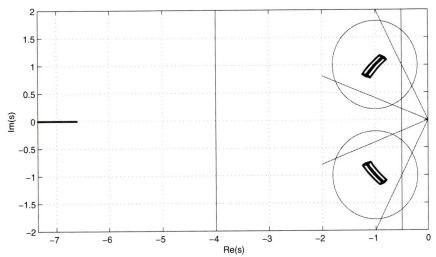

Figure 10.4   Pole spread of closed-loop system using full-rank compensator (Example 2)

10.3 General output-feedback pole assignment problem

In the real world, it is not practical to measure all the states of the system and it may be necessary to use an output-feedback compensator in many cases. Therefore, finding the general form of the output feedback compensators that would assign the nominal system poles to some desired locations is very useful. The simplest case for such a problem is when the feedback compensator is assumed to be a static one. Therefore, we examine this case first; the more general case of dynamic output-feedback is dealt with afterwards.

10.3.1 Static feedback

'The general static output-feedback pole assignment problem' can be formally stated as follows.

Given a system

$$\dot{\mathbf{x}} = A\mathbf{x} + B\mathbf{u} \tag{10.35}$$

$$\mathbf{y} = C\mathbf{x} \tag{10.36}$$

where $A \in \mathcal{R}^{n \times n}$, $B \in \mathcal{R}^{n \times m}$ and $C \in \mathcal{R}^{l \times n}$, and a set of n complex numbers, $\Gamma = \{\gamma_1, \gamma_2, \ldots, \gamma_n\}$, closed under complex conjugation, find the general form of the constant output-feedback matrix, $K_y(\mathbf{k})$, such that

$$\Gamma = \text{Eig}(A - BK_y(\mathbf{k})C) \qquad \text{for almost all } \mathbf{k} \tag{10.37}$$

Unlike the state-feedback case, there is no known linear analytical solution method for the above problem, as it stands. Under certain circumstances, however, it may be possible to find solutions to this problem. In Söylemez and Munro [17], for instance, it is shown that this problem is equivalent to an incomplete state-feedback pole assignment problem, and it imposes $n - l$ nonlinear equations to be solved. For simpler cases, solving these equations is possible with the help of symbolic algebra languages. As the number of equations and the number of unknowns increase, however, solving these equations becomes nearly impossible, and hence some more methodical solutions or some simplifications to the problem are required.

An approach that can be taken is to find the general form of the static output feedback compensator that assigns p poles of the closed-loop system, where p can be equal to or smaller than n. Here, assuming that the output-feedback matrix is divided into two parts, i.e.

$$K_y = \sum_{i \neq j} \mathbf{e}_i \mathbf{k}'_{yi} + \mathbf{e}_j \mathbf{k}'_{yj} \tag{10.38}$$

$$= \bar{K}_{yj} + \mathbf{e}_j \mathbf{k}'_{yj} \tag{10.39}$$

where $\mathbf{k}'_{yi}$ is the ith row of K_y, then the general pole assignment problem becomes that of finding $\mathbf{k}_{yj}$ such that

$$\Gamma = \mathrm{Eig}(A - B K_y C) \tag{10.40}$$

$$= \mathrm{Eig}(A - B \bar{K}_{yj} C - B \mathbf{e}_j \mathbf{k}'_{yj} C) \tag{10.41}$$

$$= \mathrm{Eig}(\bar{A}_{yj} - \beta_j \mathbf{k}'_{xj}) \tag{10.42}$$

where

$$\bar{A}_{yj} \triangleq A - B \bar{K}_{yj} C \tag{10.43}$$

$$\mathbf{k}_{xj} \triangleq C' \mathbf{k}_{yj} \tag{10.44}$$

Now, using a mapping approach type of algorithm, it is possible to show that

$$\delta_{yj} = X_{yj} \Phi'_{yj} \mathbf{k}_{xj} \tag{10.45}$$

where Φ_{yj} is the controllability matrix of $\left[\bar{A}_{yj}, \beta_j\right]$, X_{yj} is a Toeplitz matrix as in eqn. 10.14 formed by the coefficients of $p_{yj}(s) \triangleq |s I - \bar{A}_{yj}|$, and δ_{yj} is a vector obtained by subtracting the coefficients of $p_{yj}(s)$ from the coefficients of the closed-loop characteristic polynomial $p_c(s)$.

Substituting for the definition of $\mathbf{k}_{xj}$, we obtain

$$\delta_{yj} = \underbrace{X_{yj} \Phi'_{yj} C'}_{\triangleq \ Z} \mathbf{k}_{yj} \tag{10.46}$$

$$= Z \mathbf{k}_{yj} \tag{10.47}$$

Hence, $\mathbf{k}_{yj}$ has to satisfy eqn. 10.47 to yield a solution to the pole assignment problem.

Now, let us suppose that we would like to assign p poles of the closed-loop system ($p \le n$). If we define two polynomials $p_d(s)$ and $p_e(s)$, where $p_d(s)$ is the desired part of the closed-loop characteristic polynomial given by

$$p_d(s) \triangleq \prod_{i=1}^{p}(s - \gamma_i) \tag{10.48}$$

$$= s^p + \sum_{i=1}^{p} d_{p-i} s^{p-i} \tag{10.49}$$

and $p_e(s)$ is the 'residue polynomial' formed by the rest of the closed-loop characteristic polynomial poles, i.e.

$$p_e(s) \triangleq \frac{p_c(s)}{p_d(s)} \tag{10.50}$$

$$= s^t + \sum_{i=1}^{t} e_{t-i} s^{t-i} \qquad t \triangleq n - p \tag{10.51}$$

then we can write

$$p_c(s) = p_d(s)p_e(s) \tag{10.52}$$

$$= p_d(s)\left(s^t + \sum_{i=1}^{t} e_{t-i}s^{t-i}\right) \tag{10.53}$$

$$= p_d(s)\,s^t + \sum_{i=1}^{t} p_d(s)e_{t-i}s^{t-i} \tag{10.54}$$

The difference polynomial $\delta_{yj}(s)$ is now

$$\delta_{yj}(s) = p_c(s) - p_{yj}(s) \tag{10.55}$$

$$= p_d(s)\,s^t - p_{yj}(s) + \sum_{i=1}^{t} p_d(s)\,s^{t-i}e_{t-i} \tag{10.56}$$

Hence, it is possible to write

$$\delta_{yj} = \delta_0 + D_p\,\mathbf{e} \tag{10.57}$$

where δ_0 is formed from the coefficients of $p_d(s)\,s^t - p_{yj}(s)$, $\mathbf{e}$ is a vector formed from the unknown coefficients e_i, and D_p is

$$D_p \triangleq [\delta_1 \quad \delta_2 \quad \dots \quad \delta_t] \tag{10.58}$$

where the δ_i are formed from the coefficients of $p_d(s)\,s^{t-i}$.

Using eqn. 10.57 in eqn. 10.47 we obtain

$$\delta_{yj} = \mathcal{Z}\mathbf{k}_{yj} \tag{10.59}$$

$$\delta_0 + D_p\mathbf{e} = \mathcal{Z}\mathbf{k}_{yj} \tag{10.60}$$

Hence,

$$\delta_0 = \mathcal{Z}\mathbf{k}_{yj} - D_p\mathbf{e} \tag{10.61}$$

$$= \underbrace{[\mathcal{Z} - D_p]}_{\triangleq \hat{X}_j} \underbrace{\begin{bmatrix} \mathbf{k}_{yj} \\ \mathbf{e} \end{bmatrix}}_{\triangleq \hat{\mathbf{k}}_j} \tag{10.62}$$

$$= \hat{X}_j\,\hat{\mathbf{k}}_j \tag{10.63}$$

Noting that all the unknowns are in $\hat{\mathbf{k}}_j$, the desired solution is given by

$$\hat{\mathbf{k}}_j = \hat{X}_j^g\,\delta_0 \tag{10.64}$$

where g denotes the generalised Moore-Penrose inverse [18] of a matrix.
We remark that eqn. 10.64 is consistent if, and only if,

$$\hat{X}_j\,\hat{X}_j^g\,\delta_0 = \delta_0. \tag{10.65}$$

We also remark that when $p = l$, $\hat{X}_j$ is square and hence the generalised inverse given above turns into a normal inverse satisfying the consistency condition for almost all selections of $\bar{K}_{yj}$, provided that the given system is controllable and observable with B and C matrices full-rank. If $p < l$, then the equations are underdetermined and again a solution exists for almost all $\bar{K}_{yj}$. If $p > l$, however, the equations are overdetermined and eqn. 10.64 is consistent only for a subset[4] of $\bar{K}_{yj}$. Although this subset can be found with the help of symbolic algebra languages for simpler cases, the consistency condition given in eqn. 10.65 forms a nonlinear set of equations in terms of the free variables, **k**, and finding a solution to these equations is notoriously difficult in general.

Nevertheless, in practice we often want to determine the locations of only a few (dominant) poles and want to make sure that the other poles are stable and have little effect on the dynamics of the system; that is, assigning a few poles to desired locations and ensuring that the real parts of the rest of the poles are less than a predefined negative value (let us say h) is desirable in general. Hence, by keeping $\bar{K}_{yj}$ in symbolic form, eqn. 10.64 can be used to find the general form of the output-feedback compensator, $K_y(\mathbf{k})$, that assigns p poles of the closed-loop system to the desired locations and the coefficients of the general form of the residue polynomial, $p_e(\mathbf{k}, s)$, where $p_e(\mathbf{k}, s - h)$ must be Hurwitz stable.

In cases where more than l poles are required to be assigned, a possible further improvement to the above algorithm is to design a second dyadic compensator, where the poles assigned in the first step are kept and more poles are assigned. Such an algorithm is explained in Munro and Novin-Hirbod [19], and it is shown that using this approach $m + l - 1$ poles can be assigned arbitrarily. This approach, however, requires that the compensator found in the first step is fixed, and therefore restricts the degrees of freedom in the overall result, and therefore, is not preferred unless it is absolutely necessary.

Example 3

Let us again consider the system defined in Example 1. Now, let us suppose that only the first two states are available for measurement, i.e.

$$C = \begin{bmatrix} 1 & 0 & 0 \\ 0 & 1 & 0 \end{bmatrix} \tag{10.66}$$

Furthermore, suppose that the required closed-loop system specifications are the same except that we do not require the third closed-loop pole to be at -7 under nominal working conditions. Hence, $p = 2$, $t = 1$, and

$$p_p(s) = s^2 + 2s + 2 \tag{10.67}$$

---

[4] This subset can be an empty set.

If the first row of the output-feedback matrix is to be calculated in terms of the elements in the second row ($j = 1$), i.e.

$$\bar{K}_{yj} = \begin{bmatrix} 0 & 0 \\ k_1 & k_2 \end{bmatrix} \tag{10.68}$$

and hence

$$\bar{A}_{yj} = A - B\bar{K}_{yj}C = \begin{bmatrix} 1 & 0 & 4 \\ -k_1 & 1-k_2 & 2 \\ 3-k_1 & 1-k_2 & 3 \end{bmatrix} \tag{10.69}$$

then, from eqn. 10.46 we obtain

$$\mathcal{Z} = X_{yj}\,\Phi'_{yj}\,C' = \begin{bmatrix} 1 & 0 \\ k_2 & 2-k_1 \\ 3(-1+k_2) & 4-3k_1 \end{bmatrix} \tag{10.70}$$

Noting that

$$\delta_0 = \begin{bmatrix} 7-k_2 \\ 9-4k_1+2k_2 \\ -11+11k_2 \end{bmatrix} \tag{10.71}$$

and

$$D_p = \delta_1 = \begin{bmatrix} 1 \\ 2 \\ 2 \end{bmatrix} \tag{10.72}$$

then

$$\hat{\mathbf{k}}_j = [\mathcal{Z} \quad -D_p]^{-1}\delta_0 \tag{10.73}$$

$$= \begin{bmatrix} \dfrac{12k_1^2 + k_1(-26+k_2) - 10(-3+k_2)}{2+k_1-2k_2} \\[2ex] \dfrac{25 - 16k_2 + k_2^2 + 4k_1(-5+3k_2)}{2+k_1-2k_2} \\[2ex] \dfrac{16 + 12k_1^2 + 6k_2 - 2k_2^2 + k_1(-33+2k_2)}{2+k_1-2k_2} \end{bmatrix} \tag{10.74}$$

The general form of the output feedback matrix that assigns the first two poles of the closed-loop system to $\{-1+j, \ -1-j\}$ under nominal working conditions can then be given as

$$K_y = \begin{bmatrix} \dfrac{12k_1^2 + k_1(-26+k_2) - 10(-3+k_2)}{2+k_1-2k_2} & \dfrac{25 - 16k_2 + k_2^2 + 4k_1(-5+3k_2)}{2+k_1-2k_2} \\[2ex] k_1 & k_2 \end{bmatrix} \tag{10.75}$$

We note that the third nominal pole of the closed-loop system is readily calculated from the last row of $\hat{\mathbf{k}}_j$ as

$$s = -\frac{16 + 12\,k_1^2 + 6\,k_2 - 2\,k_2^2 + k_1\,(-33 + 2\,k_2)}{2 + k_1 - 2\,k_2} \tag{10.76}$$

Optimising the free variables $(k_1,\ k_2)$, the following compensator is obtained to satisfy the previously specified robust performance criteria:

$$K_y = \begin{bmatrix} 2.36216 & 6.68108 \\ -0.9 & 9.8 \end{bmatrix} \tag{10.77}$$

The pole spread of the resulting closed-loop system when this compensator is used can be seen in Figure 10.5.

10.3.2 Dynamic feedback

So far we have dealt with special cases of a larger problem, called 'the general output-feedback pole assignment problem'. In this Section we define this problem and show that it can be converted to one of the special cases we have seen previously.

First, let us define 'the general output-feedback pole assignment problem'.

Given a system

$$\begin{aligned} \dot{\mathbf{x}} &= A\,\mathbf{x} + B\,\mathbf{u} \\ \mathbf{y} &= C\,\mathbf{x} + D\,\mathbf{u} \end{aligned} \tag{10.78}$$

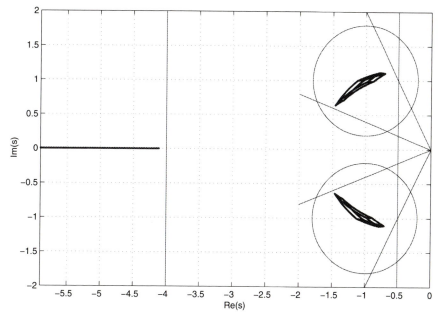

Figure 10.5   Pole spread of closed-loop system using output feedback (Example 3)

and a set of complex numbers, $\Gamma = \{\gamma_1, \gamma_2, \ldots, \gamma_{n+q}\}$, closed under complex conjugation, find the general form of the output feedback compensator, $F(\mathbf{k}')$, of order q such that

$$\begin{aligned} \dot{\mathbf{x}}_F &= A_F(\mathbf{k}')\mathbf{x}_F + B_F(\mathbf{k}')\mathbf{y} \\ \mathbf{y}_F &= C_F(\mathbf{k}')\mathbf{x}_F + D_F(\mathbf{k}')\mathbf{y} \end{aligned} \qquad (10.79)$$

where $\mathbf{x}_F \in \mathcal{R}^q$, $\mathbf{y}_F \in \mathcal{R}^m$, and A_F, B_F, C_F and D_F are matrices with appropriate dimensions, and under the feedback law

$$\mathbf{u} = \mathbf{r} - \mathbf{y}_F \qquad (10.80)$$

the closed-loop system poles are equal to Γ. Fig. 10.6 shows this arrangement.

An assumption we can make in the above problem is that the given system, G, and the compensator, F, are complete. This does not affect the generality of the problem, due to the well-known fact that only controllable and observable poles of the system can be affected by the feedback. If an incomplete system is given, then it is always possible to decompose the system into four subsystems with different controllability and observability properties. Pole assignment is (and can be) carried out only for the completely controllable and observable subsystem.

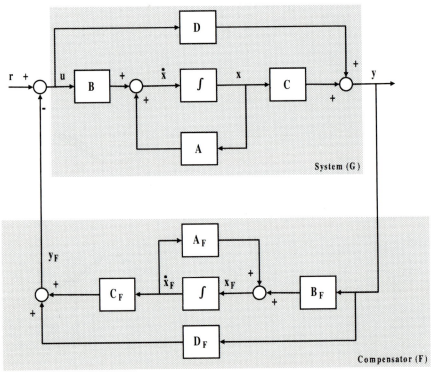

Figure 10.6   Closed-loop system block diagram in the time domain

Motivated by the fact that the closed-loop system poles are the eigenvalues of the closed-loop system A matrix, we write the closed-loop system equations as follows:

$$\dot{\mathbf{x}} = A\mathbf{x} + B\mathbf{u} \tag{10.81}$$

$$\dot{\mathbf{x}}_F = B_F C\mathbf{x} + A_F \mathbf{x}_F + B_F D\mathbf{u} \tag{10.82}$$

$$\mathbf{y}_F = D_F C\mathbf{x} + C_F \mathbf{x}_F + D_F D\mathbf{u} \tag{10.83}$$

$$\mathbf{u} = \mathbf{r} - \mathbf{y}_F \tag{10.84}$$

Hence, defining the closed-loop system state vector as

$$\mathbf{x}_c = \begin{bmatrix} \mathbf{x} \\ \mathbf{x}_F \end{bmatrix} \tag{10.85}$$

it is not difficult to show that the closed-loop system A-matrix, A_c, is given by

$$A_c = \begin{bmatrix} A & 0 \\ 0 & 0 \end{bmatrix} - \begin{bmatrix} B & 0 \\ 0 & -I_q \end{bmatrix} \begin{bmatrix} (I + D_F D)^{-1} & 0 \\ -B_F D (I + D_F D)^{-1} & I_q \end{bmatrix} \begin{bmatrix} D_F & C_F \\ B_F & A_F \end{bmatrix} \begin{bmatrix} C & 0 \\ 0 & I_q \end{bmatrix} \tag{10.86}$$

Now, defining

$$\hat{A} \triangleq \begin{bmatrix} A & 0 \\ 0 & 0 \end{bmatrix} \qquad \hat{B} \triangleq \begin{bmatrix} B & 0 \\ 0 & -I_q \end{bmatrix} \qquad \hat{C} \triangleq \begin{bmatrix} C & 0 \\ 0 & I_q \end{bmatrix} \qquad \hat{D} \triangleq \begin{bmatrix} D & 0 \\ 0 & 0 \end{bmatrix}$$

$$\hat{K} \triangleq \begin{bmatrix} D_F & C_F \\ B_F & A_F \end{bmatrix} \tag{10.87}$$

eqn. 10.86 can be rewritten as

$$A_c = \hat{A} - \hat{B}(I + \hat{K}\hat{D})^{-1}\hat{K}\hat{C} \tag{10.88}$$

We note that when $q = 0$ (constant output-feedback compensation) eqn. 10.88 reduces to

$$A_c = A - B(I + KD)^{-1}KC \tag{10.89}$$

where K is the required constant feedback compensator. However, this is in the same form as eqn. 10.88, which means that $[\hat{A}, \hat{B}, \hat{C}, \hat{D}]$ can be thought of as an *augmented system*, and many dynamic feedback pole assignment problems can be reformulated as an equivalent constant feedback pole assignment problem for the augmented system.[5] Noting also that if $[A, B, C, D]$ is complete, then so is $[\hat{A}, \hat{B}, \hat{C}, \hat{D}]$, and the general pole assignment problem can be restated as follows.

---

[5] Note that the opposite situation is not necessarily true. That is, we may not be able to find an equivalent dynamic output-feedback pole assignment problem for every given constant output-feedback problem.

Given a complete system,

$$\dot{\mathbf{x}} = \hat{A}\mathbf{x} + \hat{B}\mathbf{u} \tag{10.90}$$

$$\mathbf{y} = \hat{C}\mathbf{x} + \hat{D}\mathbf{u} \tag{10.91}$$

and a set of complex numbers, $\Gamma = \{\gamma_1, \gamma_2, \ldots, \gamma_{n+q}\}$, closed under complex conjugation, find the general form of the constant compensator, $\hat{K}(\mathbf{k})$, such that

$$\text{Eig}(A_c) = \Gamma. \tag{10.92}$$

Using eqn. 10.88 in eqn. 10.92 the pole assignment problem can now be reduced to finding a matrix, $\hat{K}$, where

$$\Gamma = \text{Eig}(\hat{A} - \hat{B}(I + \hat{K}\hat{D})^{-1}\hat{K}\hat{C}) \tag{10.93}$$

Now, by defining

$$K_y = (I + \hat{K}\hat{D})^{-1}\hat{K} \tag{10.94}$$

eqn. 10.93 simplifies to

$$\Gamma = \text{Eig}(\hat{A} - \hat{B}K_y\hat{C}) \tag{10.95}$$

We also note that when $\hat{D} = 0$ (i.e. when the system is strictly proper)

$$\hat{K} = K_y \tag{10.96}$$

and when $\hat{D} \neq 0$ then $\hat{K}$ is determined[6] using

$$\hat{K} = K_y(I - DK_y)^{-1} \tag{10.97}$$

Therefore, it is possible to assume that the given system is strictly proper, and we can solve the pole assignment problem and then calculate the required feedback matrix $\hat{K}$ from eqn. 10.97.

By further examining eqn. 10.95 it is also possible to see that

$$\text{Eig}(\hat{A} - \hat{B}K_y\hat{C}) = \text{Eig}(\hat{A}' - \hat{C}'K_y'\hat{B}') \tag{10.98}$$

Hence, the pole assignment problem for a given system $[\hat{A}, \hat{B}, \hat{C}]$ is equivalent to the pole assignment problem for the dual system $[\hat{A}', \hat{C}', \hat{B}']$. Now, under the light of this observation, it is possible to see that assuming the number of outputs is greater than or equal to the number of inputs does not change the generality of the pole assignment problem. When a system with more inputs than outputs is given, it is always possible to solve the problem for the dual system, where the above assumption holds, and then take the transpose of the resulting pole assignment matrix to find a solution to the original problem.

Another assumption we can make, and yet not change the generality of the pole assignment problem, is that the input and output matrices, B and C, and hence $\hat{B}$ and $\hat{C}$, are of full rank. This is due to the fact that a pole assignment

---

[6] This is assuming $|I - DK_y| \neq 0$. If this determinant is identically zero then a solution to the pole assignment problem probably does not exist for the given problem.

problem with the B and C matrices not full rank can be expressed as another pole assignment problem where the corresponding B and C matrices are full rank.

Considering all of the above discussions, it is now possible to see that the general output-feedback pole assignment problem can be reduced to the general static output-feedback problem as presented in the preceding Section.

Nevertheless, we should note that the dynamic output feedback pole assignment problem is, in a sense, a special case of the static output feedback pole assignment problem where the A, B and C matrices are put into a special form. Therefore, there can be (and are) methods that work only for the dynamic case. One such method was introduced in Munro and Novin-Hirbod [19], but we do not investigate such methods further here.

10.4 Conclusions

In this Chapter we have presented the possible use of symbolic algebra in the context of robust pole assignment for parametric uncertain systems. In particular, we have shown that symbolic algebra has a wide area of application when general forms of pole assignment compensators are sought.

It has also been shown that in some cases it is possible to convert the dynamic output feedback problem to a general static output-feedback pole assignment problem. Then, following the methods proposed in Section 10.3.1 it may often be possible to find general forms of the pole assignment compensators that assign $l + q$ poles of the closed-loop system. The remaining closed-loop system poles can then be readily calculated using the same formulas. This is particularly useful in reducing the search space, when robust compensators are sought.

10.5 References

1 MUNRO, N., and SÖYLEMEZ, M. T.: 'The use of symbolic algebra for uncertain systems'. Proc. Control Conf., Exeter, Vol. 9, 1996, pp. 1332–1337
2 SÖYLEMEZ, M. T., and MUNRO, N.: 'Development of a robust eigenvalue assignment toolbox using a Kharitonov based approach'. IEE Colloquium (Professional Group B1, Digest No: 97/380, Savoy Place, London, 1997)
3 WOLFRAM, S.: 'Mathematica: a system for doing mathematics by computer' (Addison-Wesley, 1991, 2nd edn.)
4 SÖYLEMEZ, M. T., and MUNRO, N.: 'Robust pole assignment in uncertain systems', *Proc. IEE*, 1997, **144**, (3), pp. 217–224
5 WONHAM, W. M.: 'On pole assignment in multi-input controllable linear systems', *IEEE Trans. Autom. Control*, 1967, **12**,(6), pp. 660–665
6 MUNRO, N.: 'Pole assignment: A review of methods', In SINGH, M. G. (Ed.): Systems and control encyclopedia' (Pergamon Press, 1987), pp. 3710–3717
7 KAILATH, T.: 'Linear systems' (Prentice Hall, Englewood Cliffs, NJ, 1980)
8 DAVISON, E. J., and WANG, S.-H.: 'Properties of linear time-invariant multivariable systems subject to arbitrary output and state feedback', *IEEE Trans. Autom. Control*, 1973, **18**,(1), pp. 24–32

9 SÖYLEMEZ, M. T.: 'Robust pole assignment using symbolic algebra'. Master's thesis, UMIST, Manchester, UK, October 1994

10 FALLSIDE, F., and SERAJI, H.: 'Direct design procedure for multivariable feedback systems', *Proc. IEE*, 1971, **118**,(6), pp. 797–801

11 SÖYLEMEZ, M. T.: 'State feedback dyadic pole assignment methods: A unification', Control Systems Centre Report 872, UMIST, Manchester, 1997

12 BASS, R. W., and GURA, I.: 'High order system design via state-space considerations'. *Proc. Joint Autom. Control Conf.*, 1965, pp. 311–318

13 SÖYLEMEZ, M. T., and MUNRO, N.: 'Pole assignment and symbolic algebra: A new way of thinking'. Proc. UKACC Control'98, to be published

14 GOLDBERG, D. E.: 'Genetic algorithms in search, optimization, and machine learning' (Addison-Wesley, Reading, MA, 1989)

15 STENDER, J. (Ed.): 'Genetic algorithms in optimization, simulation and modelling' (IOS Press, Amsterdam, Netherlands, 1994)

16 BELLETRUTTI, J.: 'A note on pole assignment'. Control Systems Centre Report 207, UMIST, Manchester, June 1973

17 SÖYLEMEZ, M. T., and MUNRO, N.: 'A note on pole assignment in uncertain systems', *Int. J. Control*, 1997, **66**,(4), pp. 487–497

18 PENROSE, R.: 'A generalized inverse for matrices', *Proc. Camb. Phil. Soc. Math. Phil. Sci.*, 1955, **51**, pp. 406–413

19 MUNRO, N., and NOVIN-HIRBOD, S.: 'Pole assignment using full-rank output-feedback compensators', *Int. J. Syst. Sci.*, 1979, **10**,(3), pp. 285–306

Chapter 11

Algebraic, algebrogeometric methods and symbolic computations in linear control problems

N. Karcanias, D. Vafiadis and J. Leventides

11.1 Introduction

Algebraic synthesis and algebrogeometric approaches in the study of control problems rely on the theory of invariants and canonical forms of polynomial matrices and use the theory of Groebner bases for the computation of solutions of polynomial equations in many variables. This chapter provides first a classification of the different types of algebraic computations, according to whether they can be performed symbolically and/or numerically, and then considers three representative families of algebraic control theory problems, which require the use of symbolic computation tools. The first area is that of Groebner bases, and it is considered in the context of the generalised cover problems. The second is that of canonical forms of polynomial matrices, which is used as the basis for canonical forms of implicit descriptor systems. The third is the algebraic framework of the global linearisation of determinantal type problems, which is used in the study of pole assignment by constant output feedback. The algorithmic procedures are implemented using MAPLE and Mathematica, and the above three problems are illustrated in terms of a number of examples.

Areas such as control theory and design, signal processing, coding theory, etc. have demonstrated the importance of computations relating to many branches of mathematics. In particular, the development of approaches in control theory, such as the algebraic [1–3], geometric [4–6], algebrogeometric [7–9], exterior algebra—algebraic geometry [10–12], graph structural methods [13–14] etc., have demonstrated the significance of a variety of mathematical tools and have created the need for special tools handling the different aspects of algebraic computations. More specifically, important areas for computations which have emerged include computational issues of polynomial, rational matrices, matrix pencils, exterior algebra, structural matrices, spectral factorisation, Riccati, Lyapunov equations, Diophantine equations, etc. The importance of computational issues in the control theory area has been realised during the past 15 years [15, 16], but the work has been mainly concentrated on problems related to state-space computations, which involve standard numerical linear algebra [17, 18]. The emergence of symbolic computation packages such as Mathematica,, MAPLE, REDUCE, DERIVE, COCOA, etc. has stimulated interest in the development of algorithms implementing theoretical computational procedures for polynomial and rational matrix computations [19]. The area of symbolic computations for control problems is an emerging subject that involves issues stemming from both the nature of the theoretical algorithms to be implemented and the characteristics of the system models on which computations are to be performed. This chapter deals first with the classification of the different types of algebraic computations and then considers the implementation of three different types of computational problems in pure symbolic form. Chapter 6 deals with issues stemming from the numerical inaccuracies of system models and examines the problem of approximate solutions to algebraic computations.

The general family of algebraic computations for linear control problems may be divided broadly into two classes. These are the problems which may be reduced to equivalent numerical linear algebra problems and those which are of a pure algebraic nature, and there is no complex matrix equivalent formulation. The question of whether numerical and/or symbolic methods are appropriate for a given problem revolves around this classification. In this chapter we elaborate on this general classification first and then we examine three algebraic problems in control theory for which symbolic tools may be used. The first involves the use of Groebner bases for the study of the generalised cover problem, to which problems of selection of inputs and outputs of systems may be reduced [20]; the derivation of the Groebner bases is a procedure which until now has relied entirely on the use of symbolic tools. The second problem deals with the derivation of echelon type canonical forms of polynomial matrices. The construction of echelon type forms is central in the derivation of canonical representations for autoregressive [21] and state-space representations and also provides means for the computation of algebraic invariants. The third problem under consideration is the development of algorithms for the global linearisation [22] of determinantal type problems using symbolic tools; the procedure involves polynomial matrix and exterior algebra computations together with numerical linear algebra and thus represents a typical problem where there is need for merging symbolic and numerical tools. The methodology is central to the development of algorithmic procedures for the solution of control problems such as the pole assignment by output feedback [23]. The different areas are illustrated in terms of examples.

11.2 Classification of computational problems

The development of algebraic, geometric and algebrogeometric approaches for linear and nonlinear systems has created the need for advanced computational tools, which may be used to transform such methods to systematic analysis and design methodologies. The algebraic computational procedure algorithms which have emerged involve basic steps which may be classified as follows:

(i)   purely numerical computations (PNC)
(ii)   numerically reducible computations (NRC)
(iii) numerically irreducible computations (NIC).

The first class contains computational issues, which in general may be handled only using numerical procedures (i.e. there is no general symbolic solution). A typical example here is the computation of eigenvalues of an $n \times n$ matrix, roots of polynomials, etc. The second and third classes correspond to computations for which there is a symbolic implementation. If such a procedure can also be formulated as an equivalent numerical computation, then it will be called *numerically reducible* (NR); otherwise, i.e. when there is no numerical linear algebra formulation for the process, then it will be called *numerically irreducible* (NI). The class of NR computations has the advantage that both symbolic and numerical versions are available; the existence of a numerical version also allows the introduction of notions such as 'approximate' algebraic computations, which are of relevance to system models with numerical inaccuracies (this issue is discussed in Chapter 6). A typical example of NR computations is the problem of finding the greatest common divisor (GCD) of a set of polynomials. The standard Euclid algorithm provides a symbolic implementation, whereas matrix-based methods such as the matrix pencil approach [24], or the ERES method [25] provide numerical procedures for the computation of GCD. The computation of Groebner bases [26], on the other hand, is an example of NI computational procedure.

The above classification has been based on the available methods for computing the desirable function. An alternative classification of the algebraic procedures may be based on the methodology–problem area, and we thus distinguish the following classes:

(*a*) state space algebraic computations (SSAC)
(*b*) rational matrices algebraic computations (RMAC)
(*c*) determinantal problem computations (DPC)
(*d*) graph type computations (GTC)
(*e*) nonlinear system computations (NSC)
(*f*) discrete event system computations (DESC).

Most of the computations within the state-space framework are reduced to numerical linear algebra. The standard algebraic machinery is provided by that of matrix pencils [27, 7, 16].

Although most matrix pencil computations can be carried out by numerical procedures [27], there are certain problems relating to parametrisations and canonical forms [29], which require symbolic tools. Some of these issues are considered later in this chapter. State-space numerical procedures have an additional significance, since they provide the means for the reduction of rational matrix-based algebraic computation problems.

Rational matrix models include as special cases the polynomial matrices [28] and matrices over specific rings of interest. This area is where symbolic computations have a special interest, since computations are based on algebraic steps [19], where a number of them are not yet numerically reduced; this is especially the case with systems over the ring of proper, proper and stable rational functions [30, 31]. A large number of rational matrix computational issues may be reduced to equivalent numerical computations using realisation theory; however, issues of reduction of matrices to canonical forms (computation of transformations) and parametrisation of families of solutions still remain hard numerically irreducible problems, for which symbolic tools are indispensable.

A family of problems of both state-space and transfer-function set-up are formulated as determinantal assignment problems (DAP) [10]; such problems are naturally reduced to the solution of linear systems and exterior equations, or equivalently to solution of homogeneous algebraic equations. The required algebraic tools involve exterior algebra computations and issues from computational algebraic geometry. An important methodology that reduces the solvability of the problem to an algebraic linear formulation is known as global linearisation [12], and a symbolic computational approach for such problems is described here. The area of DAP involves in a natural way the theory of Groebner bases; a problem related to the DAP formulation, which may be handled in terms of Groebner bases is also discussed here.

The theory of symbolic computations emerges also as important in areas such as graph theory and computations [13, 14], where most of the algorithms are of a combinatorial nature. In this area, the theory of structural matrices [32] and matroids [33] is becoming quite important and the symbolic computations are indispensable tools. Nonlinear systems methodologies involve the transformation of nonlinear differential equations, computation of invariants and testing of properties, which cannot be done numerically, but require symbolic tools. An account of the issues in this area is given in [34]. Symbolic computations are natural in the context of discrete event dynamic systems (DEDS), and hybrid systems [35], where different methodologies (such as max plus algebra, Petri nets, etc.) are of a pure algebraic nature and their only implementation is their symbolic representation.

In this chapter we concentrate on linear systems and control problems and focus mostly on numerically irreducible problems for which only the symbolic machinery is available. Three representative problems are considered next.

11.3 Groebner basis computation

In this Section we give brief background material about the multidimensional polynomials and the use of the Groebner basis in the solution of polynomial equations to provide the appropriate mathematical tools for the following Section, where the cover problem is formulated as a problem of solution of a system of multilinear equations. A detailed analysis of the Groebner basis technique may be found in textbooks of computational algebraic geometry [26].

Definition 1   Let $f_1, \ldots, f_p$ be polynomials in $\mathcal{R}[s_1, s_2, \ldots, s_q]$. Then the set of q-tuples defined by

$$\mathbf{V}(f_1, \ldots, f_p) = \{(a_1, \ldots, a_q) \in \mathcal{R}^q : f_i(a_1, \ldots, a_q) = 0, \ i = 1, \ldots, p\} \quad (11.1)$$

is called the affine variety defined by $f_1, \ldots, f_p$.   □

Now, it is clear that the affine variety $\mathbf{V}(f_1, \ldots, f_p)$ is the set of all the solutions of the system of equations $f_1 = f_2 = \cdots = f_p = 0$.

Proposition 1 [26]   Let the set of polynomials $f_1, \ldots, f_p$ be a basis of an ideal in $\mathcal{R}[s_1, s_2, \ldots, s_q]$. If $g_1, \ldots, g_r$ is another basis of the same ideal we have that

$$\mathbf{V}(f_1, \ldots, f_p) = \mathbf{V}(g_1, \ldots, g_r) \quad (11.2)$$

    □

From the above we see that given a system of polynomial equations, we are free to use another system of equations, generating the same ideal, in order to find the solution. A polynomial of one variable is a sum of monomials. The leading term is the term corresponding to the monomial with the higher degree and the ordering of the terms is obvious. In the case of polynomials in several variables, the ordering of the terms is not that obvious. A polynomial in several variables is the sum of monomials of the form $s_1^{a_1} s_2^{a_2} \ldots s_p^{a_p}$. The ordering of the monomials is determined by the p-tuple $(a_1, \ldots, a_p)$. The formal definition of the monomial ordering is the following:

Definition 2 [26]   A monomial ordering on $\mathcal{R}[s_1, \ldots, s_p]$ is any relation $>$ on the set of polynomials of the form $s_1^{a_1} s_2^{a_2} \ldots s_p^{a_p}$, $a_i \geq 0$, satisfying

(i) $>$ is a total ordering on $\mathcal{Z}_{\geq 0}^p$
(ii)   if $(a_1, \ldots, a_p) > (b_1, \ldots, b_p)$ then $(a_1, \ldots, a_p) + (c_1, \ldots, c_p) > (b_1, \ldots, b_p) + (c_1, \ldots, c_p)$
(iii)   Every subset of $\mathcal{Z}_{\geq 0}^p$ has a smallest element.   □

A special type of ordering is the lexicographic ordering defined as follows.

Definition 3   Let $a = (a_1, \ldots, a_p)$, $(b_1, \ldots, b_p) \in \mathcal{Z}^p_{\geq 0}$. We say $a >_{lex} b$ if the vector $a - b$ has its leftmost entry positive. Consider two monomials $s_1^{a_1} s_2^{a_2} \ldots s_p^{a_p}$ and $s_1^{b_1} s_2^{b_2} \ldots s_p^{b_p}$.   We   will   say   that $s_1^{a_1} s_2^{a_2} \ldots s_p^{a_p} >_{lex}, s_1^{b_1} s_2^{b_2} \ldots s_p^{b_p}$ if $a = (a_1, \ldots, a_p) >_{lex} (b_1, \ldots, b_p)$.   □

The lexicographic ordering plays an important role in the solution of systems of polynomial equations. Given a monomial ordering we may define the leading term of a polynomial as the greatest term corresponding to the ordering. Once a monomial ordering is chosen, every polynomial f has a unique leading term denoted by LT(f). Consider now an ideal $\mathcal{I}$ and a given monomial ordering. Let LT($\mathcal{I}$) denote the set of leading terms of elements of $\mathcal{I}$. This set is a set of monomials. The ideal generated by the elements of LT($\mathcal{I}$) is denoted by $\langle$ LT($\mathcal{I}$)$\rangle$. We may now give the definition of the Groebner basis.

Definition 4 [26]   Consider an ideal $\mathcal{I}$, a finite subset $\mathcal{G} = g_1, \ldots, g_t$ of $\mathcal{I}$ and fix a monomial ordering. We say that $\mathcal{G}$ is a Groebner basis of the ideal $\mathcal{I}$ if

$$\langle \mathrm{LT}(g_1), \ldots, \mathrm{LT}(g_t) \rangle = \langle \mathrm{LT}(\mathcal{I}) \rangle$$ □

Note that every nonzero ideal has a Groebner basis. The Groebner basis of an ideal is not unique. There is a special form of Groebner basis, the reduced Groebner basis, which is unique. An algorithm for finding the Groebner basis of an ideal is the Buchberger algorithm [36, 26].

The use of the Groebner basis to the solution of a system of polynomial equations is discussed below. Consider the system defined by the equations

$$f_1 = f_2 = \ldots = f_p = 0$$

The polynomials f_1, f_2, ..., f_p generate an ideal $\mathcal{I}$. Now, let $\mathcal{G}$ be the Groebner basis of $\mathcal{I}$. From proposition 1 it follows that the solutions of the given system of equations and the solutions of the system of equations defined by the polynomials of the Groebner basis are the same. When we use lexicographic monomial ordering, the use of the Groebner basis simplifies the solution considerably, because the equations we obtain have a nice form where some of the variables are eliminated from the equations in such a way that we may solve the system using the technique of 'back-substitution' in a way similar to the well known Gauss elimination procedure for linear systems. An example of the Groebner basis technique is the following.

Example 1

Consider the system of equations

$$s_1^2 + s_2 + s_3 - 1 = 0, \qquad s_1 + s_2^2 + s_3 - 1 = 0, \qquad s_1 + s_2 + s_3^2 - 1 = 0$$

A Groebner basis for the ideal generated by the left-hand-side polynomials is the following:

$$g_1 = s_1 + s_2 + s_3^2 - 1 \qquad g_2 = s_2^2 - s_2 - s_3^2 + s_3$$
$$g_3 = 2s_2 s_3^2 + s_3^4 - s_3^2 \qquad g_4 = s_3^6 - 4s_3^4 + 4s_3^3 - s_3^2$$

The system of equations that gives the same set of solutions to the original system is

$$g_1 = 0 \qquad g_2 = 0 \qquad g_3 = 0 \qquad g_4 = 0$$

The polynomial g_4 has one variable. Thus solving g_4 with respect to s_3 and substituting the roots to $g_3 = 0$ we get an equation with respect to s_2. Continuing this procedure of 'back-substitution' we obtain all the solutions of the original system of equations.

The Groebner basis technique is proven to be the appropriate tool for the parametric solution of the cover problem as shown in the following Section.

11.4 The solution of the cover problem via Groebner basis computation

Several problems in control theory were shown to be equivalent to the dynamic cover problem, i.e. the geometric problem of covering a given subspace with another subspace with special properties. In most of the cases, the covering space is required to be an (A, B)-invariant subspace. Such problems are the *linear functional observer problem* [38], the *model matching problem* [39–41], the *deterministic identification problem* [42] and the *disturbance decoupling problem* [4]. In the present chapter the cover problem is formulated as the problem of the solution of appropriate multilinear equations. Then the Groebner basis technique is used to find the families of solutions. We begin with the cover problem statement.

Let $\mathcal{S}(A, B, C)$ be the state-space system

$$\dot{x}(t) = Ax(t) + Bu(t) \qquad y(t) = Cx(t) \tag{11.3}$$

where $A \in \mathcal{R}^{n \times n}$, $B \in \mathcal{R}^{n \times l}$ and $C \in \mathcal{R}^{m \times n}$. It is assumed that both matrices B and C have full rank. If N is a left annihilator of B (i.e. a basis matrix for the $\mathcal{N}_l\{B\}$ and $B^\dagger$ is a left inverse of B, $(B^\dagger B = I_l)$, then the first part of eqns. 11.3 is equivalent to

$$N\dot{x} = NAx \qquad u = B^\dagger \dot{x} - B^\dagger Ax \tag{11.4}$$

and is a 'feedback-free' system description and the associated pencil $R(s) = sN - NA$ is known as the *input-state restriction pencil* [37] of the system. Before we proceed with the formal definition of the cover problem we give the definition of the (A, B)-invariant subspace.

Definition 5 [4]   A subspace $\mathcal{V}$ of the state space $\mathcal{X}$ is called an (A, B)-invariant subspace if

$$A\mathcal{V} \subseteq \mathcal{V} + \mathcal{B},$$

where $\mathcal{B} = \mathrm{Im}\{B\}$.   □

A matrix pencil characterisation of (A, B)-invariant subspaces is the following.

Theorem 2 [7]   *A subspace V of the state space X is (A, B)-invariant if and only if the matrix pencil*

$$R_V(s) = sNV - NAV$$

where V is a basis matrix of V, has only finite elementary divisors (FED) and/or column minimal indices (CMI) and possibly zero row minimal indices (ZRMI).   □

The above characterisation of the (A, B)-invariant subspaces via matrix pencils is the key tool for the development of the matrix pencil and subsequently the Groebner basis approach to the solution of the cover problem. A family of cover problems is defined below.

Definition 6   Let X be the state-space of $S(A, B)$ and J, W subspaces of X, where $J \subseteq W \subseteq X$. Finding all subspaces V of X such that

(i) V is (A, B)-invariant, i.e. $AV \subseteq V + B$ and $J \subseteq V \subseteq W$ is known as the **standard cover problem** [43, 42].

(ii) V is a subspace of the type (i) and $W = X$; then the problem will be called the **partial cover problem**.   □

The main idea underlying the matrix pencil approach to the study of the cover problems is the following. Let J be the basis matrix of the subspace to be covered. Since V is the covering subspace, then $V = J \oplus T$, where T is some appropriate subspace, or in matrix form

$$V = [J, T] \tag{11.5}$$

The restriction pencil of the covering subspace is then

$$R_V(s) = sNV - NAV = (sN - NA)[J, T] \tag{11.6}$$

From the above expression, it is clear that the cover problems defined above are equivalent to problems of Kronecker structure assignment defined below.

Kronecker structure assignment problem (KSAP): Given the J-restriction pencil $R_J(s) = sNJ - NAJ$, find an appropriate T-restriction pencil $R_T(s) = sNT - NAT$ such that the column augmented pencil $R_V(s)$ in eqn. 11.6 has only FED and/or CMI and possibly ZRMI.

The matrix pencil formulated problem may be further formulated as a problem of solution of multilinear equations. To proceed to this formulation we need the following.

Lemma 1 [29]   The pencil $sF - G$ does not have IED and nonzero RMI if and only if

$$\mathrm{Im}\{F\} \supseteq \mathrm{Im}\{G\} \qquad \square$$

From the above we have that $[sF - G, s\bar{F} - \bar{G}]$ has no IED and NZRMI if and only if

$$\text{Im}\{[F, \bar{F}]\} \supseteq \text{Im}\{[G, \bar{G}]\} \tag{11.7}$$

where $F = NJ$, $\bar{F} = NT$, $G = NAJ$, and $\bar{G} = NAT$. Thus, eqn. 11.7 may be written as

$$\text{Im}\{[NJ, NT]\} \supseteq \text{Im}\{[NAJ, NAT]\} \tag{11.8}$$

or equivalently

$$\text{rank}\{[NJ, NT, NAJ, NAT]\} = \text{rank}\{[NJ, NT]\} \tag{11.9}$$

Let $\text{rank}\{[NJ, NT]\} = \rho$. Then eqn. 11.9 may be written in terms of compound matrices as follows:

$$C_{\rho+1}\{[NJ, NT, NAJ, NAT]\} = 0 \tag{11.10}$$

Note that $\rho \leq n - l$, since N has $n - l$ rows. The procedure for the solution of the cover problem is as follows. Since matrix T represents a basis of the subspace $\mathcal{T}$, it must have full column rank. Let τ be the number of the columns of T. Then

$$C_\tau\{T\} \neq 0 \tag{11.11}$$

Now, to ensure that $\mathcal{J} \cap \mathcal{T} = 0$ we must have

$$C_{j+\tau}\{[J, T]\} \neq 0 \tag{11.12}$$

where $j = \dim\{\mathcal{J}\}$. In [29] it has been shown that if φ is the total number of IED and NZRMI of the restriction pencil $sNJ - NAJ$, then the dimension τ of a subspace $\mathcal{T}$ solving the cover problem must satisfy the condition

$$\tau \geq \varphi \tag{11.13}$$

Now, starting from $\tau = \varphi$, we consider the case where $\rho = j + 1$. Then, the condition $\text{rank}\{[NJ, NT]\} = \rho$ is equivalent to

$$C_\rho\{[NJ, NT]\} \neq 0 \quad \text{and} \quad C_{\rho+1}\{[NJ, NT]\} = 0 \tag{11.14}$$

Next, we solve eqn. 11.9 with respect to T. We say that there exists a solution of dimension τ to the cover problem if the solution of eqn. 11.9 does not contradict eqns. 11.10, 11.11 and 11.14. To find all the subspaces of dimension τ which solve the cover problem, we repeat the above, increasing ρ up to $p = \min\{n - l, j + \tau\}$. The next step is to increase τ and repeat the procedure until $\tau = n - j$, since then $\mathcal{J} \oplus \mathcal{T} = \mathcal{X}$.

Summarising, the procedure for the derivation of all the solutions of the partial cover problem consists in solving the following equations:

$$C_\rho\{[NJ, NT]\} \neq 0 \tag{11.15}$$

$$C_{\rho+1}\{[NJ, NT]\} = 0 \tag{11.16}$$

$$C_{\rho+1}\{[NJ, NT, NAJ, NAT]\} = 0 \tag{11.17}$$

$$C_\tau\{T\} \neq 0 \tag{11.18}$$

$$C_{j+\tau}\{[J, T]\} \neq 0 \tag{11.19}$$

for $\rho = 1, \ldots, \min\{n - l, j + \tau\}$, $\tau = \varphi + 1, \ldots, n - j$, with respect to T. The above may be considered as a homogeneous system of polynomial equations in several variables. The indeterminates of the polynomials are the entries of the matrix T. The solution of the above systems is obtained via the Groebner basis technique described in the preceding Section.

For the overall cover problem $\mathcal{J} \subset \mathcal{W} \subset \mathcal{X}$ we have

$$sNJ - NAJ = (sNV^* - NAV^*)\bar{J} \tag{11.20}$$

where V^* is a basis matrix of the maximal (A, B)-invariant subspace V^*. The (A, B)-invariance condition according to Lemma 1 is

$$\text{Im}\{[NV^*\bar{J}, NV^*\bar{T}]\} \supseteq \text{Im}\{[NAV^*\bar{J}, NAV^*\bar{T}]\} \tag{11.21}$$

where $T = V^*\bar{T}$. The algorithm of the solution procedure is the same as in the case of the partial cover problem, i.e. solve the system of equations

$$C_\rho\{[NV^*\bar{J}, NV^*\bar{T}]\} \neq 0 \tag{11.22}$$

$$C_{\rho+1}\{[NV^*\bar{J}, NV^*\bar{T}]\} = 0 \tag{11.23}$$

$$C_{\rho+1}\{[NV^*\bar{J}, NV^*\bar{T}, NAV^*\bar{J}, NAV^*\bar{T}]\} = 0 \tag{11.24}$$

$$C_\tau\{\bar{T}\} \neq 0 \tag{11.25}$$

$$C_{j+\tau}\{[\bar{J}, \bar{T}]\} \neq 0 \tag{11.26}$$

for $\rho = 1, \ldots, \min\{n - l, j + \tau\}$, $\tau = \varphi + 1, \ldots, v^* - j$, with respect to $\bar{T}$.

Remark 1   The method described above gives a complete parametrisation of the (A, B)-invariant subspaces $V = \mathcal{J} \oplus \mathcal{T}$. Eqns. 11.15–11.19 and 11.22–11.26 may have more than one set of solutions. Every set gives the parametric expressions of the basis matrices of $\mathcal{T}$.   □

Next we give an example to illustrate the Groebner basis method.

Example 2 [43]

Consider the system $\mathcal{S}(A, B)$ with controllability indices $\sigma_1 = 4$ and $\sigma_2 = 2$. The subspace $\mathcal{J}$ to be covered has a basis matrix $J = [1, -1, -1, 1, 1, 2]^T$. The Kronecker canonical form of the restriction pencil of $\mathcal{J}$ is $[0, 0, s, -1]^T$, i.e. $\mathcal{J}$ is an RMI subspace, with two ZRMI and one NZRMI $\eta = 1$. According to the algorithm presented we start with $\tau = 1$ and $\rho = 2$. Let $T = [t_{11}, t_{21}, t_{31}, t_{41}, t_{51}, t_{61}]^T$. Consider the equation

$$C_2\{[NJ, NT]\} = 0 \tag{11.27}$$

The above is equivalent to

$$t_{11} + t_{21} = 0, \qquad t_{11} + t_{31} = 0, \qquad -t_{11} + t_{51} = 0, \qquad t_{21} - t_{31} = 0,$$
$$-t_{21} - t_{51} = 0, \qquad -t_{31} - t_{51} = 0$$

The Groebner basis of the ideal generated from the left-hand side polynomials of the above equations is the following:

$$g_1 = -t_{31} - t_{51} \qquad g_2 = -t_{21} - t_{51} \qquad g_3 = -t_{11} + t_{51}$$

Thus, the solutions of the above are

$$t_{11} = t_{51} \qquad t_{21} = -t_{51} \qquad t_{31} = -t_{51}$$

Therefore the original conditions are satisfied if any of the following holds true:

$$t_{11} \neq t_{51} \qquad t_{21} \neq -t_{51} \qquad t_{31} \neq -t_{51}$$

The next step is to solve eqn. 11.17, i.e.

$$C_3\{[NJ, NT, NAJ, NAT]\} = 0 \tag{11.28}$$

or equivalently to solve the system

$$t_{11} + 2t_{31} = 0 \qquad t_{21}^2 - t_{11}t_{31} - t_{21}t_{31} - t_{31}^2 + t_{11}t_{41} + t_{21}t_{41} = 0$$
$$-2t_{21}t_{21} - 2t_{41} = 0 \qquad t_{21}^2 - t_{11}t_{31} + t_{21}t_{31} - t_{31}^2 - t_{11}t_{41} + t_{21}t_{41} = 0$$
$$t_{11} + 3t_{21} + 2t_{51} = 0 \qquad -t_{21}^2 + t_{11}t_{31} - t_{21}t_{51} - t_{31}t_{51} + t_{11}t_{61} + t_{21}t_{61} = 0$$
$$-t_{21} - 3t_{31} - 2t_{61} = 0 \qquad 2t_{21}^2 - 2t_{11}t_{31} + t_{21}t_{51} - t_{31}t_{51} - t_{11}t_{61} + t_{21}t_{61} = 0$$
$$3t_{11} + 3t_{31} = 0 \qquad -t_{21}t_{31} + t_{11}t_{41} - t_{21}t_{51} - t_{41}t_{51} + t_{11}t_{61} + t_{31}t_{61} = 0$$
$$-3t_{21} - 3t_{41} = 0 \qquad 2t_{21}t_{31} - 2t_{11}t_{41} - t_{21}t_{51} - t_{41}t_{51} + t_{11}t_{61} + t_{31}t_{61} = 0$$
$$3t_{21} - t_{31} + 2t_{51} = 0 \qquad -t_{31}^2 + t_{21}t_{41} - t_{31}t_{51} + t_{41}t_{51} + t_{21}t_{61} - t_{31}t_{61} = 0$$
$$-3t_{31} + t_{41} - 2t_{61} = 0 \qquad t_{31}^2 - 2t_{21}t_{41} - t_{31}t_{51} - t_{41}t_{51} + t_{21}t_{61} + t_{31}t_{61} = 0$$

The corresponding Groebner basis is

$$f_1 = -5t_{41} + 3t_{51} + t_{61} \qquad f_2 = 5t_{21} + 3t_{51} + t_{61} \qquad f_3 = -5t_{31} + t_{51} - 3t_{61}$$
$$f_4 = -5t_{11} - t_{51} + 3t_{61}$$

which leads to the following parametric set of solutions:

$$t_{61} = \frac{3t_{11} - t_{21}}{2} \qquad t_{51} = \frac{-t_{11} - 3t_{21}}{2} \qquad t_{41} = -t_{21} \qquad t_{31} = -t_{11}$$

The parameters t_{11} and t_{21} are free and they may be chosen arbitrarily. The basis matrix of the subspace T has the following form:

$$T = \left[t_{11}, \ t_{21}, \ -t_{11}, \ -t_{21}, \ \frac{-t_{11} - 3t_{21}}{2}, \ \frac{3t_{11} - t_{21}}{2} \right]^T$$

Now, choosing $t_{11} = t_{21} = -1$ we obtain the basis matrix of a covering space,

$$V = \begin{bmatrix} 1 & -1 & -1 & 1 & 1 & 2 \\ -1 & -1 & 1 & 1 & 2 & -1 \end{bmatrix}^T$$

The corresponding restriction pencil is

$$sNV - NAV = s \begin{bmatrix} 1 & -1 & -1 & 1 \\ -1 & -1 & 1 & 2 \end{bmatrix}^T - \begin{bmatrix} -1 & -1 & 1 & 2 \\ -1 & 1 & 1 & -1 \end{bmatrix}^T$$

The above pencil has two ZRMI, two FED at $s = j$ and $s = -j$ and has no IED, i.e. V is a fixed spectrum (A, B)-invariant subspace. The computation of compound matrices, Groebner bases as well as the solution of the systems of equations of the example was done with Mathematica. Most of the symbolic computations packages have built-in functions for Groebner basis computations using several types of orderings, solution of multilinear equations, etc.

11.5 Echelon form and canonical forms of descriptor systems

It is well known that the echelon canonical form for polynomial matrices plays an important role in several linear algebraic control problems. The direct relation of the echelon form to the canonical forms of state-space systems under similarity transformations is a fundamental result in system theory. The echelon canonical form for polynomial matrices was first introduced by Popov in [44]. A detailed treatment of the issue can be found in the classical paper of Forney [45]. In that paper the procedure of obtaining the echelon form was described in detail and its relation to several system theory and coding theory problems was investigated. A numerical approach towards the computation of the echelon form was given in [46].

In this chapter we consider the issue of symbolic computation of the echelon form. The available computational tools for symbolic algebra provide the means for implementing the original algorithm given by Forney. The polynomial matrix operations involved in this algorithm are mainly elementary column operations since the echelon form is obtained by multiplication by an appropriate unimodular matrix. The use of symbolic algebra can greatly simplify the programming work since the basic algebraic operations can be performed in a natural way and there is no need to translate the numerical results to polynomial matrix data (degree, coefficients of polynomials, etc.).

In this section the problem of canonical forms of descriptor systems under restricted system equivalence [21] is used as an example of the use of the echelon form in linear systems theory. A brief discussion on that problem is given next.

Consider the descriptor system $\mathcal{S}(E, A, B, C)$

$$\mathcal{S}: E\dot{x} = Ax + Bu \qquad y = Cx \tag{11.29}$$

where $(E, A, B, C) \in \mathcal{R}^{\tau \times n} \times \mathcal{R}^{\tau \times n} \times \mathcal{R}^{n \times l} \times \mathcal{R}^{m \times n}$. Eqn. 11.29 is assumed to be minimal under external equivalence [47]. The action of RSE transformations on the system $\mathcal{S}(E, A, B, C)$ is defined, in terms of system matrices, as follows:

$$\begin{bmatrix} P & 0 \\ 0 & I \end{bmatrix} \begin{bmatrix} sE - A & B \\ C & 0 \end{bmatrix} \begin{bmatrix} Q & 0 \\ 0 & I \end{bmatrix} = \begin{bmatrix} sPEQ - PAQ & PB \\ CQ & 0 \end{bmatrix} \tag{11.30}$$

where $(P, Q) \in \mathcal{R}^{\tau \times \tau} \times \mathcal{R}^{n \times n}$ and $\det\{P\} \neq 0$, $\det\{Q\} \neq 0$. The RSE transformations define an equivalence relation which partitions the set of all quadruples $(E, A, B, C) \in \mathcal{R}^{\tau \times n} \times \mathcal{R}^{\tau \times n} \times \mathcal{R}^{n \times l} \times \mathcal{R}^{m \times n}$ to equivalence classes or orbits. The problem of finding the CF of eqn. 11.29 is equivalent to the problem of the CF of the pencil T(s) under the transformation shown in eqn. 11.30. In [21] it has been shown that if $[N_c(s), D_c(s)] = R(s) \in \mathcal{R}^{r \times (l+m)}[s]$ is the echelon canonical form polynomial matrix corresponding to the autoregressive (AR) system with the same external behaviour to eqn. 11.29, then the canonical form is obtained as a realisation of the above AR matrix as follows.

Let $R_{hr} = [N_{hr}, D_{hr}]$ be the row high-order coefficient matrix of $R(s)$ and define the integers μ_i as follows:

$$\mu_i = \begin{cases} \partial_{r_i}(R(s)) & \text{if } \underline{n}_i = 0 \\ \partial_{r_i}(R(s)) + 1 & \text{if } \underline{n}_i \neq 0 \end{cases} \tag{11.31}$$

where $\underline{n}_i^T$ is the ith row of N_{hr} and $\partial_{r_i}(\cdot)$ denotes the degree of the ith row. Assume that $\mu_i \neq 0, i = 1, \ldots, r$, and write $N_c(s)$ and $D_c(s)$ as follows:

$$N_c(s) = S(s)B_c \qquad S(s) = \text{bl-diag}\{[1, s, \ldots, s^{\mu_i - 1}]\} \qquad i = 1, \ldots, r \tag{11.32}$$

$$D_c(s) = [\underline{d}_1(s), \ldots, \underline{d}_r(s)]^T \qquad \underline{d}_i^T(s) = (\underline{k}_{\mu_i}^i)^T s^{\mu_i} + (\underline{\lambda}_{\mu_i}^i)^T s^{\mu_i - 1} + \cdots + (\underline{\lambda}_1^i)^T$$

$$i = 1, \ldots, r \tag{11.33}$$

where B_c is obtained in an obvious way from the coefficients of the polynomial entries of $N_c(s)$. Note that $\underline{k}_{\mu_i}^i$ in eqn. 11.33 are not necessarily nonzero. Consider now the system with system matrix

$$T_c(s) = \begin{bmatrix} L(s) & sK_c - \Lambda_c & B_c \\ 0 & -I & 0 \end{bmatrix} = \begin{bmatrix} sE_c - A_c & B_c \\ C_c & 0 \end{bmatrix} \tag{11.34}$$

where

$$L(s) = \text{bl-diag}\{L_{\mu_i}(s)\} = sL_1 - L_2$$
$$L_{\mu_i}^{\mathrm{T}}(s) = s\left[I_{\mu_i-1}|0_{(\mu_i-1)\times 1}\right] - \left[0_{(\mu_i-1)\times 1}|I_{\mu_i-1}\right] \tag{11.35}$$

$$K_c = [K_1^{\mathrm{T}}, \ldots, K_r^{\mathrm{T}}]^{\mathrm{T}}, \quad K_i = [0, \ldots, 0, \underline{k}_{\mu_i}^i]^{\mathrm{T}} \tag{11.36}$$

$$\Lambda_c = [\Lambda_1^{\mathrm{T}}, \ldots, \Lambda_r^{\mathrm{T}}]^{\mathrm{T}}, \quad \Lambda_i = [\underline{\lambda}_1^i, \ldots, \underline{\lambda}_{\mu_i}^i]^{\mathrm{T}} \tag{11.37}$$

$$B_c = [B_1^{\mathrm{T}}, \ldots, B_r^{\mathrm{T}}]^{\mathrm{T}}, \quad B_i = [\underline{b}_1^i, \ldots, 0, \underline{b}_{\mu_i}^i]^{\mathrm{T}} \tag{11.38}$$

The blocks of $L(s)$ are canonical row minimal index (RMI) blocks [28]. The partitioning of K_c, Λ_c and B_c is conformable to the partitioning of $L(s)$. The following result is useful for the computation of the canonical form.

Proposition 2 [40]   Consider the system $A_1(\sigma)x(t) + B_1(\sigma)w(t) = 0$, where $A_1(s)$, $B_1(s)$ are polynomial matrices, $w(t)$ is the vector of external signals and σ is the differentiation operator. Let $U(s)$ be a unimodular matrix such that $U(s)A_1(s) = [\hat{A}_1^{\mathrm{T}}(s), 0]^{\mathrm{T}}$ with $\hat{A}_1(s)$ of full row rank. Write $U(s)B_1(s) = [\hat{B}_1^{\mathrm{T}}(s), \hat{B}_2^{\mathrm{T}}(s)]^{\mathrm{T}}$. Then the above system is externally equivalent to the autoregressive system

$$\hat{B}_2(\sigma)w(t) = 0 \tag{11.39}$$

□

The use of symbolic computations for the canonical form is illustrated in the example below.

Example 3

Consider the system with

$$T(s) = \begin{bmatrix} s+1 & 1 & 0 & 0 & 1 & -1 & 0 & 0 & 1 \\ 0 & s & 3 & 0 & 0 & -1 & 0 & 0 & -4 \\ 0 & 0 & s & -1 & 1 & 0 & 0 & 0 & 0 \\ 0 & 0 & 0 & s & 0 & 0 & 0 & 0 & 0 \\ 0 & 0 & -1 & 0 & 0 & 0 & -1 & 0 & 0 \\ 0 & -1 & 0 & 0 & 0 & -1 & 0 & -1 & 1 \\ \hline 0 & 0 & 0 & 0 & 1 & 0 & 0 & 0 & 0 \\ 0 & 0 & 0 & 0 & 0 & 1 & 0 & 0 & 0 \\ 1 & 0 & 0 & 0 & -1 & 0 & 0 & 0 & 0 \end{bmatrix}$$

First we have to find an externally equivalent autoregressive representation of the system. To do this we use Proposition 2 with

$$
A_1 = \left[\begin{array}{cccccc}
s+1 & 1 & 0 & 0 & 1 & -1 \\
0 & s & 3 & 0 & 0 & -1 \\
0 & 0 & s & -1 & 1 & 0 \\
0 & 0 & 0 & s & 0 & 0 \\
0 & 0 & -1 & 0 & 0 & 0 \\
0 & -1 & 0 & 0 & 0 & -1 \\
\hline
0 & 0 & 0 & 0 & 1 & 0 \\
0 & 0 & 0 & 0 & 0 & 1 \\
1 & 0 & 0 & 0 & -1 & 0
\end{array}\right]
\quad
B_1 = \left[\begin{array}{ccc}
0 & 0 & 1 \\
0 & 0 & -4 \\
0 & 0 & 0 \\
0 & 0 & 0 \\
-1 & 0 & 0 \\
0 & -1 & 1 \\
\hline
0 & 0 & 0 \\
0 & 0 & 0 \\
0 & 0 & 0
\end{array}\right]
$$

The unimodular matrix $U(s)$ is calculated by using the built-in function for the Hermite form. This function returns the Hermite form of the given matrix as well as the unimodular matrix that transforms the given matrix to the Hermite form and the unimodular $U(s)$ matrix as

$$
\left[\begin{array}{cccccc}
1 & 0 & 0 & 0 & 0 & 0 \\
0 & 1 & 0 & 0 & 0 & 0 \\
0 & 0 & 1 & 0 & 0 & 0 \\
0 & 0 & 0 & 1 & 0 & 0 \\
0 & 0 & 0 & 0 & 1 & 0 \\
0 & 0 & 0 & 0 & 0 & 1 \\
0 & 0 & 0 & 0 & 0 & 0 \\
0 & 0 & 0 & 0 & 0 & 0 \\
0 & 0 & 0 & 0 & 0 & 0
\end{array}\right]
$$

$$
\left[\begin{array}{ccccccccc}
0 & 0 & 0 & 0 & 0 & 0 & 1 & 0 & 1 \\
\dfrac{1}{2} & 0 & 0 & 0 & 0 & -\dfrac{1}{2} & -\dfrac{s}{2}-1 & 0 & -\dfrac{s}{2}-\dfrac{1}{2} \\
-\dfrac{s}{6}-\dfrac{1}{6} & \dfrac{1}{3} & 0 & 0 & 0 & \dfrac{s}{6}-\dfrac{1}{6} & \dfrac{(s+1)(2+s)}{6} & 0 & \dfrac{(s+1)^2}{6} \\
0 & 0 & -1 & 0 & -s & 0 & 1 & 0 & 0 \\
0 & 0 & 0 & 0 & 0 & 0 & 1 & 0 & 0 \\
-\dfrac{1}{2} & 0 & 0 & 0 & 0 & -\dfrac{1}{2} & \dfrac{s}{2}+1 & 0 & \dfrac{s}{2}+\dfrac{1}{2} \\
0 & 0 & s & 1 & s^2 & 0 & -s & 0 & 0 \\
0 & 1 & 0 & 0 & 3 & s & 0 & s+1 & 0 \\
1 & 0 & 0 & 0 & 0 & 1 & -s-2 & 2 & -s-1
\end{array}\right]
$$

Example 4

From the Hermite form of $A_1(s)$ and Proposition 2 it follows that the matrix $\hat{B}_2(s)$ is formed by the last three rows of $U(s)B_1(s)$.

The result is the following matrix, which represents the externally equivalent autoregressive system

$$
\hat{B}_2(s) = \left[
\begin{array}{ccc|ccc}
-s^2 & 0 & 0 & -s & 0 & 0 \\
-3 & -s & -s+4 & 0 & s+1 & 0 \\
0 & -1 & 2 & -s-2 & 2 & -s-1
\end{array}
\right]
$$

The echelon form of $\hat{B}_2(s)$ is computed as

$$
R(s) = \left[
\begin{array}{ccc|ccc}
3 & s & -s+4 & 0 & -s-1 & 0 \\
0 & 1 & -2 & s+2 & -2 & s+1 \\
s^2 & -1 & 2 & -2 & 2 & -s-1
\end{array}
\right]
$$

From the above we have $\mu_1 = 2$, $\mu_2 = 1$, $\mu_3 = 3$. The canonical form $T_c(s)$ of $T(s)$, which is the realisation (by inspection) of $R(s)$, is

$$
T_c(s) = \left[
\begin{array}{ccc|ccc|ccc}
s & & & 0 & -1 & 0 & 3 & 0 & 4 \\
-1 & & & 0 & -1 & 0 & 0 & 1 & -1 \\
0 & & & s+2 & -2 & s+1 & 0 & 1 & -2 \\
 & s & 0 & -2 & 2 & -1 & 0 & -1 & 2 \\
 & -1 & s & 0 & 0 & -1 & 0 & 0 & 0 \\
 & 0 & -1 & 0 & 0 & 0 & 1 & 0 & 0 \\
\hline
0 & 0 & 0 & -1 & 0 & 0 & & & \\
0 & 0 & 0 & 0 & -1 & 0 & & & \\
0 & 0 & 0 & 0 & 0 & -1 & & &
\end{array}
\right]
$$

$$
= \left[
\begin{array}{cc c}
L(s) & sK_c - \Lambda_c & B_c \\
0 & -I & 0
\end{array}
\right]
$$

The source code for the computation of the echelon form and other polynomial matrix functions useful in control are given in Reference 29.

11.6 Symbolic methods for global linearisation of pole assignment maps

One way to tackle the problem of stabilising a MIMO plant by feedback controllers of given degree is to solve the pole placement equations. These are multilinear algebraic equations with polynomial unknowns, and their exact solvability is an algebrogeometric problem. From the point of view of the computations, due to its polynomial algebraic nature, this problem can be more easily addressed with the use of symbolic computation packages.

It is well known that, for a proper system represented by the transfer function $G(s) \in R[s]^{m \times p}$ and a feedback controller represented by the transfer function $C(s) \in R[s]^{p \times m}$, the closed-loop pole polynomial $p(s)$ is given by:

$$p(s) = \det\left([D_1(s), N_1(s)]\begin{bmatrix} D(s) \\ N(s) \end{bmatrix}\right) = \det(K(s)M(s))$$

where $G(s) = N(s)D(s)^{-1}$ and $C(s) = D_1(s)^{-1}N_1(s)$ are coprime MFDs of the transfer functions of the plant and the controller, respectively, and $n = \deg(\det D(s))$, $n_1 = \deg(\det D_1(s))$ denote their corresponding McMillan degrees. The pole placement problem via output feedback may thus be reduced to a general problem, i.e. calculate a feedback (LTI) controller representation $K(s)$ so that the closed-loop system has a given pole polynomial $p(s)$. This can be formulated as a determinantal assignment problem, i.e. to solve the determinantal equation:

$$\det(K(s)M(s)) = cp(s) \tag{11.40}$$

with respect to $K(s) \in R[s]^{p \times (p+m)}$ and $\deg(K(s)) = n_1$, where $M(s)$ is a given polynomial matrix of degree n, $p(s)$ a given monic polynomial of degree less than or equal to $n + n_1$, and c is a constant number. This problem naturally induces the function:

$$\chi: R^{(n_1+p) \times (p+m)} \rightarrow R^{n+n_1+1} \tag{11.41}$$

which maps the coefficient matrix of the controller representation $K(s)$ to the coefficient vector of the closed-loop polynomial $\det(K(s)M(s))$. The problem of solving eqn. 11.40 or the equivalent,

$$\chi(K) = c\underline{p}$$

with respect to $K(s)$, is a nonlinear problem and a necessary condition for the existence of the solution for a generic $M(s)$ is [22]:

$$n_1 \geq \frac{n - mp}{m + p - 1}$$

Although an explicit solution of eqn. 11.40 for the smallest possible controller degree n_1 is not yet known, there exists a way of constructing a one-parameter family of controllers solving the problem and their degree to be very close to the optimum one. This is based on the formal power series solution of the equation

$$f(x, Y) = 0$$

where

$$f(x, Y) = \chi(Y) - x\underline{p}$$

In such a case a power series solution of the form

$$Y(x) = Y_0 + \sum_{i=1}^{\infty} x^i Y_i$$

is sought, given that $f(0, Y_0) = 0$. This solution can be constructed (see [12, 22]) by successively solving with respect to $Y_1, Y_2, \ldots$ the equations

$$f_i(Y_1, Y_2, \ldots, Y_i) = 0 \qquad \text{for} \qquad i = 1, 2, \ldots \qquad (11.42)$$

and the f_is are found from the expansion

$$f(x, Y(x)) = \sum_{i=1}^{\infty} f_i(Y_1, Y_2, \ldots, Y_i) x^i \qquad (11.43)$$

The initial solution Y_0 is the so-called degenerate controller and satisfies

$$\chi(Y_0) = \underline{0}$$

i.e. it is such that the feedback loop is not defined. In fact Y_0 corresponds to the $K_0(s)$ composite representation, for which $\det(K_0(s)M(s)) = 0$.

The great advantage of symbolic computations for this problem is evident. The problem is based on the expansion of the determinantal expression

$$\det((Y_0(s) + x Y_1(s) + x^2 Y_2(s) + \cdots) M(s))$$

first in the powers of the variable x and then in the powers of the variable s. This expansion can be done almost trivially in any symbolic language using standard symbolic commands. There are standards commands , like the 'Det' of Mathematica, that can be used to calculate the determinand of a symbolic square matrix. Then commands extracting coefficients of the resulting expression, with respect to x, can be used. This is an expression in terms of $Y_1(s)$, $M(s)$ which is linear with respect to $Y_1(s)$. This can be used, along with the pole placement equation, to calculate $Y_1(s)$, by solving a set of linear equations. Subsituting $Y_1(s)$ in the determinantal expression and applying the same idea again, taking now the coefficient of x^2, we can calculate $Y_2(s)$, and so on.

All the above can be done without the understanding of the algebra of the problem. The power series solution is produced, using the symbolic software, exactly in the same manner as in the nondeterminantal scalar case, as the computer takes the load of the difficult determinantal symbolic calculations.

Example 5

Consider the system described by a transfer function whose composite MFD description $M(s)$ is given by

$$M(s) = \begin{bmatrix} s^3 & 1 & s^2+1 & s+3 & s+1 \\ 0 & s^2 & s+1 & s & 1 \end{bmatrix}^T$$

and assume that we would like to assign the stable polynomial $p(s) = s^5 + 5s^4 + 10s^3 + \frac{117}{9}s^2 + \frac{51}{9}s + 1$ using static output feedback. Take now the composite matirix

$$Y_0 = \begin{bmatrix} 1 & 0 & 0 & 0 & 0 \\ 0 & 0 & 1 & -1 & -1 \end{bmatrix}$$

which corresponds to a degenerate compensator. The first direction

$$Y_1 = \begin{bmatrix} f_{11} & f_{12} & f_{13} & f_{14} & f_{15} \\ f_{21} & f_{22} & f_{23} & f_{24} & f_{25} \end{bmatrix}$$

of the power series can be calculated so that the coefficient of x of the expansion $\det((Y_0 + xY_1)M(s))$ is equal to $p(s)$. This determinant can be calculated using just one command of the symbolic language Mathematica to be

$$\begin{aligned}
& s^5 x f_{22} + 3xf_{13} + 3xf_{15} + xf_{13}s^2 + 3xf_{14}s + 2xf_{15}s + 3xf_{12}s^2 + 5xf_{13}s \\
& + s^4 xf_{23} + s^3 xf_{23} + s^4 xf_{24} + s^3 xf_{25} + x^2 f_{12}f_{23} + x^2 f_{12}f_{25} + 3x^2 f_{14}f_{23} \\
& + 3x^2 f_{14}f_{25} + s^5 x^2 f_{11}f_{22} + s^4 x^2 f_{11}f_{23} + s^3 x^2 f_{11}f_{23} + s^4 x^2 f_{11}f_{24} \\
& + s^3 x^2 f_{11}f_{25} + x^2 f_{12}f_{23}s + x^2 f_{12}f_{24}s + x^2 f_{13}s^4 f_{22} + x^2 f_{13}s^3 f_{24} \\
& + x^2 f_{13}f_{22}s^2 - 3x^2 f_{13}f_{24}s + x^2 f_{14}s^3 f_{22} + x^2 f_{14}s^2 f_{23} + 3x^2 f_{14}sf_{23} \\
& + 3x^2 f_{14}f_{22}s^2 + x^2 f_{15}s^3 f_{22} + 2x^2 f_{15}sf_{23} + x^2 f_{15}s^2 f_{24} + x^2 f_{15}f_{22}s^2 \\
& - xf_{12}s^4 + 2xf_{12}s^3 - xf_{13}s^3 - x^2 f_{13}f_{22} - 3x^2 f_{13}f_{24} - xf_{14}s^3 + 2xf_{14}s^2 \\
& - x^2 f_{15}f_{22} - xf_{15}s^3 - 3x^2 f_{15}f_{24} - x^2 f_{12}s^5 f_{21} - x^2 f_{12}s^4 f_{23} - x^2 f_{12}s^2 f_{23} \\
& - x^2 f_{12}s^3 f_{24} - 3x^2 f_{12}s^2 f_{24} - x^2 f_{12}s^3 f_{25} - x^2 f_{12}s^2 f_{25} - x^2 f_{13}s^4 f_{21} \\
& - x^2 f_{13}sf_{22} - x^2 f_{13}s^2 f_{24} - 2x^2 f_{13}sf_{25} - x^2 f_{13}f_{21}s^3 - x^2 f_{14}s^4 f_{21} \\
& - x^2 f_{14}sf_{22} - x^2 f_{14}s^3 f_{23} - x^2 f_{14}s^2 f_{25} - x^2 f_{15}f_{21}s^3
\end{aligned}$$

The coefficient of x of the above expression can be found again, using just one command, to be the polynomial

$$\begin{aligned}
& s^5 f_{22} + (f_{23} + f_{24} - f_{12})s^4 + (-f_{13} + f_{23} - f_{14} + 2f_{12} + f_{25})s^3 \\
& + (-f_{15} + 3f_{12} + f_{13} + 2f_{14})s^2 + (3f_{14} + 5f_{13} + 2f_{15})s + 3f_{13} + 3f_{15}
\end{aligned}$$

and the f_{ij} can be found by equating the coefficient of s^i of the above with the corresponding coefficients of $p(s)$ and solving the system of linear equations. The solution taken this way is

$$Y_1 = \frac{1}{3} \begin{bmatrix} 0 & 10 & 0 & 5 & 1 \\ 0 & 3 & 0 & 25 & 15 \end{bmatrix}$$

This idea can be used iteratively to calculate the higher-order terms of the power series, and has been implemented. The solution found here is a 'high gain' solution. It is shown in Reference 22 that by selecting Y_0 appropriately, we can obtain a 'bounded gain' solution.   □

11.7 Conclusions

This chapter has considered the problem of algebraic computations using symbolic tools and it has demonstrated the significance of symbolic packages in deriving the solution of difficult algebraic problems for which alternative means (i.e. numerical procedures) are either nonexistent or difficult to use. Although, in theory, such symbolic procedures are easy to use for small dimension problems, with well-defined models, there are many issues deserving special consideration before such methods can be used in a control design context. In fact, issues of complexity of algorithms and effect of model dimensionality on symbolic computational procedures need proper study. An alternative issue is the parameter uncertainty of the engineering models and the effects of this uncertainty on the symbolic computations results. A study of related issues, that is the approximate algebraic computations, is given in Chapter 6. The classification of computations given here to PNC, NRC and NIC is as far as the fundamental or elementary algebraic computation problems. Complex computational procedures are naturally of hybrid nature, i.e. they may involve both symbolic and numerical procedures; the optimal merging of such procedures is still an open question. The subject of symbolic computations for control problems is a challenging area, with many issues beyond the implementation of standard algebraic computations.

11.8 References

1 ROSENBROCK, H. H.: 'State space and multivariable theory' (Nelson, London, 1970)
2 KALMAN, R. E., FALB, P. I., and ARBIB, M. A.: 'Topics in mathematical systems theory' (McGraw-Hill, 1969)
3 KAILATH, T.,: 'Linear systems' (Prentice Hall, 1980)
4 WONHAM, W. M.: 'Linear multivariable control: a geometric approach' (Springer, New York 1979)
5 WILLEMS, J. C.: 'Almost invariant subspaces: an approach to high gain feedback design–Part I, almost controlled invariant subspaces', *IEEE Trans. Autom. Control*, 1981, **AC-26**, pp. 235–252

6 BASILE, G., and MARRO, G.: 'Controlled and conditioned invariant subspaces in linear system theory' (Prentice Hall, 1969)

7 JAFFE, S., and KARCANIAS, N.: 'Matrix pencil characteristation of almost (A, B)-invariant subspaces: a classification of geometric concepts', *Int. J. Control*, 1981, **33**, pp. 51–93

8 KARCANIAS, N., and KALOGEROPOULOS, G.: 'Geometric theory and feedback invariants of generalized linear systems: a matrix pencil approach', *Circuits Syst. Signal Process.*, 1989, **8**, (3), pp. 375–397

9 KARCANIAS, N.: 'Minimal bases of matrix pencils, algebraic, Toeplitz structure and geometric properties', *Lin. Algebr. Appl.* 1994, **205,206**, pp. 831–868

10 KARCANIAS, N., and GIANNACOPOULOS, C.: 'Grassmann invariants, almost zeros and the determinantal zero, pole assignment problems of linear systems', *Int. J. Control*, 1984, **40**, pp. 673–698

11 KARCANIAS, N., and GIANNACOPOULOS, C.: 'Necessary and sufficient conditions for zero assignment by constant squaring down', *Lin. Algebr. Appl.*, Special issue on control theory, 1989, **122/123/124**, pp. 415–446

12 LEVENTIDES, J., and KARCANIAS, N.: 'Global asymptotic linearisation of the pole placement map: a closed form solution for the constant output feedback problem', *Automatica*, 1995, **31**, (9), pp. 1303–1309

13 REINSCHKE, K. J.: 'Multivariable control: a graph-theoretic approach', *Lect. Notes Control Inf. Sci.*, 1988, **108**

14 SILJAK, D. D.: 'Decentralised control of complex systems' (Academic Press, *Math. Sci. Eng.*, London), vol. 184

15 MOORE, B., and LAUB, A.: 'Computation of supremal (A, B)-invariant and controllability subspaces', *IEEE Trans. Autom. Control*, 1978, **AC-23**, pp. 873–792

16 VAN DOOREN, P.: 'The generalized eigenstructure problem in linear system theory', *IEEE Trans. Autom. Control*, 1981, **AC-26**, pp. 111–129

17 GOLUB, G., and VAN LOAN, P.: 'Matrix computations' (North Oxford Academic, Oxford, 1983)

18 WILKINSON, J.: 'The algebraic eigenvalue problem' (Clarendon Press, Oxford, 1988)

19 MUNRO, N.: 'The development of computer aided control system design tools', *Trans. Inst. Meas. Control*, 1992, **25**, pp. 134–136

20 KARCANIAS, N., and VAFIADIS, D.: 'On the cover problems of geometric theory', *Kybernetika*, 1993, **29**, (6), pp. 547–562

21 VAFIADIS, D., and KARCANIAS, N.: 'Canonical forms for descriptor systems under restricted system equivalence', *Automatica*, 1997, **33**, (5), pp. 955–958

22 LEVENTIDES, J., and KARCANIAS, N.: 'Dynamic pole assignment using global blow up linearisation: low complexity solution', *J. Optim. Theory Appl.*, 1996, **96**, pp. 57–86

23 MITROULI, M., KARCANIAS, N., and GIANNAKOPOULOS, C.: 'The computational framework of the determinantal assignment problem'. Proc. of 1st ECC, Genoble, 2–5 July 1991, Vol. 4, pp. 98–103

24 KARCANIAS, N., and MITROULI, M., 'A matrix pencil based numerical method for the computation of GCD of polynomials', *IEEE Trans. Autom. Control*, 1994, **AC-39**, 977–981

25 MITROULI, M., and KARCANIAS, N.: 'Computation of the GCD of polynomials using Gaussian transformations and shifting', *Int. J. Control*, 1993, **58**, pp. 211–228

26 COX, D., LITTLE, J., and O'SHEA, D.: 'Ideals varieties and algorithms' (Springer-Verlag, New York, 1992)

27 VAN DOOREN, P.: 'The computation of Kronecker's canonical form of a singular pencil', *Lin. Algreb. Appl.*, 1979, **27**, pp. 103–140

28 GANTMACHER, F. R.: 'The theory of matrices', vol. I, II (Chelsea, New York, 1959)

29 VAFIADIS, D.: 'Algebraic and geometric methods and problems for implicit linear systems'. PhD thesis, City University, London, 1995

30 VIDYASAGAR, M.: 'Control system synthesis: a factorization approach' (MIT Press, Cambridge, 1985)

31 VARDULAKIS, A. I. G.: 'Linear multivariable control: algebraic analysis and synthesis methods' (John Wiley & Sons, Chichester, 1991)
32 BRUALDI, R. A., and RYSER, H. J.: 'Combinatorial matrix theory', *Encycl. Math. Applic.*, 1991, **39**
33 MUROTA, K.: 'Systems analysis by graphs and matroids', in 'Algorithms and combinatorics', Vol. 3 (Springer-Verlag, Berlin)
34 AMIRAT, Y. A., and DIOP, S.: 'Towards an algebraic system theory toolbox. Proc. 13th Trien. World Congress of IFAC, San Francisco, 2b-24 4, pp. 299–304
35 PLANTIN, J., GUNNARSON, J., and GERMUNDSSON, R.: 'Symbolic algebraic discrete systems theory applied to a fighter aircraft'. Proc 34th CDC, New Orleans, TA 10, pp. 1863–1864
36 BUCHBERGER, B.: 'Gröbner basis: an algorithmic method in polynomial ideal theory', in BOSE, N. K. (Ed.), 'Multidimensional systems theory' (Reidel, Dordrecht, 1985), pp. 184–232
37 KARCANIAS, N., and KOUVARITAKIS, B.: 'The output zeroing problem and its relationship to the invariant zero structure: A matrix pencil approach', *Int. J. Control*, 1979, **30**, pp. 395–415
38 WONHAM, W. M., and MORSE, A. S.: 'Feedback invariants for linear multivariable systems', *Automatica*, 1972, **8**, pp. 93–100
39 MORSE, A. S.: 'Structure and design of linear model following systems', *IEEE Trans. Autom. Control*, 1973, **AC-18**, (4), pp. 346–354
40 MORSE, A. S.: 'Minimal solutions to transfer matrix equations', *IEEE Trans. Autom. Control*, 1976, **AC-21**, pp. 131–133
41 EMRE, E., and HAUTUS, M. L. J.: 'A polynomial characterization of (A, B)-invariant and reachability subspaces', *SIAM J. Control*, 1980, **18**, pp. 420–436
42 EMRE, E., SILVERMAN, L. M., and GLOVER, K.: 'Generalised dynamic covers for linear systems with applications to deterministic identification problems', *IEEE Trans. Autom. Control*, 1977, **AC-22**, pp. 26–35
43 ANTOULAS, A. C.: 'New results on the algebraic theory of linear systems: the solution of cover problems', *Lin. Algebr. Appl.*, 1983, **50**, pp. 1–43
44 POPOV, V. M.: 'Some properties of the control systems with irreducible matrix transfer functions', *Lec. Notes Math.*, 1969, **144**, pp. 250–261
45 FORNEY, G. D.: 'Minimal bases of rational vector spaces, with applications to multivariable linear systems', *SIAM J. Control*, 1975, **13**, pp. 493–520
46 KUNG, S., KAILATH, T., and MORF, M.: 'Fast and stable algorithms for minimal design problems', Proc. 4th IFAC Internat. Symp. on *Multivariable Technological Systems*, ATHERTON, D. P. (Ed.) (Pergamon Press, 1977), pp. 97–104
47 KUIJPER, M.: 'Descriptor representations without direct feedthrough term', *Automatica*, 1992, **28**, (3), pp. 633–639
48 WILLEMS, J. C.: 'Input–output and state space representations of finite dimensional linear time invariant systems', *Lin. Alg. Appl.*, 1983, **22**, pp. 561–580

## Part IV

Nonlinear systems

Chapter 12

Symbolic aids for modelling, analysis and synthesis of nonlinear control systems

Bram de Jager

12.1 Introduction

This Chapter describes the role of symbolic computation in several phases of the control system design cycle, namely modelling, analysis and synthesis. It provides some mathematical background of nonlinear system theory and presents a guided tour through those phases using typical examples, both contrived and real life. To name them, the symbolic computation of (i) the zero dynamics, used for different purposes, (ii) the input-output exact linearising state feedback, and (iii) the state-space exact linearising state feedback, all for nonlinear dynamic systems, are discussed. The examples come from a diverse range of applications and provide a picturesque landscape of this area of research. The possibilities and limitations of the so-called NON$\frac{z}{5}$CON package, based on the symbolic computation program MAPLE, are outlined with these examples. This package was developed to aid with analytical computations in the area of nonlinear control systems.

Owing to increased consumer expectations and to the accumulated knowledge and experience of producers, there is a tendency to demand for and provide higher quality goods and better services. Control systems play an important role in some of these goods and services: for example, in engine control, drive line control, (semi) active suspension control, hydraulic steering, and automatic brake systems for vehicles, laser beam focusing and radial servos for compact disc players, controllers that stabilise inherently unstable air planes (fly-by-wire) and automated highway systems for the benefit of transport services; and in the systems that produce these goods or components—for example, advanced tracking controllers for robotic manipulators in automated production lines, and advanced process control in chemical plants like refineries.

To increase the performance of control systems, to ensure the required product quality and speed of production, and to shrink the control system design cycle, for a reduced time-to-market, use is made of computer-aided design tools for all phases of the control design cycle, specifically in modelling, analysis, and synthesis.

For linear systems an abundant number of theories, algorithms and design tools is available. The standard tools employ programs like MATLAB. However, to produce high quality goods with low expenditure, these tools are not always sufficient, because they cannot cope well with nonlinear control systems. They are not adequate because they can only manipulate numbers, *i.e.*, data entry, number crunching, and data visualisation are their main forte. Therefore, symbolic computation is an alternative or complement for these programs because it can manipulate symbols, *i.e.*, manipulate mathematical expressions, and so may replace tedious manual paper and pencil work.

To effectively use tools based on symbolic computation they must be:

1   cost-effective, so they can be used widely
2   easy to use, without a steep learning curve, to get up to speed with the tool in a short time
3   powerful and able to solve most problems, preferably without much user programming
4   flexible, adaptable and expandable, to fit users' needs closely.

When a tool does not fulfil these requirements, it is not of much value for the practising engineer and will not find widespread acceptance. The question is: Do current symbolic computation programs and the tools based thereon satisfy these requirements? In trying to answer this question, several algorithms were implemented in the symbolic computation program MAPLE and evaluated in the three phases of the control design cycle mentioned before.

The use of symbolic computation programs for control purposes has been investigated by several researchers. Some control problems are handled by REDUCE, see [1, 2]. REDUCE has also been used for modelling of systems and model-based control, see [3].

The use of MACSYMA is discussed by Barker *et al.* [4] for a specific class of control problems. Zeitz *et al.* [5, 6] employ the program MACNON that is based on MACSYMA. MACNON is used to analyse observability and reachability, and to design observers and controllers for nonlinear systems. They discuss ten observer design methods ranging from a working point observer to an extended Kalman-filter observer. These methods are implemented in MACNON, with some controller design methods. Results' for larger scale problems are published in [7, 8]. For linear systems some work using MACSYMA is reported in [9]. Blankenship [10] also used MACSYMA to design output tracking controllers for nonlinear systems via feedback linearisation and (left) invertibility with his implementation CONDENS, but has switched to Mathematica [11]. A control toolbox for this platform is provided [12]. Some Mathematica notebooks, e.g., COSY_PAK, are developed to mimic MATLAB toolboxes, primarily with the aim of obtaining a more powerful visualisation and a possible integration of symbolic capabilities, although those capabilities are not fully exploited yet. A symbolic toolbox for MATLAB, using the OEM kernel of MAPLE, is commercially available and employed in [13].

To study stability properties of a restricted class of single-input single-output nonlinear systems MAPLE was used in [14]. The use of MAPLE for several control problems is reported in [15], and in that paper some problems were reported, with solving partial differential equations, for example, that are partly resolved in the research reported in [16]. They describe a MAPLE package called NON$^Z_\xi$CON (a successor of the ZERODYN package presented and used in [15, 17]), that will compute the zero dynamics and provide solutions to the exact linearisation problem, for example. In Reference 17, attention is focused on the computation of the zero dynamics, while in Reference 18 the focus is on exact linearisation. In the meantime this package has been extended (see Reference 19).

Here, some key components of, and algorithms implemented in, the package are highlighted. Its use is illustrated by some textbook and practice oriented problems that were solved with it. As in [17, 18, 20], attention is focused on the computation of the zero dynamics and on the solution of the input-output exact linearisation and state-space exact linearisation problems. The examples display the advantages and limitations of the package.

The main contribution of this work is an overview and assessment of the suitability of symbolic computation programs for the analysis and design of control systems. Other goals are to supply feedback to the developers of these programs, to solve some difficult nonlinear control problems, and to document some applications of NON$^Z_\xi$CON. The chapter also assesses explicitly the areas in the implementation of the package and the underlying symbolic computation program that are problematic and limit its usefulness. The results presented here are available in scattered form in the literature already [18–21]. The conclusions made there are still valid.

The Chapter is structured as follows. First, Section 2 presents and makes remarks on the specific setup that is used, and provides mathematical details of basic procedures and algorithms. Section 3 follows with a discussion of the implementation in the symbolic computation program MAPLE, in the form of the NON$^Z_\xi$CON package. It also investigates which of the mathematical tools needed can be implemented easily and which cannot. Sections 4, 5 and 6 give applications in the areas of modelling, analysis, and synthesis, respectively, using the NON$^Z_\xi$CON package. Finally, Section 7 presents conclusions and discusses possibilities for future research.

12.2 Theory and tools

Several areas in control may profit from symbolic computation. Here, only nonlinear control is treated. The main argument is that for linear control numerical tools are available in sufficient number. This is not quite true, because some problems in linear control cannot be solved readily with a numerical approach. For instance, the observability check by means of the rank of the observability matrix is numerically tricky, often giving a wrong answer for systems with badly damped or wide-ranging eigenvalues, due to the numerically bad conditioning of the observability matrix. A symbolic computation is not hampered by this problem. Also, for studies of the influence of system parameters it is sometimes more convenient to obtain an expression depending explicitly on these parameters, than it is to get a numerical result in the form of a table or graph. The last example is the area of factorisation methods in optimal control. Therefore, symbolic computations are also helpful in linear control.

In the following the computation of the zero dynamics, the input-output exact linearisation and the state-space exact linearisation problem for nonlinear systems are discussed. To be able to pose and solve these problems, a system definition and some transformations of the system are needed.

12.2.1   Problem setup

In our setup, the system itself is not allowed to be changed—the only possible modifications are state transformations and the judicious manipulation of control signals, i.e. signals that act on the system and can be influenced from the outside.

Examples of control signals are valve settings, that can influence flow rates or heat inputs, and electrical currents, that can influence the torque exerted by motors. Most of these control inputs are generated by control devices that are manipulated by sending standardised electrical signals (current or voltage).

The generation of control signals will be done by a control law, where information of the system is used to generate the control input. To solve the problems with an appropriate nonlinear state feedback based control law the following model for the system to be controlled (the *plant*):

$$\dot{x} = f(x) + g(x)u \qquad y = h(x) \tag{12.1}$$

is used. The *state* vector $x \in \mathbb{R}^n$, containing all necessary information of the plant, the *input* vector $u \in \mathbb{R}^m$, and the *output* vector $y \in \mathbb{R}^m$ are introduced. The number of inputs is equal to the number of outputs, i.e. the plant is square. This squareness assumption mostly is for convenience only and makes a simplified presentation possible. The vector field f is a smooth one, g has m columns g_i of smooth vector fields, and h is a column of m scalar-valued smooth functions h_i. The implicit assumptions that a model can be used that is affine in the input u, i.e. u enters linearly in eqn. 12.1, and with no direct feed-through from u to y, i.e. $h(x)$ is not an explicit function of u, can often be circumvented by an appropriate redefinition of the state x, the input u or the output y, i.e. by the introduction of additional states and using derivatives of the original signals. Therefore it is not very restrictive.

Not all systems can be described with differential equations of the type of eqn. 12.1, e.g. sometimes it is convenient to include derivatives $u^{(k)}$ of the input u in the model equations. Then, a more general model is needed. For an illustration on how this can be done, see [22], where a general controller canonical form is introduced,

The possible modifications are *state feedback*, i.e. a feedback based on the explicit knowledge of the value of the state vector $x(t)$ of the plant, and a (local) *change of co-ordinates* in the state-space. Although for linear systems a linear change of co-ordinates $z = Tx$ in the state space $\mathbb{R}^n$, with T a nonsingular matrix, is usually adequate, for nonlinear systems it is more appropriate to allow for a nonlinear change of co-ordinates

$$z = \Phi(x) \tag{12.2}$$

It is required that the change of co-ordinates is a diffeomorphism, so there is an inverse mapping $\Phi(x)^{-1}$. It is sufficient that the Jacobian $\partial\Phi/\partial x$ of the transformation vector Φ is locally invertible for Φ to qualify as a local change of co-ordinates.

The type of control law used is restricted to static or dynamic state feedback. In a *static* state feedback the input vector u at time t depends on the state $x(t)$ and a new *reference* input vector $v(t)$. This dependency is assumed to be of the form

$$u = \alpha(x) + \beta(x)v \quad v \in \mathbb{R}^m \tag{12.3}$$

because it preserves the structure of eqn. 12.1. Here α and β contain smooth functions. In a *dynamic* state feedback the value of the input vector u at time t depends on the state $x(t)$, a new *reference* input vector $v(t)$, and an *auxiliary state* vector $\zeta(t) \in \mathbb{R}^k$. This dependence is of the form

$$u = \alpha(x, \zeta) + \beta(x, \zeta)v$$
$$\dot{\zeta} = \gamma(x, \zeta) + \delta(x, \zeta)v \tag{12.4}$$

because, again, it preserves the structure of eqn. 12.1.

See Figure 12.1 for an overview of the setup of a static state feedback loop.

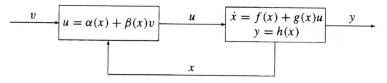

Figure 12.1   Control structure for static state feedback

12.2.2   Basic notions

In the following some basic notions in nonlinear system theory are required. The relative degree and the normal form are introduced. More advanced algorithms, like the structure algorithm and the zero dynamics algorithm, are mentioned in passing, while the conditions needed to solve the state-space exact linearisation problem are given a more thorough treatment. Other algorithms, like the controlled invariant distribution algorithm are not covered here.

To introduce the relative degree, first the *Lie derivative* $L_f\lambda$ for a scalar function $\lambda(x)$ along a vector field f is defined as $L_f\lambda(x) = (\partial\lambda/\partial x)f(x)$. Now the *relative degree vector* $\{r_1, \ldots, r_m\}$ is defined at a point x° if

(i)   the scalar

$$L_{g_j}L_f^k h_i(x) = 0$$

for all $1 \le j \le m$, for all $1 \le i \le m$, for all $k < r_i - 1$, and for all x in a neighbourhood of x°

(ii)   the matrix

$$\begin{bmatrix} L_{g_1}L_f^{r_1-1}h_1(x) & \cdots & L_{g_m}L_f^{r_1-1}h_1(x) \\ \vdots & \ddots & \vdots \\ L_{g_1}L_f^{r_m-1}h_m(x) & \cdots & L_{g_m}L_f^{r_m-1}h_m(x) \end{bmatrix}$$

is nonsingular at x°.

Remark 1   It is possible that there does not exist a set of integers $r = \{r_1, \ldots, r_m\}$ fulfilling the two conditions given above. In that case the system does not have a well-defined relative degree r, otherwise the relative degree r is well defined.

Using the outputs of the system, their derivatives and the values of the relative degree r, it is possible to introduce a co-ordinate transformation

$$z = \begin{bmatrix} \xi \\ \eta \end{bmatrix} = \Phi(x)$$

which brings the system in the following *normal form*:

$$y_i = \xi_1^i$$
$$\dot{\xi}_1^i = \xi_2^i$$
$$\vdots$$
$$\dot{\xi}_{r_1-1}^i = \xi_{r_i}^i$$
$$\dot{\xi}_{r_i}^i = b_i(\xi, \eta) + \sum_{j=1}^m a_{ij}(\xi, \eta)u_j$$

for $1 \leq i \leq m$ and with

$$\xi = \{\xi_1^1, \ldots, \xi_{r_1}^1, \xi_1^2, \ldots, \xi_{r_2}^2, \ldots, \xi_1^m, \ldots, \xi_{r_m}^m\}$$

while for the remaining states, the *internal dynamics* holds:

$$\dot{\eta} = q(\xi, \eta) + p(\xi, \eta)u$$

If the distribution spanned by $g_1(x), \ldots, g_m(x)$ is involutive near x° (see later), a change of co-ordinates is possible such that $p(\xi, \eta) = 0$ and, hence, the internal dynamics has an especially simple form,

$$\dot{\eta} = q(\xi, \eta)$$

The zero dynamics for nonlinear systems has been introduced as a generalisation of the concept of zeros of the transfer function for linear systems, although there are more interpretations of the zero dynamics of a nonlinear system.

The zero dynamics of a nonlinear system can be characterised as the dynamics of the system, i.e. its time evolution or behaviour, when it is constrained so that the output of the system is required to be 0 for all times. In other words, it is the dynamics of the system on the largest unobservable submanifold that can be obtained by judiciously manipulating the input to the system by a control law, and by choosing appropriate initial conditions. A formal definition of the problem is as follows (see also [23, 24]).

The *zero dynamics* problem: what is the dynamics of the system when the output y is required to be 0 for all t, achieved by a proper choice of initial state $x(0)$ and control input $u(t)$?

The zero dynamics can be computed in at least two ways: first, if the relative degree of the system is well defined, by use of the normal form; second, independent of the relative degree, but restricted by some regularity conditions, by application of the *zero dynamics algorithm* [23, Section 6.1].

Our aim is to implement these solutions and, especially, to compute the zero dynamics by using the zero dynamics algorithm. This algorithm proceeds as follows:

Step 0   Set $M_0 = \{x \in \mathbb{R}^n : h(x) = 0\}$.
Step k   If for some neighbourhood U_{k-1} of x° the submanifold $M_{k-1} \cap U_{k-1}$ is smooth, then

$$M_k = \{x \in M_{k-1}^c : f(x) \in \text{span}\{g_1(x), \ldots, g_m(x)\} + T_x M_{k-1}^c\}$$

with M_{k-1}^c the connected component of $M_{k-1} \cap U_{k-1}$ which contains x° and $T_x M_{k-1}^c$ the tangent space to M_{k-1}^c at x.

The zero dynamics algorithm is an iterative procedure, that, under certain regularity conditions, converges in a finite number of steps to a locally output zeroing submanifold. So, for a certain k^* it holds that $Z^* = M_{k^*}^c$, with Z^* the locally maximal output zeroing submanifold. A manifold M is called an output zeroing submanifold if it contains x°, if for each $x \in M$, $h(x) = 0$, and if M is locally controlled invariant at $x°$, that is, there is an input u and a neighbourhood $U°$ of $x°$ so that M is locally invariant under the vector field $f(x) + g(x)u$. Characteristic for an output zeroing submanifold M is that—for a suitable choice of feedback $u(x)$—the trajectories of the closed-loop system that start in M stay in M for a certain time, and the corresponding output is zero.

Remark 2   For linear systems the regularity conditions guarantee invertibility of the transfer function matrix of the system.

The co-ordinate-free algorithm presented above is not directly suitable for implementation in a symbolic computation program, because that lacks the high level operations used to describe the algorithm. It is possible to give a co-ordinate bound description of the algorithm [23]. An analysis of such a description of the zero dynamics algorithm shows that a symbolic computation program should be able to compute the Lie derivative (in the computation of the tangent space), the Jacobian, the rank, the Gauss Jordan decomposition, and the determinant of a matrix. The main problems are the computation of solutions for sets of nonlinear equations and the computation of the kernel of a symbolic matrix.

The structure algorithm is treated in [23]. Development of this algorithm was needed to solve the input-output exact linearisation problem for systems without a well-defined relative degree. For systems with a well-defined relative degree a simpler solution by means of the normal form is possible.

An analysis of the structure algorithm shows that a symbolic computation program should be able to compute the Lie derivative, perform matrix-vector and matrix-matrix multiplication, check if some expression is equal to 0, perform rank tests, compute the kernel of a mapping, etc. The computations are much like those needed in an implementation of the zero dynamics algorithm. Some computations are relatively straightforward. Here, also, the main problem is to find solutions for sets of nonlinear equations.

The *input-output exact linearisation* problem is the following: under which conditions is it possible to transform the system of eqns. 12.1 into a linear one by the state feedback of eqn. 12.3? The linearity property should be established between the new input v and the output y.

The problem is treated in [23], and its solution is given with the help of the structure algorithm. For all systems that have a well-defined relative degree the input-output exact linearisation problem can be solved, only the normal form is needed to do that and it suffices to give an explicit expression of the state feedback needed. For systems without a well-defined relative degree a solution is possible with the structure algorithm. For all systems for which the structure algorithm converges in a finite number of steps the problem has a solution.

Now it is time to state our final problem, which is the *state-space exact linearisation* problem: Under which conditions is it possible to transform the system of eqn. 12.1 to a linear one by the state feedback of eqn. 12.3 and a change of co-ordinates, eqn. 12.2? The linearity property should be established between the new input v and the transformed state z. This problem has been solved (see, for example, Reference 23), and our goal is to test these conditions and to derive explicit expressions for the feedback and the change of co-ordinates for specific plants.

First, the conditions for solving the exact linearisation problem are given, and then some remarks are made on how the state feedback and the change of co-ordinates can be computed. To state the solution more easily, so-called distributions are introduced. To do this, first the definition of the adjoint ad, using the Lie product $[f, g_i] = L_f g_i - L_{g_i} f$, is given. Recall that the Lie derivative is defined as $L_f \lambda = (\partial \lambda / \partial x) f$. In terms of the Lie product, ad is defined recursively as $\mathrm{ad}_f^k g_i = [f, \mathrm{ad}_f^{k-1} g_i]$ with $\mathrm{ad}_f^0 g_i = g_i$. Next, introduce the distributions

$$G_0 = \mathrm{span}\{g_1, \ldots, g_m\}$$

$$\vdots$$

$$G_i = \mathrm{span}\{\mathrm{ad}_f^k g_j : 0 \le k \le i, 1 \le j \le m\}$$

for $i = 0, \ldots, n - 1$.

The conditions for a solution of the state-space exact linearisation problem to exist can be stated as follows [23, Theorem 5.2.3].

Theorem 1   *Suppose a system*

$$\dot{x} = f(x) + g(x)u; \qquad x \in \mathbb{R}^n; \qquad u \in \mathbb{R}^m$$

with rank $g(x^\circ) = m$ *is given. There exists a solution for the state-space exact linearisation problem if and only if*

(i) G_i *has constant dimension near x° for each $0 \le i \le n - 1$*
(ii) G_{n-1} *has dimension n*
(iii) G_i *is involutive for each $0 \le i \le n - 2$.*

Here, involutive means that the distribution is closed under the Lie or bracket product, i.e. the Lie product of each two vector fields in G_i is contained in G_i.

Remark 3   For $m = 1$ the situation is easier, because condition (ii) implies (i), and then condition (iii) for $i = n - 2$ implies (iii) for $0 \le i \le n - 3$.

Remark 4   For linear systems the conditions are equivalent with conditions for the controllability of the system.

When the conditions for the exact linearisation problem are fulfilled, the state feedback and change of co-ordinates that realise the linearisation are still to be determined. It can be shown that, if the problem is solvable, there exist solutions $\lambda_i(x)$, $i = 1, \ldots, m$, for the following partial differential equations:

$$L_{g_j} L_f^k \lambda_i(x) = 0, \quad \text{for} \quad 0 \le k \le r_i - 2 \quad \text{and} \quad 0 \le j \le m \qquad (12.5)$$

It holds that $\sum_{i=1}^{m} r_i = n$, where the set of integers $\{r_1, \ldots, r_m\}$ is again the relative degree vector. The m functions λ_i can be computed, based on a constructive proof of Theorem 1. This is based on a result already obtained by Frobenius.

Using the functions λ_i, the change of co-ordinates and state feedback that solve the state-space exact linearisation problem are given by

$$z = \Phi(x) = \begin{bmatrix} \lambda_1(x) \\ \vdots \\ L_f^{r_1-1}\lambda_1(x) \\ \vdots \\ \lambda_m(x) \\ \vdots \\ L_f^{r_m-1}\lambda_m(x) \end{bmatrix}$$

$$\alpha(x) = -A^{-1}(x)b(x)$$

$$\beta(x) = A^{-1}(x)$$

with the $m \times m$ nonsingular matrix A and the $m \times 1$ column b given by

$$A(x) = \begin{bmatrix} L_{g_1}L_f^{r_1-1}\lambda_1(x) & \cdots & L_{g_m}L_f^{r_1-1}\lambda_1(x) \\ \vdots & \ddots & \vdots \\ L_{g_1}L_f^{r_m-1}\lambda_m(x) & \cdots & L_{g_m}L_f^{r_m-1}\lambda_m(x) \end{bmatrix}$$

$$b(x) = \begin{bmatrix} L_f^{r_1}\lambda_1(x) \\ \vdots \\ L_f^{r_m}\lambda_m(x) \end{bmatrix}$$

Using the state feedback eqn. 12.3, the transformed state $z = \Phi(x)$ depends linearly on the new input v.

An analysis of these formulas shows that a symbolic computation program should be able to compute the Lie derivative and Lie product, perform matrix-vector multiplication, compute a matrix inverse, etc. These computations are relatively easy. The main problem is in the computation of the functions λ_i, where partial differential equations have to be integrated. This is discussed in the following Section.

Remark 5 A completely different algorithm to solve the exact linearisation problem, totally avoiding the integration step, does also exist (see [25]). This algorithm is not implemented in NonᶻCon. It is unclear whether or not this algorithm provides any computational advantages.

12.3 Computer algebra implementation

The algorithms described in the preceding Section, to compute the relative degree, the normal form, the zero dynamics and the structure algorithm, together with the solution for the input-output exact linearisation problem, for systems with and without a well-defined relative degree, and solutions for the state-space exact linearisation problem, for systems with a well-defined relative degree only, are included in NonᶻCon. Algorithms for other problems, like the dynamic extension algorithm and the controlled invariant distribution algorithm, are also included in this package.

The NonᶻCon package is documented in detail in [26]. There is no need for repetition here. The focus is on some specific problems that appeared during the implementation of the algorithms indicated in Section 2. These problems present the major bottlenecks for the symbolic solution of the design problems discussed here.

One of the main problems was to compute solutions for sets of nonlinear equations. In MAPLE a facility is available to solve these sets, namely the `solve` function, but this facility is not always sufficient. No attempt has been made to improve this, but a more powerful `solve` function in future releases of MAPLE would be advantageous. Even for problems where a solution is known to exist, it is not always practical to compute, because the complexity of some underlying algorithms—for example, to compute solutions for sets of nonlinear equations—is double exponential in a measure of the problem size, see Reference 27.

Another problem was to compute the functions λ_i, and solving sets of partial differential equations is an intermediate step in this computation. In MAPLE, and also in other general purpose symbolic computation programs, no extensive facilities are available to solve partial differential equations. Therefore the following route was chosen.

The partial differential equations that need to be solved are from the 'completely integrable' type; in other words, based on Frobenius' theorem, it is known that a solution for the partial differential equations exists. To compute the solutions, Frobenius' theorem itself is of no help. This problem was solved by computing the solutions with an algorithm based on a constructive proof of Frobenius' theorem. The procedure is as follows.

The solution of a partial differential equation can be constructed from the solutions of related sets of ordinary differential equations. Because MAPLE provides a facility for this type of problem, the `dsolve` command, the problem seems solved. For more details of this procedure see Reference 28. However, the `dsolve` command is not very powerful, and is often unable to present a solution, although this solution is known to exist. Therefore the `dsolve` procedure was extended in an ad hoc way, so a larger class of problems could be handled. Nevertheless, the computation of the functions λ_i is often unsuccessful, especially for more complicated systems, so NON$^{\text{Z}}_{\equiv}$CON is then unable to finish the computations. This part of NON$^{\text{Z}}_{\equiv}$CON should therefore be considered as experimental. It seems unlikely that another symbolic computation program will improve this situation. A possible solution could be to use the alternate algorithm of Gardner and Shadwick [25].

For some other minor problems, that were encountered during the implementation of the algorithms, workarounds are applied. Some of these workarounds are discussed. The problem was that some of the standard MAPLE functions in the `linalg` package are only suitable for a restricted class of functions, namely rational polynomials. This was too limited for our purposes. Therefore, the `rank` and, implicitly, the `gausselim`, and `gaussjord` procedures were extended, so a larger class of problems could be handled. The modification consisted simply in removing the check for the type of entries of the matrix for which the rank, Gauss elimination or Gauss Jordan form should be computed, and adding an additional call to the `simplify` function to ensure that the detection of zero expressions was guaranteed for a larger class of problems. The new functions `extrank`, `extgausselim`, and `extgaussjord` are therefore extended in a rather ad hoc way, and this part of NON$^{\text{Z}}_{\equiv}$CON should therefore be considered as experimental, because the results of the modified procedures are only valid in a generic sense.

The facilities of NON$^{\text{Z}}_{\equiv}$CON used in the examples are summarised in Table 12.1.

12.4 Modelling

In the standard books on nonlinear control [23, 24, 29], the zero dynamics mainly play a role in the characterisation of the possibilities of input-output

Table 12.1   Overview of facilities of NON$^{\text{Z}}_{\equiv}$CON and related functions

	Functions	
Facility	Well defined relative degree	No well defined relative degree
Relative degree	`reldeg`	—
(Extended) normal form	`normform`	`extnormform`
Zero dynamics	`normform`	`extnormform`
Input-output linearisation	`normform`	`inoutlin`
State space linearisation	`outputfunc`	—
	`statelin`	
	`transform`	

linearisation and in other applications in control. There are more applications where the zero dynamics can be employed, not only in analysis and synthesis, but also in modelling. Two applications of this kind will be treated: systematic modelling of large scale systems and characterising nonlinear models.

12.4.1   Systematic modelling

During systematic modelling of large scale systems, due to the restrictions introduced by the requirement of general applicability and modularity, one always ends up with a model of differential and algebraic equations (DAE). This type of equation has two disadvantages: first, they are difficult to simulate when their index is >1—the higher the index the more difficult the problem will be; secondly, the number of analysis and design tools for these systems is limited compared to those available for systems in state-space form, i.e. given by a set of explicit differential equations for the system states. One of the aims of the application of the zero dynamics to DAE is to reduce the index (but not necessarily to totally remove the algebraic equations) to make the problem simpler. Index reduction is possible by solving algebraic equations, whose solutions have to be substituted in the differential equations.

One way to do that is by computing the zero dynamics of a suitably defined system. The reasoning is as follows. The zero dynamics represents the dynamics of a system that is compatible with the requirement that outputs of the system are kept at zero, by a suitable control input and initial conditions. For this application area, the output is formed by equating it to one or more of the algebraic equations, written in such a form that they are satisfied when a 'slack' variable is zero; the output is then defined as this variable. Computing the zero dynamics then yields the dynamics of the differential equations that is compatible with the algebraic constraints. This approach can be taken for DAE in semi-explicit form, where the algebraic relations are separated out.

Two examples are presented. The zero dynamics is computed symbolically. The first example uses the model of a drive line, where an algebraic restriction is introduced when a clutch, that was formerly open, is closed. In general this will reduce the number of degrees of freedom of the system by one, so the structure of the dynamics is changed. Given the original model of the system, the zero dynamics for a suitable defined output then represents the model for the system with a closed clutch.

The second example is the model of a double pendulum. This example falls in the class of multibody systems with position constraints, which have DAE models of index three. Writing the equations of motion for both parts of the pendulum separately (as is done in a systematic and modular modelling approach), the fact that endpoints of the pendulums are attached to each other is equivalent to algebraic relations that equate the position, and therefore also the speed, of the connecting ends of the pendulums. The resulting DAE is of higher index, and so is difficult to simulate. The zero dynamics of this system, using the derivative of the position constraints to define an output, is computed. It has a reduced index, and so is easier to simulate than the original DAE.

Example 1

A simple drive line model is given by two rotating masses with a clutch in between (see Figure 12.2). When the clutch is open the system has two degrees of freedom. If it is closed, one degree of freedom (DOF) is removed, because both inertias then have the same rotational speed. This effect can be accounted for by computing the zero dynamics of the two DOF system and requiring that an output, in this case the speed difference between the two inertias, is to be kept zero. The zero dynamics in this case represents only the one DOF of the combined inertias. The input u required to keep the output, i.e., the difference in velocity, equal to zero is precisely the torque transmitted through the clutch. A more detailed treatment of this example is given in [21].

In this example the zero dynamics could easily be computed by hand, because the model is linear. However, if the system is higher dimensional or completely parametrised, or if effects like nonlinear friction or springs are introduced in the model, solving the problem by paper and pencil becomes harder. A package for symbolic computation can be an asset in such a situation.

Example 2

The double pendulum example starts by proposing a general model of two links with concentrated unit mass at the far end of the links, moving freely in a two-dimensional space, and then requiring that the near end of a link is attached to a fixed mounting, while its far end is attached to the near end of the other link at a fixed distance of the mounting. An illustration of such a set-up is given in Figure 12.3.

The two joints allow free rotation but no relative translations, so both the Cartesian co-ordinates and the velocities of the parts of the links that connect at the joints should be equal, and the links are of fixed length. This gives a set of algebraic conditions. Stacking the Cartesian e_1-co-ordinates of the joint between the two links and of the endpoint of the second link and their derivatives first in x, (i.e. x_i, $i = 1, \ldots, 4$), and the e_2-co-ordinates and their derivatives last, (i.e. x_i,

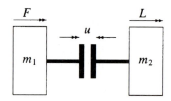

Figure 12.2   Two rotating masses with clutch

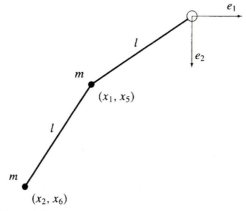

Figure 12.3   Double pendulum

$i = 5, \ldots, 8$), and taking the origin in the base joint and neglecting gravity, the differential equations for the model are

$$\dot{x} = \begin{bmatrix} x_3 \\ x_4 \\ 0 \\ 0 \\ x_7 \\ x_8 \\ 0 \\ 0 \end{bmatrix} + \begin{bmatrix} 0 \\ 0 \\ 2x_1 \\ 0 \\ 0 \\ 0 \\ 2x_5 \\ 0 \end{bmatrix} \lambda_1 + \begin{bmatrix} 0 \\ 0 \\ 2(x_1 - x_2) \\ 2(x_2 - x_1) \\ 0 \\ 0 \\ 2(x_5 - x_6) \\ 2(x_6 - x_5) \end{bmatrix} \lambda_2$$

$$= f(x) + g_1 u_1 + g_2 u_2$$

while the position constraints are

$$l^2 = x_1^2 + x_5^2$$
$$l^2 = (x_2 - x_1)^2 + (x_6 - x_5)^2$$

The derivatives of the constraints, i.e. constraints on the velocity, are

$$0 = x_1 \dot{x}_1 + x_5 \dot{x}_5 = x_1 x_3 + x_5 x_7$$
$$0 = (x_2 - x_1)(\dot{x}_2 - \dot{x}_1) + (x_6 - x_5)(\dot{x}_6 - \dot{x}_5)$$
$$= (x_2 - x_1)(x_4 - x_3) + (x_6 - x_5)(x_8 - x_7)$$

Here, λ_i are Lagrange multipliers, related to the joint forces, and l is the length of the links. Eliminating the velocity constraints is possible by keeping

$$y = \begin{bmatrix} x_1 x_3 + x_5 x_7 \\ (x_2 - x_1)(x_4 - x_3) + (x_6 - x_5)(x_8 - x_7) \end{bmatrix}$$

equal to zero. Computing the zero dynamics, in this case via the normal form, is possible and has been carried out with NON$\frac{z}{=}$CON. To obtain the especially simple normal form with $p(\xi, \eta) = 0$, a complete set of independent solutions $\varphi(x)$ for the partial differential equations

$$\frac{\partial \varphi}{\partial x} [g_1 \quad g_2] = 0$$

is needed. The equations are completely integrable, and so solvable according to Frobenius' theorem, if the distribution spanned by g_1 and g_2 is involutive, as is the case here, because the Lie product $[g_1, g_2] = 0$. The solutions as computed by NON$\frac{z}{=}$CON are

$$z = \Phi(x) = \begin{bmatrix} x_2 \\ x_1 \\ x_6 - 2l \\ x_5 - l \\ \dfrac{x_8(x_1 + x_2) + x_4(x_5 + x_6)}{x_2 - x_1} \\ \dfrac{x_1(x_7(x_2 - x_1) + x_4 x_6) - x_5(x_3(x_2 - x_1) + x_2 x_4)}{x_1(x_2 - x_1)} \end{bmatrix}$$

Using this result the zero dynamics is rather involved. The solutions of the partial differential equations are not unique, and the most simple solution is not always found. Another solution, due to M. Weiss, in this case leading to simpler expressions for the zero dynamics, is

$$z = \begin{bmatrix} x_1 \\ x_2 \\ x_5 \\ x_6 \\ x_5(x_3 + x_4) - x_1(x_7 + x_8) \\ x_4(x_6 - x_5) - x_8(x_2 - x_1) \end{bmatrix}$$

The main simplification is that now the new co-ordinates do not have a denominator term. Nevertheless, the expressions for the zero dynamics are still too involved to be included. Note that the input u needed to keep the output y equal to zero is related to the connecting force between the links.

12.4.2 *System characterisation*

Systems can be characterised by properties that are of interest and that are not the same for all systems. In the past, this characterisation has often been performed for systems of differential equations, with or without control input. In control one often considers systems whose characteristics can be modified, because one has a control input available to do so. If one allows, for example, state feedback, systems that previously had different characteristics can be made equivalent.

It is also possible to characterise systems with respect to their input-output behaviour. This is relatively unexplored for nonlinear systems. It is known that for linear systems state feedback does not influence the zeros of the system. It is possible to change the zeros by a precompensator, but only if the zeros are in the left half-plane; otherwise the system with state feedback and a precompensator will not be internally stable. This leads to the conclusion that controllable, and in some sense also stabilisable, systems of the same order and same relative degree have equivalent input-output characteristics under the operations of state feedback and precompensation if their unstable zeros impose a dynamics that is equivalent.

It is possible, with some restrictions, to generalise this idea to nonlinear systems, where the unstable part of the zero dynamics then characterises the system. An example is given.

Example 3

Two models are compared: a cart-pendulum and a monocycle model. It can be shown that part of their zero dynamics is unstable and is of different form. The comparison is performed by symbolically computing the zero dynamics of both models and looking at the results. A symbolic routine that analyses the dynamics and then states whether or not they are equivalent has not yet been developed. An extensive treatment of this example can be found in [21].

12.5 Analysis

The characterisation of a system by properties of its zero dynamics is of importance for some design goals; for example, if the zero dynamics is unstable, a certain type of control law is guaranteed to be unable to stabilise the system, so this control law should be avoided. An unstable zero dynamics is also an indication of inherent restrictions for the control system performance. It has the same role as right-hand-plane zeros for linear systems. This marks the interest in the zero dynamics of a system. The zero dynamics can be computed based on the normal form (but then the system must have a well defined relative degree) or with the zero dynamics algorithm. Both ways are possible in NON≦CON, but the computation with the normal form, if applicable, is often easier to complete.

To illustrate the use of NON≦CON an example is considered which computes the zero dynamics for a system with two inputs and two outputs. This example is a contrived one. It is taken from [23, Example 6.1.5]. The system has no well-defined relative degree, so the zero dynamics algorithm is employed.

Example 4

The model of the system is

$$\dot{x} = \begin{bmatrix} 0 \\ x_4 \\ \lambda x_3 + x_4 \\ 0 \end{bmatrix} + \begin{bmatrix} 1 & 0 \\ x_3 & 0 \\ 0 & 0 \\ 0 & 1 \end{bmatrix} u, \qquad y = \begin{bmatrix} x_1 \\ x_2 \end{bmatrix}$$

From the functions supplied by NON$\stackrel{z}{=}$CON, only `extnormform` is needed to compute the zero dynamics. Part of the output from `extnormform` is displayed as follows:

* the zero dynamics $\dot{x}_3 = \lambda x_3$

* the zeroing input $[0, 0]$

Both components of the input u that keep $y = 0$ valid are just zero in this case. The results coincide with those in [23].

12.6 Synthesis

To synthesise controllers, several tools from nonlinear control theory can be used. Only the input-output and state-space exact linearisation problems and related tools are discussed and illustrated by some examples.

12.6.1   Input-output exact linearisation

The input-output exact linearisation problem is of long-standing interest in control theory. In essence, it is the problem of modifying a nonlinear dynamic system such that, after the modification, it behaves like a linear one, so powerful design methods for linear systems can again be employed. The goal is to get a linear (dynamical) relation between the new input v and the output y of the plant.

In a complete control system design the input-output exact linearisation is only a subordinate goal, to make it possible to use other design methods for attaining additional goals.

The input-output exact linearisation problem has been solved (see, for example, Reference 23, Sections 5.6 and 12.2). Our goal is to test the conditions under which the problem can be solved and to derive explicit expressions α and β for the feedback eqn. 12.3.

This static state feedback can be computed with the structure algorithm, or be based on the normal form. An example is given, for a system with two inputs and two outputs without a well-defined relative degree, from [23, Example 5.6.4].

Example 5

The model of the system is

$$\dot{x} = \begin{bmatrix} x_1^2 + x_2 \\ x_1 x_3 \\ -x_1 + x_3 \\ 0 \\ x_5 + x_3^2 \end{bmatrix} + \begin{bmatrix} 0 & 0 \\ 0 & 1 \\ 1 & 0 \\ 1 & 0 \\ x_2 & 0 \end{bmatrix} u, \qquad y = \begin{bmatrix} x_3 \\ x_4 \end{bmatrix}$$

The following part of a MAPLE session shows that NON$\leqslant$CON can compute the input-output linearising state feedback. From the functions supplied by NON$\leqslant$CON only `inoutlin` is needed. The computed state feedback agrees with the result in [23].

The exact linearising feedback u is generated as

$$[v_1 + x_1 - x_3, v_2 - 2x_1 x_2 - 2x_1^3 - x_1 x_3]$$

and is a function of the new input v. The linearised system dynamics of the closed loop is of the form

$$\dot{x} = \tilde{f}(x) + \tilde{g}(x)v$$

with $\tilde{f}(x)$ computed as

$$[x_2 + x_1^2, -2x_1 x_2 - 2x_1^3, 0, x_1 - x_3, x_5 + x_3^2 + x_1 x_2 - x_2 x_3]$$

and $\tilde{g}(x)$ as

$$\begin{bmatrix} 0 & 0 & 1 & 1 & x_2 \\ 0 & 1 & 0 & 0 & 0 \end{bmatrix}^T$$

The functions in $\tilde{f}(x)$ are nonlinear in x and $\tilde{g}(x)$ is not constant. Nevertheless, the operator from v to y is linear.

12.6.2   *State-space exact linearisation*

The state-space exact linearisation problem has been discussed extensively in Section 12.2. To illustrate the use of NON$\leqslant$CON, two examples are considered: (i) for a system with $m = 1$, i.e. a system with one input and one output signal; (ii) for a system with multiple input and output signals, and this is more complicated. Both examples are contrived ones: the first example is taken from [23, Example 4.2.5], the second is from [23, Example 5.2.6].

Example 6

The model of the system is

$$\dot{x} = \begin{bmatrix} x_3(1 + x_2) \\ x_1 \\ x_2(1 + x_1) \end{bmatrix} + \begin{bmatrix} 0 \\ 1 + x_2 \\ -x_3 \end{bmatrix} u$$

To check whether this system can be transformed into a linear and controllable one via state feedback and a change of co-ordinates, we have to compute the functions $\mathrm{ad}_f g(x)$ and $\mathrm{ad}_f^2 g(x)$ and test the conditions of Theorem 1 (p. 305). See also Remark 4.

Calculations given in [23] show that

$$\mathrm{ad}_f g(x) = \begin{bmatrix} 0 \\ x_1 \\ -(1 + x_1)(1 + 2x_2) \end{bmatrix}$$

$$\mathrm{ad}_f^2 g(x) = \begin{bmatrix} (1 + x_1)(1 + x_2)(1 + 2x_2) - x_1 x_3 \\ x_3(1 + x_2) \\ -x_3(1 + x_2)(1 + 2x_2) - 3x_1(1 + x_1) \end{bmatrix}$$

At $x = 0$, the matrix

$$\begin{bmatrix} g(x) & \mathrm{ad}_f g(x) & \mathrm{ad}_f^2 g(x) \end{bmatrix}_{x=0} = \begin{bmatrix} 0 & 0 & 1 \\ 1 & 0 & 0 \\ 0 & -1 & 0 \end{bmatrix}$$

has rank 3 and therefore condition (ii) is satisfied. The Lie product $[g, \mathrm{ad}_f g](x)$ is of the form

$$[g, \mathrm{ad}_f g](x) = \begin{bmatrix} 0 \\ * \\ * \end{bmatrix}$$

so condition (iii) is also satisfied, because the matrix

$$\begin{bmatrix} g(x) & \mathrm{ad}_f g(x) & [g, \mathrm{ad}_f g](x) \end{bmatrix}$$

has rank 2 for all x.

A function $\lambda(x)$ that solves the equation

$$\frac{\partial \lambda}{\partial x} \begin{bmatrix} g(x) & \mathrm{ad}_f g(x) \end{bmatrix} = 0$$

is

$$\lambda(x) = x_1.$$

This function is not unique.

Around $x = 0$, the system is transformed into a linear one by the state feedback

$$u = \frac{v - x_3^2(1 + x_2) - x_2 x_3(1 + x_2)^2 - x_1(1 + x_1)(1 + 3x_2)}{(1 + x_1)(1 + x_2)(1 + 2x_2) - x_1 x_3}$$

and the change of co-ordinates

$$z = \begin{bmatrix} z_1 \\ z_2 \\ z_3 \end{bmatrix} = \Phi(x) = \begin{bmatrix} x_1 \\ x_3(1 + x_2) \\ x_1 x_3 + x_2(1 + x_1)(1 + x_2) \end{bmatrix}$$

The next, edited, excerpt of a MAPLE session shows that these results can be reproduced by NON$\frac{z}{z}$CON. From the functions supplied by NON$\frac{z}{z}$CON the following are needed: **outputfunc** for the computation of the λ_i, **statelin** for the linearising state feedback u, and `transform` for the change of co-ordinates $\Phi(x)$.

The results of `outputfunc`, generating the functions λ_i, are

the function(s) lambda that fulfil the demands are: x_1

The exact linearising feedback u is presented by `statelin` as

$$(-v_1 + x_3^2 + x_3^2 x_2 + x_3 x_2 + 2x_3 x_2^2 + x_2^3 x_3 + x_1 + x_1^2 + 3x_1 x_2 + 3x_1^2 x_2)/$$
$$(x_3 x_1 - 1 - 3x_2 - x_1 - 3x_1 x_2 - 2x_2^2 - 2x_1 x_2^2)$$

It is a nonlinear function of the state x and contains the new input v. The input u is not well defined when its denominator becomes 0. The transformation to the normal form is derived by `transform`. The new co-ordinates z are given as

$$x_1, x_3 + x_3 x_2, x_3 x_1 + x_2 + x_1 x_2 + x_2^2 + x_1 x_2^2$$

It is easy to check that all these results coincide with the previous ones.

Example 7

The model of the system is

$$\dot{x} = \begin{bmatrix} x_2 + x_2^2 \\ x_3 - x_1 x_4 + x_4 x_5 \\ x_2 x_4 + x_1 x_5 - x_5^2 \\ x_5 \\ x_2^2 \end{bmatrix} + \begin{bmatrix} 0 & 1 \\ 0 & 0 \\ \cos(x_1 - x_5) & 1 \\ 0 & 0 \\ 0 & 1 \end{bmatrix} u$$

This system satisfies the conditions of Theorem 1. Two functions λ_i need to be constructed. This is worked out in [23] and the result is

$$\lambda_1 = x_1 - x_5 \qquad \lambda_2 = x_4$$

The following part of a MAPLE session, with some edits, shows that these results can also be obtained by NON$\frac{z}{=}$CON.

First the new output functions λ_i are computed in `outputfunc` as

$$[-x_1 + x_5, x_4]$$

The exact linearising feedback u is computed in `statelin` as

$$(v_1 - v_2 + x_2^2)/\cos(x_1 - x_5), v_2 - x_2^2$$

while the new co-ordinates z are found as follows by `transform`

$$x_1 - x_5, x_2, x_3 - x_1 x_4 + x_4 x_5, x_4, x_5$$

The computed state feedback and change of co-ordinates complete the results given in [23]. Recall that the functions λ_i are not unique—at least their sign is not defined. The linearising state feedback is not always well defined, as can be seen from the expression for u. The denominator is not allowed to become 0.

12.7 Conclusions

Computing the zero dynamics and solving the input-output and state-space exact linearisation problems can be automated by using symbolic computation, e.g. by using the NON⊵CON package. This means that it is now easier to design controllers based on the linearisation approach, that can fully take into account the nonlinearities in real systems. An enhancement of the performance of some control systems, for a large set of operation conditions, can therefore be expected.

At the moment, the computations cannot be performed for complicated systems, due to the limited capability to solve sets of nonlinear equations or sets of partial differential equations. Therefore, the designers of control systems cannot yet routinely compute solutions for these problems, using tools that are based on symbolic computation programs. To remedy this, a possibility is to extend the capabilities of symbolic computation programs for solving large and intricate sets of nonlinear and (partial) differential equations. Another possibility is to try to partition the problem into smaller ones using, eventually, intrinsic structure.

12.8 Acknowledgments

Harm van Essen implemented the zero dynamics and structure algorithms. Gose Fisher made several extensions to the NON⊵CON package. The idea of using the zero dynamics algorithm for the index reduction problem was suggested by Martin Weiss.

12.9 References

1 SAITO, O., KANNO, M., and ABE, K.: 'Symbolic manipulation CAD of control engineering by using REDUCE', *Int. J. Control*, 1988, **48**, (2), 781–790

2 UMENO, T., ABE, K., YAMASHITA, S., and SAITO, O.: 'A software package for control design based on the algebraic theory using symbolic manipulation language REDUCE'. Proc. TENCON 87, Vol. 3, IEEE, Seoul, Korea, 1987, pp. 1102–1106

3 GAWTHROP, P. J., and BALLANCE, D. J.: 'Symbolic algebra and physical-model-based control'. Proc. of UKACC Internat. Conf. CONTROL'96, Vol. 2, IEE, Exeter, UK, pp. 1315–1320

4 BARKER, H. A., KO, Y. W., and TOWNSEND, P.: 'The application of a computer algebra system to the analysis of a class of nonlinear systems', 'Nonlinear control systems design: science papers of the IFAC Symposium' IFAC, Oxford: Pergamon Press, Capri, Italy, 1989, pp. 131–136

5 BIRK, J., and ZEITZ, M.: 'Anwendung eines symbolverarbeitenden Programmsystems zur Analyse und Synthese von Beobachtern für nichtlineare Systeme', *MSR, Mess. Steuern Regeln*, 1990, **33**, (12), pp. 536–543

6 ROTHFUSS, R., SCHAFFNER, J., and ZEITZ, M.: 'Rechnergestützte Analyse und Synthese nichtlinearer Systeme', 'Nichlineare Regelung: Methoden, Werkzeugen, Anwendungen', *VDI Ber.* 1993, **1026**, pp. 267–291

7 HAHN, H., LEIMBACH, K.-D., and ZHANG, X.: 'Nonlinear control of a spatial multi-axis servo-hydraulic test facility'. Preprints of 12th IFAC World Congress, Vol. 7, IFAC, Sydney, Australia, 1993, pp. 267–270

8 ALLGÖWER, F., and GILLES, E. D.: 'Nichtlinearer Reglerentwurf auf der Grundlage exakter Linearisierungstechniken', in 'Nichtlineare Regulung: Methoden, Werkzeugen, Anwendungen', *VDI-Ber.* 1993, **1026**, pp. 209–234

9 HO, D. W. C., LAM, J., TIN, S. K., and HAN, C. Y.: 'Recent applications of symbolic computation in control system design'. Preprints of 12th IFAC World Congress, Vol. 9, IFAC, Sydney, Australia, 1993, pp. 509–512

10 AKHRIF, O., and BLANKENSHIP, G. L.: 'Computer algebra for analysis and design of nonlinear control systems'. Proc. of 1987 American Control Conf., Vol. 1, IEEE, Minneapolis, MN, pp. 547–554

11 KWATNY, H. G., and BLANKENSHIP, G. L.: 'Symbolic tools for variable structure control system design: the zero dynamics'. Proc. of IEEE Workship on Robust Control via Variable Structure and Lyapunov Techniques, IEEE, Benevento, Italy, 1994

12 BLANKENSHIP, G. L., GHANADAN, R., KWATNY, H. G., LA VIGNA, C. and POLYAKOV, V.: 'Tools for integrated modeling, design, and nonlinear control', *IEEE Control Syst. Mag.*, 1995, **15**, (2), pp. 65–79

13 SAITO, O., HINO, H. and XU, L.: 'Symbolic CAD system for algebraic approach on MATLAB—Symbolic Control Toolbox—', Proc. of UKACC Internat. Conf. CONTROL'96, Vol. 2, IEE, Exeter, UK, 1996, pp. 1338–1343

14 WANG, R.: 'Symbolic computation approach for absolute stability', *Int. J. Control*, 1993, **58**, (2), pp. 495–502

15 DE JAGER, B.: 'Nonlinear control system analysis and design with MAPLE', in HOUSTIS, E. N. and RICE, J. R. (Eds.), 'Artificial intelligence, expert systems and symbolic computing'. Selected and revised papers from the IMACS 13th World Congress, IMACS, Amsterdam, 1992, pp. 155–164

16 VAN ESSEN, H. and DE JAGER, B.: 'Analysis and design of nonlinear control systems with the symbolic computation system MAPLE', in NIEUWENHUIS, J. W., PRAAGMAN, C. and TRENTELMAN, H. L. (Eds.), Proc. of Second European Control Conf., Vol. 4, Groningen, The Netherlands, 1993, pp. 2081–2085

17 DE JAGER, B.: 'Symbolic calculation of zero dynamics for nonlinear control systems', in S. M. WATT (Ed.). Proc. of 1991 Internat. Symp. on *Symbolic and Algebraic Computation*, ISSAC'91, 1991, pp. 321–322

18 DE JAGER, B.: 'Symbolics for control: MAPLE used in solving the exact linearization problem', in COHEN, A. M., VAN GASTEL, L., and VERDUYN LUNEL, S. (Eds.), 'Computer algebra in industry 2: Problem solving in practice' (John Wiley & Sons, Ltd., Chichester, 1995), pp. 291–311

19 DE JAGER, B.: 'Generalized normal forms and disturbance rejection: A symbolic approach'. Proc. of UKACC Internat. Conf. CONTROL'96, Vol. 2, IEE, Exeter, UK, 1996, pp. 1321–1325

20 DE JAGER, B.: 'Zero dynamics and input-output exact linearization for systems without a relative degree using MAPLE'. Proc. of 1993 Internat. Symp. on *Nonlinear Theory and its Applications*, Vol. 4, IEICE, Honolulu, Hawaii, 1993, pp. 1205–1208

21 DE JAGER., B.: 'Applications of zero dynamics with symbolic computation'. Proc. of UKACC Internat. Conf. CONTROL'98, IEE, Swansea, UK

22 FLIESS, M.: 'Generalized controller canonical forms for linear and nonlinear dynamics', *IEEE Trans. Autom. Control*, 1990, **35**, (9), pp. 994–1001

23 ISIDORI, A.: 'Nonlinear control systems' (Springer-Verlag, Berlin, 1995, 3rd edn.)

24 NIJMEIJER, H. and VAN DER SCHAFT, A. J.: 'Nonlinear dynamical control systems' (Springer-Verlag, New York, 1990)

25 GARDNER, R. B., and SHADWICK, W. F.: 'The GS algorithm for exact linearization to Brunovsky normal form', IEEE Trans. Autom. Control, 1992, **37**, (2), pp. 224–230

26 VAN ESSEN, H.: 'Symbols speak louder than numbers: Analysis and design of nonlinear control systems with the symbolic computation system MAPLE'.

Master's thesis, Eindhoven University of Technology, Faculty of Mechanical Engineering. Report WFW 92.061, 1992.

27 BAYER, D. and STILLMAN, M.: 'On the complexity of computing syzygies', in ROBBIANO, L. (Ed.), 'Computational aspects of commutative algebra' (Academic Press, London, 1989), pp. 1–13

28 DE JAGER, B. and VAN ASCH, B.: 'Symbolic solutions for a class of partial differential equations', *J. Symb. Comput.*, 1996, **22**, (4), pp. 459–468

29 KHALIL, H. K.: 'Nonlinear systems' (Prentice-Hall, New York, 1996, 2nd edn.)

Chapter 13

Symbolic methods for global optimisation

E. M. B. Smith and C. C. Pantelides

13.1 Introduction

The use of formal optimisation techniques for the determination of optimal designs and operating strategies is increasing rapidly in all branches of engineering. Many of the problems of interest require the solution of nonlinear programming (NLP) problems involving the minimisation or maximisation of nonlinear objective functions $\Phi(\cdot)$ subject to nonlinear equality and inequality constraints $g(\cdot)$ and $h(\cdot)$, respectively:

$$\left. \begin{array}{c} \min_{x} \Phi(x) \\[2mm] \text{subject to} \\[2mm] g(x) = 0 \\[2mm] h(x) \leq 0 \\[2mm] \text{and} \\[2mm] x^l \leq x \leq x^u \end{array} \right\} \qquad (13.1)$$

The variables x can take any value between specified lower and upper bounds x^l and x^u, respectively.

If the objective function and/or the constraints are nonconvex, then there is a possibility that the above problem has multiple local minima. Most of the currently available numerical methods aim to determine one of these local solutions. However, in many applications, it is important to determine *globally* optimal solutions. One reason for this could be that there is substantial economic incentive in obtaining the absolutely best design that is achievable. In other cases, *only* the global solution is physically meaningful or useful; such problems include those of determining the equilibrium state of a system via energy mini-misation, or for verifying its safety or stability by identifying the worst possible combination of values of uncertain parameters or events that may affect the system.

In recent years, methods for the determination of global solutions to the problem in eqn. 13.1 have begun to emerge. Most of these may be classified as either stochastic or deterministic. The first category (see review by [1]) encompasses a wide variety of methods such as adaptive random search, simulated annealing and genetic algorithms. All of them employ an element of random search and, consequently, can guarantee global optimality only as the search time goes to infinity. Ensuring constraint satisfaction as random changes are made to variable values is also not straightforward.

Deterministic global optimisation methods have the advantage that they provide a rigorous guarantee of the global optimum being determined within a specified ε-tolerance[1] and within a finite amount of computation. Deterministic techniques include generalised Benders' decomposition [2], cutting plane algorithms [3, 4], interval analysis techniques [5, 6], and spatial branch-and-bound (sBB) algorithms [7–10].

The symbolic manipulation techniques described in this paper aim at allowing the application of sBB-type algorithms to a wide range of problems. The following Section outlines the key steps of the standard sBB algorithm. Then Section 13.3 describes a symbolic manipulation algorithm that automatically reformulates a wide class of nonlinear programming problems of the type of eqn. 13.1 to a form amenable to the application of sBB algorithms. Section 13.4 presents some examples of the application of the algorithm to some standard global optimisation test problems. Finally, some concluding remarks on the applicability and limitations of the algorithm are made in Section 13.5.

13.2 The spatial branch-and-bound algorithm

This Section provides an overview of spatial branch-and-bound algorithms to the extent necessary for understanding the material presented in this chapter. More detailed algorithmic descriptions have recently been provided by several authors (see, for instance, [11–13])).

13.2.1 Outline of spatial branch-and-bound algorithm

Given a hyperrectangular subdomain defined by upper and lower bounds on the variables x, spatial branch-and-bound algorithms compute lower and upper bounds, Φ^{min} and Φ^{max}, respectively, on the objective function value Φ over this subdomain. The upper bound also corresponds to a feasible point of the original

---

[1] This refers to the maximum difference between the value of the objective function at the solution obtained and the true optimum.

optimisation problem (eqn. 13.1). If, in any subdomain, the lower and upper bounds differ by less than ε, then Φ^{max} (and the corresponding point x) may be taken as the globally ε-optimal point in this particular subdomain.

The best point (i.e. the one with the lowest value of the objective function) determined from all the subdomains that have been considered up to any stage in the execution of the algorithm constitutes an upper bound Φ^* on the globally optimal value of Φ. Obviously, if, in a given subdomain, we find a Φ^{max} that is smaller than the current Φ^*, then this value becomes the new Φ^*.

If the lower bound Φ^{min} in a particular subdomain exceeds the current value of Φ^*, then the subdomain may be discarded from further consideration as it cannot possibly contain the global solution of eqn. 13.1.

If, on the other hand, $\Phi^{min} < \Phi^* - \varepsilon$, then there is a possibility that the global optimum lies within the subdomain under consideration. In such a case, we partition the subdomain into two smaller ones, and apply the above procedure to each one of them.

The standard sBB algorithm starts by considering the entire domain of the x variables as defined by the hyperrectangle $[x^l, x^u]$, with the global upper bound Φ^* being set at $+\infty$. It then applies the above procedure to this domain, partitioning it into subdomains as necessary according to the ideas presented above.

13.2.2 Upper and lower bounds for spatial branch-and-bound algorithms

As described above, sBB algorithms rely on the generation of upper and lower bounds on the objective function over a given subdomain of the variables. The generation of a rigorous upper bound Φ^{max} in a given subdomain is, in principle, simple as *any* feasible point of eqn. 13.1 can provide an upper bound on the objective function value at the global optimum. In practice, the upper bound feasible point is usually chosen to be a local minimiser of eqn. 13.1.

The most common way of obtaining the lower bound Φ^{min} is via the minimisation of a *convex relaxation* of the original problem, eqn. 13.1. This is generally of the form

$$
\left.
\begin{aligned}
&\min_x \Phi^L(x) \\[4pt]
\text{subject to} \quad & \\
&g^L(x) \leq 0 \leq g^U(x) \\
&h^L(x) \leq 0 \\[4pt]
\text{and} \quad & \\
&x^l \leq x \leq x^u
\end{aligned}
\right\} \tag{13.2}
$$

where $\Phi^L(x)$ is a convex underestimator of the objective function $\Phi(x)$. Also $g^L(x)$ and $h^L(x)$ are convex underestimators of $g(x)$ and $h(x)$, respectively, and $g^U(x)$ a concave overestimator of $g(x)$. It can be shown that the sBB algorithm will converge to an ε-optimal solution after examining a finite number of sub-domains provided the gap between the convex relaxation and the original nonconvex problem decreases monotonically as the size of the subdomain is reduced.

13.2.3 Underestimators and overestimators of nonconvex functions

The main technical requirement for the implementation of the sBB algorithm is the construction of the estimators $\Phi^L(x)$, $g^L(x)$, $g^U(x)$ and $h^L(x)$. Such estimators are known for simple nonconvex functions such as bilinear products and linear fractional terms. For instance, McCormick [14] proposed the following convex relaxation for a bilinear function $x_i x_j$:

$$
\begin{aligned}
x_i x_j &\geq x_i^l x_j + x_j^l x_i - x_i^l x_j^l \\
x_i x_j &\geq x_i^u x_j + x_j^u x_i - x_i^u x_j^u \\
x_i x_j &\leq x_i^l x_j + x_j^u x_i - x_i^l x_j^u \\
x_i x_j &\leq x_i^u x_j + x_j^l x_i - x_i^u x_j^l
\end{aligned}
\tag{13.3}
$$

We note that all the expressions on the right hand sides of the above constraints are linear and involve the bounds of the variables x_i and x_j.

It is also possible to derive the following bounds for the quotient of two variables [15]:

$$
\begin{aligned}
\frac{x_i}{x_j} &\geq \frac{x_i^u}{x_j} + \frac{x_i}{x_j^l} - \frac{x_i^u}{x_j^l} \dots (A) \\[2mm]
\frac{x_i}{x_j} &\geq \frac{x_i^l}{x_j} + \frac{x_i}{x_j^u} - \frac{x_i^l}{x_j^u} \dots (B) \\[2mm]
\frac{x_i}{x_j} &\leq \frac{x_i^u}{x_j} + \frac{x_i}{x_j^u} - \frac{x_i^u}{x_j^u} \dots (C) \\[2mm]
\frac{x_i}{x_j} &\leq \frac{x_i^l}{x_j} + \frac{x_i}{x_j^l} - \frac{x_i^l}{x_j^l} \dots (D)
\end{aligned}
\tag{13.4}
$$

However, only two of these four estimators are actually convex for any combination of bounds of x_i and x_j. For instance, it can easily be shown that, if $x_i^l \geq 0$ and $x_j^l > 0$ (i.e. if x_i is non-negative and x_j is positive), then estimators (A) and (B) are convex while (C) and (D) are nonconvex. Clearly, only convex estimators may be used in forming the convex relaxation, eqn. 13.2.

Finally, simple estimators can be proposed for many unary transcendental functions [13]. One such class of functions $f(\cdot)$ comprises those that are convex over their entire range of definition (e.g. e^x or x^n, where n is a constant exceeding unity). In such cases, the function itself provides a convex 'underestimator' while the secant,

$$
S(x) \equiv f(x^l) + \frac{f(x^u) - f(x^l)}{x^u - x^l}(x - x^l)
\tag{13.5}
$$

is a concave (linear) overestimator of $f(x)$ at any point $x \in [x^l, x^u]$. On the other hand, for functions that are concave over their domain of definition (e.g. $\sqrt{x}$ or $\ln(x)$), the secant (eqn. 13.5) provides a convex underestimator while the function itself can act as a concave 'overestimator'.

The problem with all of the above estimators is that the functions that appear in practical applications are often much more complex than those considered above. Consider, for instance, the problem of proving or disproving diagonal dominance in a multiple-input multiple-output (MIMO) transfer function at a given frequency in the presence of uncertainty in the values of various system parameters. In such an application, we have to consider the relative magnitudes of the elements of each row of the transfer function matrix. The expression for the magnitude of a first-order transfer function of the form

$$G_{ij}(s) \equiv \frac{K}{\tau s + 1} \tag{13.6}$$

is given by

$$|G_{ij}(j\omega)| = \frac{K}{\sqrt{\tau^2 \omega^2 + 1}} \tag{13.7}$$

Here K and τ are functions of one or more independent parameters that can take any value within given ranges; the problem of determining diagonal dominance involves two or more such expressions and aims to determine the worst possible combination of the uncertain parameter values. We return to consider this problem again in Section 13.4.5. However, for the moment, suffice it to say that the expression on the right-hand side of eqn. 13.7 is already outside the scope of the simple functions for which convex underestimators and concave overestimators are available.

Techniques which allow the formation of the necessary estimators for *any* continuous twice differentiable function have been proposed by Androulakis *et al.* [16] and Adjiman and Floudas [17]. These generally involve the augmentation of the relevant function by a new convex term that is sufficiently large to render the entire combination convex. However, the relaxations obtained in this fashion are usually not as tight as the specific ones mentioned above.

In the following Section we consider a symbolic manipulation procedure that allows the automatic construction of the relaxed problem eqn. 13.2 for a wide range of problems of the type of eqn. 13.1.

13.3 A symbolic reformulation algorithm

13.3.1 The standard form for a nonlinear programming problem

The procedure considered in this Section comprises two main steps. In the first one, any nonlinear programming problem of the form of eqn. 13.1 is reformulated to the following *standard form*:

$$\min_{w} w_{obj}$$

$$Aw = b$$

$$w^l \leq w \leq w^u$$

$$
\begin{aligned}
w_k &\equiv w_i w_j & \forall (i, j, k) \in \mathcal{T}_{bt} \\
w_k &\equiv \frac{w_i}{w_j} & \forall (i, j, k) \in \mathcal{T}_{lft} \\
w_k &\equiv w_i^n & \forall (i, k, n) \in \mathcal{T}_{et} \\
w_k &\equiv fn(w_i) & \forall (i, k) \in \mathcal{T}_{uft}
\end{aligned}
$$

$$(13.8)$$

where w is a vector comprising the original continuous variables x together with other auxiliary continuous variables introduced during the reformulation. Also A and b are, respectively, a matrix and vector of real constant coefficients. The index 'obj' denotes the position of a single variable corresponding to the objective function value within the vector w.

The symbolic reformulation is exact, i.e. problems in eqns. 13.1 and 13.8 are completely equivalent. However, in the latter, all the nonlinearities (and potential nonconvexities) are described by the sets $\mathcal{T}_{bt}$, $\mathcal{T}_{lft}$, $\mathcal{T}_{et}$ and $\mathcal{T}_{uft}$, corresponding to bilinear product, linear fractional, simple exponentiation and univariate function terms,[2] respectively. It should be noted that each element of these sets is either a triplet or a pair of indices that relate the corresponding elements of the vector w in the appropriate manner. For instance,

$$(2, 5, 13) \in \mathcal{T}_{bt} \Leftrightarrow w_{13} = w_2 \times w_5$$

13.3.2 Basic ideas of symbolic reformulation algorithm

The reformulation algorithm is based on the simple observation that *any* algebraic expression is made up of binary operators corresponding to the five basic operations of arithmetic (addition, subtraction, multiplication, division and exponentiation), and unary operators corresponding to a relatively small number of transcendental functions (e.g. logarithms, exponentials, etc.). Thus, by introducing an appropriate set of new variables, it is possible to break any expression into sets of linear constraints and simple nonlinearities.

---

[2] In reality, we form a different set $\mathcal{T}_{uft}$ for each type of univariate function (e.g. $\ln(\cdot)$, $\exp(\cdot)$, etc.), but here we represent them all by a single set for simplicity of notation.

Consider, for instance, the expression on the right-hand side of eqn. 13.7. If we define the following variables:

$$z_1 = \tau^2 \tag{13.9a}$$

$$z_2 = \omega^2 \tag{13.9b}$$

$$z_3 = z_1 \times z_2 \tag{13.9c}$$

$$z_4 = z_3 + 1 \tag{13.9d}$$

$$z_5 = \sqrt{z_4} \tag{13.9e}$$

$$z_6 = K \div z_5 \tag{13.9f}$$

then the original expression is simply equivalent to z_6. Moreover, we have broken it down into one bilinear term (eqn. 13.9c), one linear fractional term (eqn. 13.9f), two simple exponentiation terms (eqns. 13.9a and 13.9b), a univariate function term (eqn. 13.9e), and a linear constraint (eqn. 13.9d). Thus, if such an expression were to occur in the original nonlinear programming problem, its reformulation could become part of the standard form, eqn. 13.8. In this case, the augmented variable vector w would include both the original variables K and τ and the newly introduced variables z_1 to z_6.

13.3.3 An automatic symbolic reformulation algorithm

In the example considered in Section 13.3.2, the reformulation was easily obtained by inspection. In an earlier article [18], the authors proposed a new *automatic* reformulation procedure that applies symbolic manipulation to the standard binary tree representation of algebraic expressions [19]. This type of representation for the expression on the right-hand side of eqn. 13.7 is shown in Figure 13.1. Each node in such a tree has zero, one or two descendants. Nodes without any descendants are known as the 'leaves' of the tree and correspond to constants or variables. On the other hand, nodes with one or two descendants correspond to unary or binary operators, respectively.

Binary tree structures are naturally recursive in that they can be defined in terms of simpler binary trees. More precisely, *any* binary tree can be viewed as one of the following:

- a leaf node
- a unary operator with a binary tree as a descendant
- a binary operator with left and right binary trees as descendants.

Because of this recursive nature, symbolic manipulation algorithms applied to a binary tree can also be defined recursively (see, for instance, [20]). In this particular case, the symbolic reformulation of a binary tree could be expressed as follows:

1. If the root node of the tree is a unary operator (a transcendendal function), then:

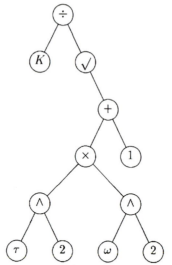

Figure 13.1   Binary tree representation of algebraic expression in eqn. 13.7

(a)   reformulate the descendant of the tree, introducing a new variable z_i equivalent to its value

(b)   introduce a new variable, z_k, to represent the value of the binary tree

(c)   add the pair (i, k) to the set T_{uft}.

2. If the root node of the tree is a binary operator, then:

(a)   reformulate the left descendant of the tree, introducing a new variable z_i equivalent to its value

(b)   reformulate the right descendant of the tree, introducing a new variable z_j equivalent to its value

(c)   introduce a new variable, z_k, to represent the value of the binary tree

(d)   add the triplet (i, j, k) to the appropriate triplet set T_{bt} or T_{lft} or T_{et} depending on the nature of the operator in the root node.

It can be seen that recursive invocations of the procedure occur at steps $1(a)$, $2(a)$ and $2(b)$.

Although the above procedure certainly achieves the required aim of breaking any complex expression into simple functions, it may introduce a number of unnecessary new variables. For instance, in the reformulation described by eqns. $13.9a - 13.9f$, the definition of a new variable z_2 to represent ω^2 is unnecessary as ω is a given constant. The definition of z_3 is also unnecessary as it can be combined directly with those of z_1 and z_2 to define the linear constraint eqn. $13.9d$. Overall, the reformulation of the right-hand side of eqn. 13.7

to a form conforming to that of problem eqn. 13.8 can be written more compactly as:

$$z_1 = \tau^2 \tag{13.10a}$$

$$z_4 = \omega^2 z_1 + 1 \tag{13.10b}$$

$$z_5 = \sqrt{z_4} \tag{13.10c}$$

$$z_6 = K \div z_5 \tag{13.10d}$$

The symbolic manipulation procedure formally described by Smith and Pantelides [18] is designed to produce the reformulation that introduces the *minimum* number of new variables. This is achieved by identifying both constant subexpressions and maximal linear subexpressions within the expression being reformulated. No new variable needs to be introduced to represent the former while a single variable is used to represent a linear expression of any length. Moreover, the procedure detects identical nonlinear expressions occurring in *different* constraints in the original problem, eqn. 13.1, thereby introducing the necessary variables once only for each such expression.

13.3.4 Implementation of the spatial branch-and-bound algorithm

As mentioned at the start of this Section, our global optimisation procedure comprises two main steps. We have already considered how nonlinear programming problems of the form of eqn. 13.1 can be reformulated to the standard form of eqn. 13.8.

A spatial branch-and-bound algorithm of the type outlined in Section 13.2 can now be applied to the problem eqn. 13.8. It will be recalled that a fundamental requirement for the application of sBB algorithms is the ability to construct convex relaxations of the original nonconvex problem. Because eqn. 13.8 is an exact reformulation of eqn. 13.1, any convex relaxation of the former is also a valid convex relaxation of the latter. Hence, the reformulation procedure allows us to construct a convex relaxation for any optimisation problem described by eqn. 13.1 provided we restrict ourselves to univariate functions for which the appropriate estimators are available. As indicated in Section 13.2.3, the latter class includes functions that are convex or concave over their entire domain of definition. This comprises many commonly occurring functions, such as $\exp(\cdot)$, $\ln(\cdot)$, $\sqrt{(\cdot)}$, but excludes trigonometric ones.

Overall, then, applying an sBB procedure to the problem described by eqn. 13.8 will locate a global optimum which is also the global optimum of the original problem, eqn. 13.1. However, the efficiency of sBB algorithms depends crucially on the tightness of the convex relaxation in any given subdomain, i.e. the difference between the value of the optimal objective function of the convex relaxation and that of the global minimum in this subdomain. This tightness, in turn, depends on the size of the subdomain as determined by the bounds x^l and x^u on the variables x. It is, therefore, very important to tighten these bounds as much as possible, both before the start of the sBB algorithm and during it when considering any particular subdomain. Ways of achieving this aim have recently been reviewed by Smith [13].

Of course, for the symbolic reformulation procedure to be practically implementable, the objective function and constraints appearing in the nonlinear programming problem eqn. 13.1 must be made available in binary tree form. The latter can be obtained automatically provided a high level symbolic representation of these expressions is available. In our case, this has been achieved by implementing our algorithm within the *gPROMS* process modelling environment [21]. This essentially allows the application of our procedure to a wide variety of problems described by algebraic equations and also partial differential equations in one or more spatial dimensions. *gPROMS* automatically discretises the latter to algebraic equations [22] before applying to them the symbolic reformulation procedure described in this chapter.

13.4 Numerical experiments

In this Section, we examine the application of the techniques presented in this paper to some typical small-scale global optimisation problems. More examples, including some that are much larger than those considered here, have been presented by Smith [13], together with a critical comparison with other global optimisation algorithms.

13.4.1 Konno's bilinear program

This example was used by Konno [23] to illustrate his cutting plane algorithm:

$$\min_{x,y} \ \Phi = x_1 - x_2 - y_1 - (x_1 - x_2)(y_1 - y_2)$$

subject to

$$x_1 + 4x_2 \leq 8$$
$$4x_1 + x_2 \leq 12$$
$$3x_1 + 4x_2 \leq 12$$
$$2y_1 + y_2 \leq 8$$
$$y_1 + 2y_2 \leq 8$$
$$y_1 + y_2 \leq 5$$

where

$$x_1, x_2, y_1, y_2 \geq 0$$

$$(13.11)$$

The nonconvexity of the problem is due to the presence of the expression $(x_1 - x_2)(y_1 - y_2)$ in the objective function. The symbolic reformulation algorithm introduces two new variables representing $(x_1 - x_2)$ and $(y_1 - y_2)$, respectively, and a third one representing their product.

In this case, it is possible to deduce *a priori* from the problem constraints (cf. Section 13.3.4) the upper bounds $\mathbf{x} = [3.0, 0.0]^T$ and $\mathbf{y} = [4.0, 0.0]^T$.

The solution of the first convex relaxation (cf. eqn. 13.2) is $\mathbf{x} = [3.0, 0.0]^T$ and $\mathbf{y} = [4.0, 0.0]^T$, with an objective function value of -13. Since all variables happen to lie on their bounds, the bilinear term estimators (eqn. 13.3) are exact in this case. Consequently, the solution obtained is immediately confirmed to be the global optimum without the need for any further iteration. This is in accordance with the observations of other researchers who have considered this problem [24, 25].

13.4.2 The doughnut slice problem

The problem is to minimise the sum of the co-ordinates of a point that lies between two circles within a banded region (see Figure 13.2). The problem is defined as:

$$\min_{y} \ \Phi = y_1 + y_2$$

subject to

$$
\left.
\begin{aligned}
y_1^2 + y_2^2 &\leq 4 \\
y_1^2 + y_2^2 &\geq 1 \\
y_1 - y_2 - 1 &\leq 0 \\
y_2 - y_1 - 1 &\leq 0
\end{aligned}
\right\}
\quad (13.12)
$$

where

$$-2 \leq \{y_1, y_2\} \leq 2$$

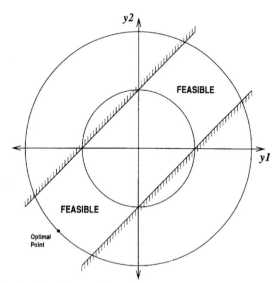

Figure 13.2   The doughnut slice problem

The nonconvexity arises from the second constraint. The global optimum is $\Phi = -2\sqrt{2}$ at $\mathbf{y} = [-\sqrt{2}, -\sqrt{2}]^T$.

As the nonconvex inner circle constraint is *not* active at the global optimum, its relaxation does not affect the feasibility of the relaxed problem solution when considered in the context of the original problem. On the other hand, it will be recalled from Section 13.2.3 that the convex underestimators for unary functions (such as y_1^2 and y_2^2) which are convex everywhere are the functions themselves. Hence, the 'relaxation' of the outer circle constraint $y_1^2 + y_2^2 \leq 4$ is, in fact, the constraint itself.

Under such circumstances, the solution of the relaxed problem will be the global optimum and the algorithm should terminate in one iteration. This general property is indeed observed when applying the algorithm to this particular problem.

13.4.3 Haverly's pooling problem

The pooling problem by Haverly [26] is illustrated graphically in Figure 13.3. Three feeds *A*, *B* and *C* with different sulphur contents are combined to make two products *X* and *Y* with specified maximum sulphur contents. The problem is modelled as follows:

$$\text{min } \Phi = 6A + \beta B + 10(Cx + Cy) - 9x - 15y$$

subject to

$$Px + Py - A - B = 0$$
$$x - Px - Cx = 0$$
$$y - Py - Cy = 0$$
$$pP(x + y) - 3A - B = 0$$
$$pPx + 2Cx - 2.5x \leq 0$$
$$pPy + 2Cy - 1.5y \leq 0$$
$$x \leq x^u$$
$$y \leq y^u$$

(13.13)

where β is a cost coefficient and x^u and y^u are, respectively, the specified upper bounds for x and y. All other symbols correspond to unknown variables. Nonconvexity is present in both the objective function and the constraints in the form of bilinear terms. Our algorithm reformulates these automatically by introducing one new variable for each distinct bilinear term.

Three instances of this problem are solved, details of which are presented with their corresponding solutions in Table 13.1.

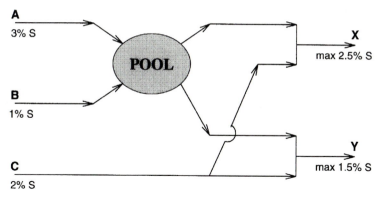

Figure 13.3 Haverly's pooling problem flowsheet

13.4.4 A six-hump camel back problem

This is a well-known test for global optimisation algorithms, having been used as such since the mid 1970s [27]. The function is

$$\min_{x} \Phi = 4x_1^2 - 2.1x_1^4 + \tfrac{1}{3}x_1^6 + x_1x_2 - 4x_2^2 + 4x_2^4$$

where

$$-4 \leq \{x_1, x_2\} \leq 4$$

$$(13.14)$$

Table 13.1 Parameters and solutions for three cases of Haverly's pooling problem

	β	x^u	y_u	Φ
Case 1	16	100	200	−400
Case 2	16	600	200	−600
Case 3	13	100	200	−750

Solution	Case 1	Case 2	Case 3
Φ	−400.0	−600.0	−750.0
A	0.0	300.0	50.0
B	100.0	0.0	150.0
C	0.5	0.5	0.0
P	0.5	0.5	1.0
x	0.0	600.0	0.0
y	200.0	0.0	200.0
p	1.0	3.0	1.5
Iterations	7	21	11

Two views of the objective function surface for this problem are shown in Figure 13.4. The reformulated problem involves one bilinear and five exponentiation terms.

After 147 iterations, two points are located as global optima. These are $\mathbf{x} = [0.08984, -0.71266]$ and $\mathbf{x} = [-0.08981, 0.71265]$ with the same objective function value of -1.0316.

Byrne and Bogle [28] solved this problem using an interval analysis algorithm based on high order inclusion functions and located the same global optima.

13.4.5 *MIMO control system diagonal dominance problem*

The recent work of Kontogiannis and Munro [29] has led to the extension of the direct Nyquist stability test to multiple-input multiple-output (**MIMO**) parametric systems. A necessary condition of this extended controller stability test is diagonal dominance of the transfer function matrix being tested [30].

Consider the application of this new stability test to a candidate rate and composition control scheme for a solution copolymerisation reactor. The process transfer function has been taken from the work of Congalidis *et al.* [31] on the design of a single-input single-output feedforward and feedback control scheme for such a reactor. The first row of this transfer function is

$$G_1 = \left[\frac{0.98711(0.12011s + \alpha)}{0.065948s^2 + 0.36662s + 1} \quad \frac{0.20527\beta}{0.4195s + 1} \quad \frac{0.49882(0.49849s + 1)}{0.12424\gamma s^2 + 0.40242s + \alpha} \right]$$

(13.15)

where α, β and γ are independent uncertain parameters whose values are bounded such that

$$0.9 \leq \{\alpha, \beta, \gamma\} \leq 1.1$$

(13.16)

The above row of the transfer function matrix is diagonally dominant if

$$|G_{1,1}(s)| > |G_{1,2}(s)| + |G_{1,3}(s)|$$

(13.17)

For a given frequency ω, we substitute $s = j\omega$ in each element of the transfer function matrix and compute the magnitude of the resulting complex number,

$$\left| \frac{a + bs}{cs^2 + ds + e} \right| \equiv \left| \frac{a + bj\omega}{(c - e\omega^2) + dj\omega} \right| \equiv \sqrt{\frac{a^2 + b^2\omega^2}{(c - e\omega^2)^2 + d^2\omega^2}}$$

These magnitudes may then be used to formulate the diagonal dominance criterion

$$D = \min_{\alpha, \beta, \gamma} \left(|G_{1,1}(i\omega)| - |G_{1,2}(i\omega)| - |G_{1,3}(i\omega)| \right)$$

(13.18)

If the globally minimum value of D is positive, then the transfer function row is diagonally dominant for *any* combination of values of the uncertain parameters. However, if this value is negative, then there exist combinations of the uncertain

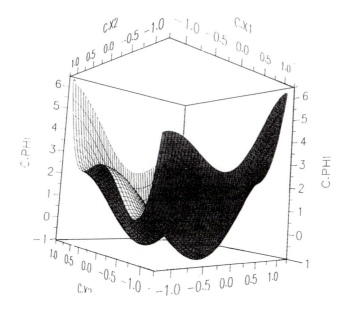

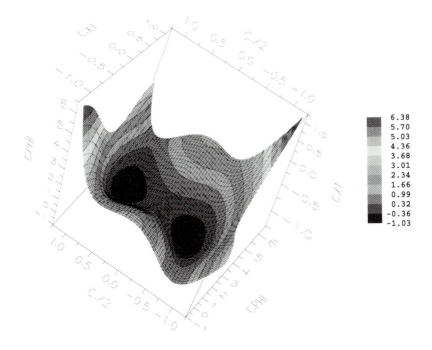

Figure 13.4 Objective function of the six-hump camel back problem

parameter values for which the diagonal dominance test fails, i.e. the overall multi-input multi-output control scheme may be unstable.

The symbolic reformulation of the above problem results in a problem with 22 linear constraints, three linear fractional terms (one for each element of the transfer function row), and 15 terms corresponding to the various squares and square roots. The total number of variables in the reformulated problem is 43.

The test is to be performed at the frequencies of 1 and 3 rad/s. The global optimisation carried out for these frequencies yields optimal objective function values of 0.0391 and −0.0648, respectively. Therefore, the control scheme cannot be guaranteed to be stable at the latter frequency as there are combinations of values of the uncertain parameters under which diagonal dominance does not hold.

13.5 Concluding remarks

This chapter has described some recent work for the global solution of nonconvex nonlinear programming problems. The algorithm has two strands. First, a symbolic manipulation procedure is applied to convert the original problem (eqn. 13.1) into a standard form (eqn. 13.8). Once this is done, a spatial branch-and-bound algorithm that takes advantage of the features of the standard form is applied to it, solving it to global optimality.

Our algorithm is, in principle, applicable to *any* nonlinear programming problem where the objective function and the constraints can be expressed in terms of the five basic operations of arithmetic and univariate functions that are convex or concave over the entire domain of interest.

In practice, as with other branch-and-bound algorithms (such as those used to solve mixed integer linear programming problems), it is difficult to predict *a priori* the computational effort that will be necessary to solve a particular problem with our algorithm. Nevertheless, practical experience indicates reasonable behaviour with problems involving up to several tens of variables. Again, it must be emphasised that problem size is only one factor that affects computational cost; the convexity gap of the problem (i.e. the difference between the objective function values at its optimal solution and at the solution of its convex relaxation) is perhaps a more important characteristic in this context, albeit one that is much more difficult to quantify *a priori*. In any case, the interested reader is referred to Smith [13] for a much more extensive set of test problems and a discussion on the significance of the results obtained.

Finally, it is worth mentioning that a recent extension of the algorithm presented here has allowed it to deal with mixed integer nonlinear programming (MINLP) problems where a subset of the variables is allowed to take integer values only [32]. This increases its applicability to problems involving discrete decisions.

13.6 References

1 SCHOEN, F.: 'Stochastic techniques for global optimization: a survey of recent advances', *J. Glob. Optim.*, 1991, **1**, pp. 207–228

2 FLOUDAS, C. A., and VISWESWARAN, V.: 'A global optimization algorithm (GOP) for certain classes of nonconvex NLP's—I. Theory', *Comput. Chem. Eng.*, 1990, **12**, pp. 1397–1417

3 BALAS, E.: 'Integer programming and convex analysis: Intersection cuts from the outer polars', *Math. Program.*, 1972, **2**, pp. 330–382

4 BALAS, E.: 'Nonconvex quadratic programming via generalized polars', *SIAM J. Appl. Math.*, 1975, **28**, pp. 335–349

5 ICHIDA, K., and FUJII, Y.: 'An interval arithmetic method of global optimization', *Computing*, 1979, **23**, pp. 85–97

6 HANSEN, E. R.: 'Global optimization using interval analysis—the multidimensional case', *Numer. Math.*, 1980, **34**, 247–270

7 FALK, J. E., and SOLAND, R. M.: 'An algorithm for separable nonconvex programming problems', *Manage. Sci.*, 1969, **15**, pp. 550–569

8 SOLAND, R. M.: 'An algorithm for separable nonconvex programming problems II: Nonconvex constraints', Manage. Sci., 1971, **17**, pp. 759–773

9 HORST, R.: 'An algorithm for nonconvex programming', *Math. Program.*, 1976, **10**, 213–321

10 TUY, H., and HORST, R.: 'Convergence and restart in branch-and-bound algorithms for global optimization. Application to concave minimization and D.C. optimization problems', *Math. Program.*, 1988, **41**, pp. 161–183

11 RYOO, H. S., and SAHINIDIS, N. V.: 'Global optimization of nonconvex NLPs and MINLPs with applications in process design', *Comput. Chem. Eng.*, 1995, **19**, pp. 551–566

12 RYOO, H. S., and SAHINIDIS, N. V.: 'A branch-and-reduce approach for global optimization', *J. Glob. Optim.*, 1996, **8**, pp. 107–138

13 SMITH, E. M. B.: 'On the optimal design of continuous processes'. PhD thesis, University of London, 1996

14 McCORMICK, G. P.: 'Computability of global solutions to factorable nonconvex programs: Part I—Convex underestimating problems', *Math. Program.*, 1976, **10**, pp. 146–175

15 QUESADA, I., and GROSSMANN, I. E.: 'A global optimization algorithm for linear fractional and bilinear programs', J. Glob. Optim., 1995, **6**, pp. 39–76

16 ANDROULAKIS, I. P., MARANAS, C. D., and FLOUDAS, C. A.: 'αBB: a global optimization method for general constrained nonconvex problems', *J. Glob. Optim.*, 1995, **7**, pp. 337–363

17 ADJIMAN, C. S., and FLOUDAS, C. A.: 'Rigorous convex underestimators for general twice-differentiable problems', *J. Glob. Optim.*, 1996, **7**, pp. 337–363

18 SMITH, E. M. B., and PANTELIDES, C. C.: 'Global optimisation of general process models', in GROSSMANN, I. E. (Ed.): 'Global optimization in engineering design' (Kluwer Academic Publichers Inc., Dordrecht, 1996), Chap. 12, pp. 355–386

19 KNUTH, D. E.: 'The art of computer programing—1. Fundamental algorithms' Computer Science and Information Processing (Addison-Wesley, Reading, MA, USA, 1973, 2nd edn.)

20 PANTELIDES, C. C.: 'SpeedUp—recent advances in process simulation', *Comput. Chem. Eng.*, 1988, **12**, pp. 745–755

21 BARTON, P. I., and PANTELIDES, C. C.: 'Modeling of combined discrete/continuous processes', *AIChE J.*, 1994, **40**, pp. 966–979

22 OH, M., and PANTELIDES, C. C.: 'A modelling and simulation language for combined lumped and distributed parameter systems', *Comput. Chem. Eng.*, 1996, **20**, pp. 745–755

23 KONNO, H.: 'A cutting plane algorithm for solving bilinear programs', *Math. Program.*, 1976, **11**, pp. 14–27

24 SAHINIDIS, N. V.: 'BARON: an all-purpose global optimization software package'. Technical report UILU-ENG-95-4002, University of Illinois, Urbana-Chamgaign, March 1995

25 EPPERLY, T. G. W.: 'Global optimization of nonconvex nonlinear programs using parallel branch and bound', PhD thesis, University of Wisconsin, Madison, WI, USA, 1995

26 HAVERLY, C. A.: 'Studies of the behaviour of recursion for the pooling problem', *SIGMAP Bull.*, 1978, **25**, p. 19

27 DIXON, L. C. W., GOMULKA, J., and SZEGO, G. P.: 'Towards a global optimisation technique', in DIXON, L. C. W. and SZEGO, G. P. (Eds.), 'Towards global optimisation'. (North-Holland, Amsterdam, 1975), pp. 29–54

28 BYRNE, R. P., and BOGLE, I. D. L.: 'Solving nonconvex process optimisation problems using interval subdivision algorithms', in GROSSMANN, I. E. (Ed.), 'Global optimization in engineering design'. (Kluwer Academic Publishers Inc., Dordrecht, 1996), Chap. 5, pp. 155–174

29 KONTOGIANNIS, E., and MUNRO, N.: 'The application of symbolic computation in the analysis of systems having parametric uncertainty'. IEEE Colloquium on Symbolic Computation for Control, April 1996

30 KONTOGIANNIS, E., and MUNRO, N.: 'Multi-input multi-output systems with parametric uncertainty'. Presented at IFAC Workshop on *Robust Control*, San Fransisco, USA, June 1996

31 CONGALIDIS, J. P., RICHARDS, R. R., and RAY, W. H.: 'Feedforward and feedback control of a solution copolymerization reactor', *AIChE J.*, 1989, **35**, pp. 891–907

32 SMITH, E. M. B., and PANTELIDES, C. C.: 'Global optimisation of non-convex MINLPs', *Comput. Chem. Eng.*, 1997, **21**, pp. S791–S796

## Chapter 14

Solving strict polynomial inequalities by Bernstein expansion

J. Garloff and B. Graf

Dedicated to Professor Dr Karl Nickel on the occasion of his 75th birthday

14.1 Introduction

Many interesting control system design and analysis problems can be recast as systems of inequalities for multivariate polynomials in real variables. In particular, for linear time-invariant systems, important control issues such as robust stability and robust performance can be reduced to such systems. Typically, the variables in the (multivariate) polynomials come from plant (controlled system) and compensator (controller) parameters. In this Chapter, we describe a method for solving such systems of inequalities. By solving we mean that we end up with a collection of axis-parallel boxes in the parameter space whose union provides an inner approximation of the solution set, i.e. the polynomial inequalities are fulfilled for each parameter vector taken from such a box. This method is based on the expansion of a multivariate polynomial into Bernstein polynomials. It provides an alternative to symbolic methods like quantifier elimination, for which the application to control problems was demonstrated in [1]. The number of operations required by quantifier elimination methods is still doubly exponential in the number of variables, so that only relatively small problems can actually be solved, whereas Bernstein expansion has been applied to larger robust stability problems [2–4]. However, it should be noted that in contrast to symbolic methods Bernstein expansion requires *a priori* bounds on the design parameter range. This is not a hard restriction since the designer can often estimate the interesting parameter range.

A third approach, the *probabilistic approach* (see, for example [5, 6]), is to solve problems in control theory which can be formulated as systems of strict inequalities. Here again bounds on the parameter range must be known. This approach is applicable to very complex systems but it provides only 'probabilistic' answers.

14.1.1 *Notation*

For compactness, we define a *multi-index I* as an ordered *l*-tuple of non-negative integers $(i_1, \ldots, i_l)$. We will use multi-indices, for example, to shorten power products for $x = (x_1, \ldots, x_l) \in \mathbf{R}^l$ we set $x^I = x_1^{i_1} x_2^{i_2} \cdot \ldots \cdot x_l^{i_l}$. For simplicity, we sometimes suppress the brackets in the notation of multi-indices. We write $I \leq N$ if $N = (n_1, \ldots, n_l)$ and if $0 \leq i_k \leq n_k$, $k = 1, \ldots, l$. Further, let $S = \{I : I \leq N\}$. Then we can write an *l*-variate polynomial p in the form

$$p(x) = \sum_{I \in S} a_I x^I, \qquad x \in \mathbf{R}^l \tag{14.1}$$

and refer to N as the *degree* of p and to

$$\hat{n} = \max\{n_i : i = 1, \ldots, l\} \tag{14.2}$$

as the *total degree* of p. Also, we write I/N for $(i_1/n_1, \ldots, i_l/n_l)$ and $\binom{N}{I}$ for

$$\binom{n_1}{i_1} \cdot \ldots \cdot \binom{n_l}{i_l}.$$

14.1.2 *Problem statement*

Let $p_1, \ldots, p_n$ be *l*-variate polynomials and let an axis-parallel box Q in the $\mathbf{R}^l$ be given. We want to find

$$\Sigma := \{x \in Q : p_i(x) > 0, i = 1, \ldots, n\} \tag{14.3}$$

The set Σ is called the *solution set* of the system of polynomial inequalities.

This Chapter is organised as follows: Section 14.2 contains a short review of quantifier elimination methods and their application to control problems. In Section 14.3 we recall the Bernstein expansion, and apply it to the solution of systems of polynomial inequalities in Section 14.4. Our algorithm is explained in Section 14.5, numerical examples are presented in Section 14.6 and conclusions are given in Section 14.7.

14.2 Quantifier elimination

In many practical control problems some of the polynomial variables are quantified by the logic quantifier $\forall$ (for all) or $\exists$ (there exists). Typically, $\forall$ quantifies the plant parameters (for robust design) and $\exists$ quantifies the controller parameters (to define the feasible controller-parameter set). In addition, the polynomial inequalities are combined by the Boolean operators $\wedge$ (and) and $\vee$ (or). Examples from control theory can be found in [1, 7, 8].

The problem to find an equivalent expression involving only unquantified variables is called the *quantifier elimination* (QE) *problem*. In 1948, Tarski [9] showed that there is a procedure that solves this problem in a finite number of steps. Although Tarski gave a constructive proof, the resulting algorithm is inpractical even with the power of the today's computers. One of the first attempts to use QE methods to solve control design problems was made in 1975 by Anderson *et al.* [10] to solve the static output-feedback stabilisation problem. However, the computational complexity and lack of software severely limited the interest in their results. In 1975, Collins [11] introduced a more efficient approach, the *cylindrical algebraic decomposition* (CAD). For an introduction to CAD see the excellent exposition in [8]. Given a set of *l*-variate polynomials, the CAD algorithm decomposes the $\mathbf{R}^l$ into components over which the polynomials have constant signs. For systems of polynomial inequalities (and even equations) the CAD method has value of its own: once a decomposition is found, a solution (if one exists) to any system determined by the given polynomials can be found. However, utilising this method the QE problem could be solved only for very small problems.

In [12–14] significantly more efficient QE algorithms based on partial CAD were presented. Software packages have been written for the implementation of the new algorithms, e.g. the software package QEPCAD for *quantifier elimination by partial cylindrical algebraic decomposition* by Hoon Hong from the Research Institute for Symbolic Computation in Linz (Austria), with contributions by G. E. Collins, J. R. Johnson and M. J. Encarnación.

The CAD algorithms always completely solve any QE problem. However, the number of operations required is still doubly exponential (for details cf. Reference 15) so that only problems with modest size can be handled. We mention two papers [16, 17] treating the special case of systems of strict polynomial inequalities—the proper subject of this Chapter. In [16, 17] algorithms are described which allow us to decide whether such a system has a solution and which are much faster than the general CAD algorithm. The simplified CAD algorithm in [17] finds a finite set of solutions such that any other solution can be connected by a continuous path of solutions with one of the solution set. We were told from Wolfram Research, Inc., that this algorithm is included in the 3.0 version of Mathematica [18] in the Standard Add-on Package [19] covering manipulating and solving algebraic inequalities and that the upcoming version of Mathematica will contain in the kernel algorithms for deciding the existence of solutions of systems of polynomial equations and inequalities based on CAD and QE methods.

14.3 Bernstein expansion

14.3.1 Bernstein transformation of a polynomial

In this sub-section we expand a given multivariate polynomial (eqn. 14.1) into Bernstein polynomials to obtain bounds for its range over an *l*-dimensional box.

This approach was used in the univariate case for the first time in [20] and subsequently in a series of papers, e.g. [21, 22]. Generalisations to the multivariate case were given in [23–26]. For a nearly complete bibliography see [27]. Without loss of generality we consider the unit box $U = [0, 1]^l$ since any nonempty box of $\mathbf{R}^l$ can be mapped affinely onto this box.

The ith Bernstein polynomial of degree n is defined as

$$b_{n,i}(x) = \binom{n}{i} x^i (1 - x)^{n-i}, \qquad 0 \le i \le n$$

for an arbitrary $x \in \mathbf{R}$. In the multivariate case, the Ith Bernstein polynomial of degree N is defined by

$$B_{N,I}(x) = b_{n_1,i_1}(x_1) \cdot \ldots \cdot b_{n_l,i_l}(x_l), \qquad x = (x_1, \ldots, x_l) \in \mathbf{R}^l \tag{14.4}$$

The transformation of a polynomial from its power form (eqn. 14.1) into its Bernstein form results in

$$p(x) = \sum_{I \in S} b_I(U) B_{N,I}(x) \tag{14.5}$$

where the Bernstein coefficients $b_I(U)$ of p on U are given by

$$b_I(U) = \sum_{J \le I} \frac{\binom{I}{J}}{\binom{N}{J}} a_J, \qquad I \in S \tag{14.6}$$

We collect the Bernstein coefficients in an array $B(U)$, i.e. $B(U) = (b_I(U))_{I \in S}$. A similar notation will be employed for other sets of related coefficients. In [23] a method was presented for calculating the Bernstein coefficients efficiently by a difference table scheme (which is similar to the sweep procedure; cf. Section 14.3.2) that avoids the binomial coefficients and products appearing in eqn. 14.6.

In the following, we use a special subset of the index set S comprising those indices which correspond to the indices of the vertices of the array $B(U)$, i.e.

$$S_0 = \{0, n_1\} \times \cdots \times \{0, n_l\}$$

We list two useful properties of the Bernstein coefficients, e.g. [22, 23, 28].

Lemma 1 *Let p be a polynomial (eqn. 14.1) of degree N. Then the following properties hold for its Bernstein coefficients $b_I(U)$ (eqn. 14.6):*

(*i*) *Sharpness of special coefficients:*
$$\forall I \in S_0 : b_I(U) = p(I/N) \tag{14.7}$$

(*ii*) *Convex hull property:*
$$\forall x \in U : \min_{I \in S} b_I(U) \le p(x) \le \max_{I \in S} b_I(U) \tag{14.8}$$

with equality in the left (resp., right) inequality if and only if $\min_{I \in S} b_I(U)$ *(resp.,* $\max_{i \in S} b_I(U)$*) is assumed at a Bernstein coefficient* $b_I(U)$ *with* $I \in S_0$.

Eqn. 14.7 follows immediately from eqn. 14.6. The property eqn. 14.8 relies on two fundamental properties of the Bernstein polynomials, namely their non-negativity on the unit box U and the fact that they form a partition of unity.

14.3.2 Sweep procedure

In this sub-Section we follow the exposition in [4]. We define a sweep in the rth direction $(1 \leq r \leq l)$ similarly to de Casteljau's algorithm in Computer-Aided Geometric Design, e.g. [28], as recursively applied linear interpolation. Let D be any sub-box of U generated by sweep operations (at the beginning we have $D = U$, then subsequently D is obtained by successively halving). Starting with $B^{(0)}(D) = B(D)$ we set for $k = 1, \ldots, n_r$,

$$
b^{(k)}_{i_1,\ldots,i_r,\ldots,i_l}(D) = \begin{cases} b^{(k-1)}_{i_1,\ldots,i_r,\ldots,i_l}(D) & i_r = 0, \ldots, k-1 \\ \frac{1}{2}(b^{(k-1)}_{i_1,\ldots,i_r-1,\ldots,i_l}(D) + b^{(k-1)}_{i_1,\ldots,i_r,\ldots,i_l}(D)) & i_r = k, \ldots, n_r \end{cases}
$$

(14.9)

To obtain the new coefficients, we apply eqn. 14.9 for $i_j = 0, \ldots, n_j$, $j = 1, \ldots, r-1, r+1, \ldots, l$.

Then the Bernstein coefficients on D_0^l, where the sub-box D_0 is given by

$$
D_0 = [\underline{d}_1, \bar{d}_1] \times \cdots \times [\underline{d}_r, (\underline{d}_r + \bar{d}_r)/2] \times \cdots \times [\underline{d}_l, \bar{d}_l]
$$

are obtained as $B(D_0) = B^{(n_r)}(D)$. At no extra cost we get as intermediate values the Bernstein coefficients $B(D_1)$ on the neighbouring sub-box D_1:[1]

$$
D_1 = [\underline{d}_1, \bar{d}_1] \times \cdots \times [(\underline{d}_r + \bar{d}_r)/2, \bar{d}_r] \times \cdots \times [\underline{d}_l, \bar{d}_l]
$$

since for $k = 0, \ldots, n_r$ the following relation holds [24]: $b_{i_1,\ldots,n_r-k,\ldots,i_l}(D_1) = b^{(k)}_{i_1,\ldots,n_r,\ldots,i_l}(D)$.

It is important to note that by the sweep procedure the explicit transformation of the sub-boxes generated by the sweeps back to U is avoided. Let $\hat{n}$ denote the total degree (eqn. 14.2) of polynomial eqn. 14.1. Since we have to perform eqn. 14.9 $n_r(n_r + 1)/2$ times, we need altogether $O(\hat{n}^{l+1})$ additions and multiplications.

14.3.3 Selection of the sweep direction

The definition of the sweep procedure shows that we are free in choosing the sweep direction. To increase the probability of finding a nonpositive sharp Bernstein coefficient proving that the polynomial under consideration is not positive, we suggest sweeping in that co-ordinate direction in which the first partial derivative is largest. Our selection rule profits from the easy calculation of the partial derivatives of a polynomial in Bernstein form, e.g. [28, 29].

---

[1] That is, the Bernstein coefficients of the polynomial shifted from this sub-box to U.

To shorten some expressions in the sequel we associate with an index $I = (i_1, \ldots, i_r, \ldots, i_l)$ the index $I_{r,k} = (i_1, \ldots, i_r + k, \ldots, i_l)$, where $0 \le k + i_r \le n_r$. Then the first partial derivative with respect to x_r of p (eqn. 14.5) is given by the following formula $(1 \le r \le l)$:

$$\frac{\partial p}{\partial x_r}(x) = n_r \sum_{I \le N_{r,-1}} [b_{I_{r,1}}(D) - b_I(D)] B_{N_{r,-1},I}(x)$$

To decide which sweep direction to choose we estimate

$$\max_{x \in D} \left| \frac{\partial p}{\partial x_r}(x) \right|$$

from the above by

$$\tilde{I}_r = \max_{I \le N_{r,-1}} n_r |b_{I_{r,1}}(D) - b_I(D)|$$

We choose the r_0 with maximum value

$$\tilde{I}_{r_0} = \max_{j=1,\ldots,l} \tilde{I}_j \tag{14.10}$$

14.4 Approximation of the solution set

In this Section we use a similar approach to that in [2, 30]. The algorithm which we describe in the following Section was applied in [31] to approximate the stability region of a polynomial family with polynomial parameter dependency. Since we are able to describe the solution set Σ only in the simplest cases, we are seeking a good approximation to it. We obtain an inner approximation of Σ by the union of some sub-boxes of Q on which all polynomials p_i are positive. Similarly, an outer approximation is given by the union of some sub-boxes of Q with the property that on each there is a polynomial p_i being nonpositive there. The boundary $\partial\Sigma$ of Σ can be approximated by the union of some sub-boxes of Q on which each polynomial p_i assumes positive as well as nonpositive values, cf. Figure 14.1.

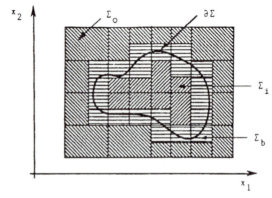

Figure 14.1   Inner and outer approximation of Σ and approximation of $\partial\Sigma$ [32]

By Σ_i and Σ_o we denote the set of sub-boxes which provide the inner and outer approximation, respectively. For a fixed positive number ε, $\Sigma_b(\varepsilon)$ is the list of sub-boxes with volume less than ε the union of which approximates the boundary $\partial\Sigma$. In our algorithm we check the (non)positivity of a polynomial by the sign of its Bernstein coefficients. From Lemma 1 immediately follows Lemma 2.

Lemma 2 *Positivity test of a multivariate polynomial*

Let p be an l-variate polynomial and let b_I be its Bernstein coefficients on Q. Then the following holds:

$$\min_{I \in S} b_I > 0 \Rightarrow p(x) > 0 \ \forall x \in Q \tag{14.11}$$

$$\max_{I \in S} b_I \leq 0 \Rightarrow p(x) \leq 0 \ \forall x \in Q \tag{14.12}$$

$$\exists I \in S_0 : b_I > 0 \Rightarrow \exists x \in Q : p(x) > 0 \tag{14.13}$$

$$\exists I \in S_0 : b_I \leq 0 \Rightarrow \exists x \in Q : p(x) \leq 0 \tag{14.14}$$

According to Lemma 2, the sets Σ_i, Σ_o and $\Sigma_b(\varepsilon)$ consist of the sub-boxes $\tilde{Q}$ generated by sweeps which fulfil the following conditions:

Σ_i: The Bernstein coefficients $(b_I^{(i)})_{I \in S}$ of all polynomials p_i, $i = 1, 2, \ldots, n$, are positive; then by property 14.11 all polynomials p_i, $i = 1, \ldots, n$, are positive on $\tilde{Q}$.

Σ_o: The Bernstein coefficients $(b_I^{(i^*)})_{I \in S}$ of a polynomial p_{i^*} are all nonpositive; then by property 14.12 the polynomial p_{i^*} is nonpositive on $\tilde{Q}$.

$\Sigma_b(\varepsilon)$: The volume of $\tilde{Q}$ is less than ε and each polynomial p_i possesses sharp positive and nonpositive Bernstein coefficients; then, according to eqns. 14.13 and 14.14, each polynomial p_i assumes on $\tilde{Q}$ positive and nonpositive values. [2]

Of course, if it turns out that a polynomial p_i is positive over a sub-box $\tilde{Q}$ of Q we can discard this polynomial from the list of polynomials to be checked further for positivity on any sub-box of $\tilde{Q}$.

14.5 Algorithm

The procedure **Test** checks a sub-box $\tilde{Q}$ of Q to which list this box will be appended. This procedure returns AP (for *all positive*) if $\tilde{Q}$ will be added to Σ_i, EN (for *exists a nonpositive polynomial*) if it will belong to Σ_o, and UD (for *undecided*) if $\tilde{Q}$ will be appended to $\Sigma_b(\varepsilon)$.

The procedure **TerminateSearch** terminates the search for sub-boxes of Q on which the polynomials p_i are positive or nonpositive if all remaining sub-boxes have a volume less than ε. For a fixed recursion depth d (i.e. a fixed number of sweeps performed on the initial box), we obtain boxes with volume $2^{-d} \text{vol}(Q)$.

---

[2] To avoid introducing a fourth list, we collect for simplicity in $\Sigma_b(\varepsilon)$ all sub-boxes generated by sweeps having volume less than ε which belong neither to Σ_i nor to Σ_o.

On the other hand, to achieve boxes with volume less than ε we have to choose the depth d as the smallest integer number greater than

$$\frac{\ln(\text{vol}(Q)) - \ln \varepsilon}{\ln 2}$$

In our algorithm we use a maximum recursion depth d which leads to parameter boxes with volume less than a given ε.

The main procedure **SolutionSet** (Figure 14.2) returns a collection of sub-boxes on which all polynomials p_i are positive. These sub-boxes are listed in a stack Σ_i. The stack BC consists of the Bernstein coefficients of the polynomials $p_i, i = 1, \ldots, n$, denoted by $B(D)$. We assume that the standard operations *MakeStack, Push, Pop* and *Isempty* are implemented, e.g. [33, 34].

Procedure 1　SolutionSet$(B(D))$

---

begin $BC = \text{MakeStack}()$; $\Sigma_i = \text{MakeStack}()$;
　　　　$\text{Push}(BC, B(D))$;
　　while $(\neg \text{ Isempty}(BC))$ **do**
　　　　$B(D) = \text{Pop}(BC)$;
　　　　$\{B(D_0), B(D_1)\} = \text{Sweep}(B(D))$;
　　　　$t0 = \text{Test}(B(D_0)); t1 = \text{Test}(B(D_1))$;
　　　　if $(t0 = UD)$ **then**
　　　　　　if$(\neg \text{ TerminateSearch}())$ **then**
　　　　　　　　$\text{Push}(BC, B(D_0))$;
　　　　　　end if
　　　　else if $(t0 = AP)$**then**
　　　　　　$\text{Push}(\Sigma_i, D_0)$;
　　　　end if
　　　　if$(t1 = UD)$ **then**
　　　　　　if $(\neg \text{TerminateSearch}())$ **then**
　　　　　　　　$\text{Push}(BC, B(D_1))$;
　　　　　　end if
　　　　else if $(t1 = AP)$ **then**
　　　　　　$\text{Push}(\Sigma_i, D_1)$;
　　　　end if
　　end while
end

Figure 14.2　Solution Set

14.6 Examples

All examples were run (on a PC equipped with a Pentium 133) with the maximum recursion depth $d = 15$. For simplicity, we apply the sweep selection rule (eqn. 4.10) only to the first of the list of polynomials so that the direction of the sweeps is completely determined by this polynomial. We denote by $\Box\Sigma$ the smallest box in Q containing Σ.

The first two examples utilise one of the Liénard-Chipart stability criteria, cf. p. 155 in [35], which states that a polynomial

$$p(s) = a_0 s^m + a_1 s^{m-1} + \cdots + a_{m-1}s + a_m \qquad \text{with } a_0 > 0$$

is stable, i.e. all its zeros have negative real parts, if and only if

$$a_m > 0, \qquad a_{m-2} > 0, \qquad a_{m-4} > 0, \ldots$$

and

$$\Delta_1 > 0, \qquad \Delta_3 > 0, \qquad \Delta_5 > 0, \ldots$$

where Δ_j is the leading principal minor of order j of the Hurwitz matrix, i.e.

$$\Delta_j = \det(h_{ik})_{i,k=1,\ldots,j}$$

with

$$h_{ik} = a_{2k-i} \qquad i, k = 1, \ldots, m$$

where by convention $a_r = 0$ if $r < 0$ or $r < m$.

Example 1

Our first example involves only two parameters, i.e. $l = 2$, so that we are able to visualise the approximations Σ_i, Σ_o and $\Sigma_b(\varepsilon)$ obtained by our algorithm. We consider the static output-feedback problem presented in [10], cf. [1], which leads to the problem to find the set of all parameters v, w such that the closed-loop characteristic polynomial

$$p(s) = s^3 + vs^2 + (w - 5v - 13)s + w$$

is stable. The Liénard-Chipart criterion gives us the conditions

$$v, w > 0$$

$$-5v^2 - 13v + vw - w > 0$$

The results for $v \in [2, 10]$, $w \in [40, 50]$ are presented in Figure 14.3. The white region is the inner approximation and the grey domain is the outer approximation to Σ. The approximation of $\partial\Sigma$ is given by the black region. The algorithm finds $\Box\Sigma = [2, 5.59375] \times [41.9922, 50]$ in 0.27 s.

Example 2

The problem taken from [1] is to find a (stable) second-order compensator with three parameters which simultaneously stabilise the three different plants with the following transfer functions:

$$\frac{2-s}{(s^2 - 1)(s + 2)} \qquad \frac{2-s}{s^2(s + 2)} \qquad \frac{2-s}{(s^2 + 1)(s + 2)}$$

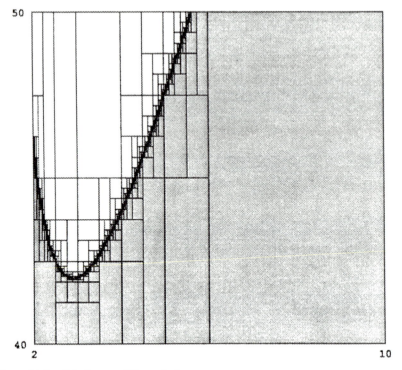

Figure 14.3   Solution set of Example 1

To reduce the number of parameters to be considered we assume a second-order compensator of the form

$$\frac{A(s+B)^2}{(s+D)^2}$$

with $D > 0$. To achieve stability, we utilise the Liénard-Chipart criterion. After some simplifications, we obtain the following set of inequalities in the parameters A, B and D:

$$A, B, D > 0$$
$$AB^2 - D^2 > 0$$
$$-AB + A + D^2 - D - 1 > 0$$
$$AB - AD - 2A + D^3 + 4D^2 + 4D > 0$$
$$AB^3 - AB^2D - 4AB^2 + 2ABD + 4AB + 2BD^3 + 5BD^2 + 2BD - D^3 - 4D^2 - 4D > 0$$
$$AB - 2A - BD^2 - 4BD - 4B + 2D^2 + 3D - 2 > 0$$

We have chosen $A \in [100, 120]$, $B \in [0, 2]$, $D \in [10, 20]$. Note that the software package QEPCAD, cf. Section 14.2, already needs 2 h of CPU time to solve the existence problem, i.e. to show that there is a solution of the above system of inequalities [1]. In 1.3 s our algorithm finds $\square\Sigma = [100, 120] \times [1.15625, 1.60938] \times [12.1875, 17.3438]$. Figure 14.4 shows the set of acceptable values of B and D (white region) for fixed $A = 110$ obtained in 0.5 s.

Example 3 [2]

We consider a plant which is assumed to be an unstable first-order system with transfer function

$$\frac{p_1}{1 - s/p_2}$$

where p_1 and p_2 are uncertainty parameters. To control the plant a proportional plus integral (PI) compensator is used with transfer function

$$C(s, q) = q_1 \frac{1 + s/q_2}{s}$$

where q_1 and q_2 are the design parameters. The controller to be designed has to meet some performance specifications represented by the following polynomial inequalities:

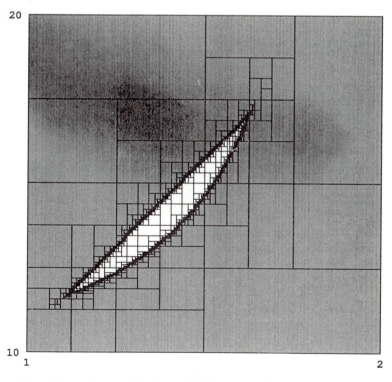

Figure 14.4 The set of acceptable values of B, D and A = 110

(i)   closed-loop stability: $q_2 > 0$, $-q_1 > 0$, $-p_1 q_1 - q_2 > 0$

(ii)   tracking error (at steady state) for unit ramp: $p_1^2 q_1^2 - 2500 > 0$

(iii)   closed-loop bandwith: $aw^4 + bw^2 + c$, where

$$a = -q_2^2$$
$$b = p_1^2 p_2^2 q_1^2 - 2p_1 p_2 q_1 q_2^2 - 2p_1 p_2^2 q_1 q_2 - p_2^2 q_2^2$$
$$c = p_1^2 p_2^2 q_1^2 q_2^2$$

(iv)   resonance peak of the closed-loop transfer function: $aw^4 + bw^2 + c$, where

$$a = 1.96 q_2^2$$
$$b = 0.96 p_1^2 p_2^2 q_1^2 + 1.96 p_2^2 q_2 + 3.92 p_1 p_2^2 q_1 q_2 + 3.92 p_1 p_2 q_1 q_2^2$$
$$c = 0.96 p_1^2 p_2^2 q_1^2 q_2^2$$

(v)   control effort: $aw^4 + bw^2 + c$, where

$$a = 400 q_2^2 - q_1^2$$
$$b = 400 p_1^2 p_2^2 q_1^2 + 800 p_1 p_2^2 q_1 q_2 + 800 p_1 p_2 q_1 q_2^2 - p_2^2 q_1^2 + 400 p_2^2 q_2^2 - q_1^2 q_2^2$$
$$c = 400 p_1^2 p_2^2 q_1^2 q_2^2 - p_2^2 q_1^2 q_2^2$$

The design parameters (q_1, q_2) are taken from $[-300, 0] \times [0, 15]$, the plant parameters p_1, p_2 are chosen from $[0.8, 1.25]$, and the variable w varies in $[0, 300]$. Our algorithm found the following parameter intervals in 8.436 s: $q_1 \in [-300, -56.25]$, $q_2 \in [0.11, 15]$, $p_1, p_2 \in [0.8, 1.25]$, $w \in [0, 18.75]$. We note that the software package QEPCAD was not able to solve this problem [7]. For the solution using QEPCAD of the simplified model involving only the design parameter q_1 with the simple output feedback $C(s, q) = q_1$ see [7].

14.7 Conclusions

Bernstein expansion provides a method for testing a multivariate polynomial for positivity over a box and therefore for finding an inner approximation of the solution set of a system of strict polynomial inequalities. Compared to quantifier elimination methods, Bernstein expansion is not so widely applicable:

• Only strict inequalities can be handled. However, many problems in linear control theory can be reduced to such systems.
• Bernstein expansion requires *a priori* bounds on the parameter range. However, the designer often has a region of special interest.

The applicability of quantifier elimination methods is severely limited by the number of variables. So many problems of practical importance are beyond the capabilities of these methods. The development of both better algorithms and of fast algorithms for special classes of problems is a very active area for research; it is hoped that the solution of significantly more complicated problems will be possible in the near future. Bernstein expansion can currently handle more complex problems, but its efficiency drastically decreases if the number of parameters exceeds about seven. Quantifier elimination provides an explicit description of the solution set, which in general is complicated. From the point of view of the designer the description of the entire solution set is often not necessary. What the designer really wants is a good inner approximation of the solution set or even only a large box inside this set, but that is what Bernstein expansion provides.

14.8 Acknowledgment

Support from the Ministry of Science and Research Baden-Württemberg and from the Ministry of Education, Science, Research, and Technology of the Federal Republic of Germany under contract 1704596 is gratefully acknowledged.

14.9 References

1 ABDALLAH, C., DORATO, P., LISKA, R., STEINBERG, S., and YANG, W.: 'Applications of quantifier elimination theory to control theory'. Technical report EECE95-007, University of New Mexico, Department of Electrical and Computer Engineering, Albuquerque, 1995, cf. Proceedings of the 4th IEEE Mediteranean Symposium on *Control and Automation*, Maleme, Crete, Greece, 1996, pp. 340–345

2 FIORIO, G., MALAN, S., MILANESE, M., and TARAGNA, M.: 'Robust performance design of fixed structure controllers for systems with uncertain parameters'. Proc. of 32nd Conference on *Decision and Control*, San Antonio, Texas,1993, pp. 3029–3031

3 GARLOFF, J., GRAF, B., and ZETTLER, M.: 'Speeding up an algorithm for checking robust stability of polynomials'. Proceedings of 2nd IFAC Symposium on *Robust Control Design* (Elsevier Science, Oxford, 1998), pp. 183–188

4 ZETTLER, M., and GARLOFF, J.: 'Robustness analysis of polynomials with polynomial parameter dependency using Bernstein expansion', *IEEE Trans. Autom. Control*, 1998, **43**, (3), pp. 425–431

5 BAI, E.-W., TEMPO, R., and FU, M.: 'Worst-case properties of the uniform distribution and randomized algorithms for robustness analysis'. Proc. of American Control Conference, Albuquerque, NM, 1997

6 YOON, A., and KHARGONEKAR, P.: 'Computational experiments using randomized algorithms for robust stability analysis', internal report, University of Michigan (a brief version appeared in Proceedings of 36th Conference on Decision and Control)

7 DORATO, P., YANG, W., and ABDALLAH, C.: 'Robust multi-objective feedback design by quantifier elimination', *J. Symbol. Comput.*, 1997, **24**, (2), pp. 153–159

8 JIRSTRAND, M.: 'Algebraic methods for modeling and design in control', *Linköping Studies in Science and Technology* Thesis No. 540, Linköping University, Linköping, Sweden, 1996

9 TARSKI, A.: 'A decision method for elementary algebra and geometry' (University of California Press, Berkeley, 1951, 2nd edn.)

10 ANDERSON, B. D. O., BOSE, N. K., and JURY, E. I.: 'Output feedback stabilization and related problems—solution via decision methods', *IEEE Trans. Autom. Control*, 1975, **AC-20**, (1), pp. 53–65

11 COLLINS, G. E.: 'Quantifier elimination for real closed fields by cylindrical algebraic decomposition', *in* BRAKHAGE, H. (Ed.): '2nd GI conf. automation theory and formal languages', *Lect. Notes Comput. Sci.*, 1975, **33**, pp. 134–183

12 HONG, H.: 'Improvements in CAD-based quantifier elimination'. PhD Dissertation, Ohio State University, Ohio, 1990

13 COLLINS, G. E., and HONG, H.: 'Partial cylindrical algebraic decomposition for quantifier elimination', *J. Symbol. Comput.*, 1991, **12**, (3), pp. 299–328

14 HONG, H.: 'Simple solution formula construction in cylindrical algebraic decomposition based quantifier elimination'. Proceedings of ISSAC'92, Intern. Symp. *Symbolic and Algebraic Computation* (ACM Press, New York, 1992), pp. 177–188

15 BASU, S., POLLACK, R., and ROY, M.-F.: 'On the combinatorial and algebraic complexity of quantifier elimination'. Proceedings 35th Annual Symposium of the IEEE Computer Soc. on *Foundations of Computer Science*, Santa Fe, NM, 1994, pp. 632–641

16 McCALLUM, S.: 'Solving polynomial strict inequalities using cylindrical algebraic decomposition', *Comput. J.*, 1993, **36**, (5), pp. 432–438

17 STRZEBONSKI, A. W.: 'An algorithm for systems of strong polynomial inequalities', *Mathematica J.*, 1994, **4**, (4), pp. 74–77

18 WOLFRAM, S.: 'The Mathematica book' (Cambridge University Press, Cambridge, 1996, 3rd edn.)

19 WOLFRAM RESEARCH, INC.: 'Mathematica 3.0 Standard Add-on Packages (Cambridge University Press, Cambridge, 1996)

20 CARGO, G.T., and SHISHA, O.: 'The Bernstein form of a polynomial', *J. Res. Nat. Bur. Standards Sect. B*, 1966, **70B**, pp. 79–81

21 LANE, J.M., and RIESENFELD, R. F.: 'Bounds on a polynomial', *BIT*, 1981, **21**, pp. 112-117

22 RIVLIN, T. J.: 'Bounds on a polynomial', *J. Res. Nat. Bur. Standards Sect. B*, 1970, **74B**, pp. 47–54

23 GARLOFF, J.: 'Convergent bounds for the range of multivariate polynomials', *in* NICKEL, K. (Ed.): 'Interval mathematics 1985', *Lect. Notes. Comput. Sci.*, **212**, (Springer, Berlin, 1986) pp. 37–56

24 GARLOFF, J.: 'The Bernstein algorithm', *Interval Comput.*, 1993, **2**, pp. 154–168

25 LANE, J. M., and RIESENFELD, R. F.: 'A theoretical development for the computer generation and display of piecewise polynomial surfaces', *IEEE Trans. Pattern Anal. Mach. Intel.*, 1980, **2**, pp. 35–46

26 MALAN, S., MILANESE, M., TARAGNA, M., and GARLOFF J.: 'B^3 algorithm for robust performances analysis in presence of mixed parametric and dynamic perturbations'. Proc. of 31st Conference on *Decision and Control*, Tucson, AZ, 1992, pp. 128–133

27 HUNGERBÜHLER, R., and GARLOFF, J.: 'Bounds for the range of a bivariate polynomial over a triangle', *Reliab. Comput.*, 1998, **4**, (1), pp 3–13

28 FARIN, G.: 'Curves and surfaces for CAGD. A practical guide' (Academic Press, New York, 1993, 3rd edn.)

29 FAROUKI, R. T., and RAJAN, V. T.: 'Algorithms for polynomials in Bernstein form', *Comput. Aided Geometr. Des.*, 1988, **5**, pp. 1–26

30 CANALE, M., FIORIO, G., MALAN, S., MILANESE, M., and TARAGNA, M.: 'Robust tuning of PID controllers via uncertainty model identification'. Proc. 1997 European Control Conference, Bruxelles, Belgium, 1997

31 GRAF, B.: 'Computation of stability regions'. Technical report 9801, Fachhochschule Konstanz, Fachbereich Informatik, Konstanz, Germany, 1998

32 WALTER, E., and PRONZATO, L.: 'Identification of parametric models from experimental data' (Springer, London, 1997)

33 KNUTH, D. E.: 'The art of computer programming, vol. 1' (Addison-Wesley, Reading, MA, 1973, 2nd edn.)

34 SEDGEWICK, R.: 'Algorithms' (Addison-Wesley, Reading, MA, 1988)

35 BARNETT, S.: 'Polynomials and linear control systems' (Marcel Dekker, New York, 1983)

Chapter 15

Computational methods for control of dynamical systems

G.L. Blankenship and H.G. Kwatny

15.1 Background—design and control of nonlinear systems

Computational methods for the design and analysis of controlled, nonlinear systems are at an early stage of development in control engineering. For the most part developers of design tools for control systems have concentrated on finite-dimensional, linear systems where computational linear algebra and numerical integration of linear ordinary differential equations provide natural, effective frameworks for the expression of algorithms. For nonlinear systems the situation is different. A much broader range of behaviour is possible (for example, 'chaotic dynamics'), and a greater variety of tools is necessary for effective design and synthesis. Optimal control methods, which one might expect to provide a natural framework for computational design of nonlinear systems, have not had a significant impact, aside from a few important special cases.

The need for design tools for nonlinear systems has always been evident; however, advances in computer and control technology are increasing the urgency. New opportunities for innovation are being realised by integration of microprocessor controlled actuators into dynamical systems of all types, e.g. spacecraft and aircraft, ground vehicles, robots, machine tools, fluid flow systems, etc. Often these systems are interconnected, multibody dynamical systems with rigid and elastic substructures. Their operational range is substantial, and their behaviour is intrinsically nonlinear—linear small displacement models are not useful. Effective design of such systems and their controls relies on computer analysis for composing and screening alternative design concepts before constructing expensive prototypes. There has been work on computational tools to support the development of models for systems with embedded control elements. Experience has shown that it is critical to integrate the design of the system structures and the embedded control architecture and laws to achieve optimal performance.

Our work is intended to contribute to the development of tools to support integrated design of controlled structural mechanical systems. We have developed and released two systems:[1]

- **TSiDynamics**, a package for automatic generation of explicit models (in C or Matlab code) for multibody dynamical systems composed of rigid and flexible bodies interconnected by simple or compound joints, including embedded actuators and sensors
- **TSiControls**, a package for design of nonlinear (and linear) control systems.

In this chapter we describe the **TSiControls** package and its predecessors and samples of its applications. The **TSiDynamics** software is described elsewhere [34].

Both packages include functions for generation of simulation models in Matlab/Simulink or C. They provide functions for the manipulation of dynamical system models into standard formats and for the basic mathematical operations commonly encountered in the analysis of control systems. In addition to standard algorithms for linearisation and manipulation of linear systems, they implement algorithms for adaptive and approximate nonlinear control and provide flexible numerical simulation of the closed-loop systems via C or Matlab code. We have used the packages to generate models for complex systems including a tracked, multiwheel vehicle (15 degrees of freedom) and to design adaptive, stabilising controllers for several systems including a conical magnetic bearing (18 states, one uncertain parameter) and helicopter and tank weapons systems.

Our technical approach in developing both packages combines integrated symbolic and numerical computing with graphics pre- and postprocessing (see Fig. 15.1). Computer algebra and mathematical symbolic manipulation systems have matured substantially in recent years. Advances in this field provide an opportunity for a new approach to the assembly of models for integrated design. Model building software based on computer algebra need not constrain systems to be composed of rigidly defined sets of components. The system architecture itself can become a design vector. This capability greatly expands the design engineer's ability to devise and experiment with new types of elements and configurations. Equally important is that access to analytical tools for nonlinear design (feedback linearisation, adaptive control etc.) and analysis (nonlinear bifurcation, sliding mode etc.) is natural in this setting.

---

[1] Developed with our colleagues R. Ghanadan, V. Polyakov, C. LaVigna, and C. Teolis, based on earlier work with O. Akhrif and J.P. Quadrat. We are grateful to our colleagues and students for their contributions to this work

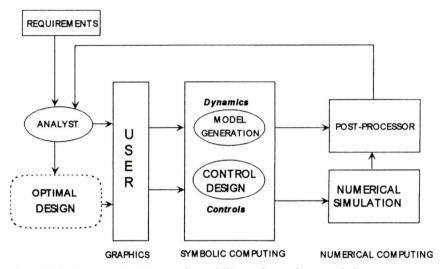

Figure 15.1   Integrated architecture for modelling and control systems design

15.2 Design of linear control systems

Although our focus is on nonlinear systems, certain basic tools are provided in **TSiControls** for linear systems in state-space or transfer-function form.[2] State-space models are:

$$\dot{x} = Ax + Bu$$
$$y = Cx + Du$$

(15.1)

where $x \in R^n$, $u \in R^m$, $y \in R^l$ and A, B, C, D are constant matrices; the transfer function models are

$$Y(s) = G(s)U(s)$$

(15.2)

where $U(s) \in C^m$, $Y(s) \in C^l$, $s \in C$ and the elements of the transfer matrix $G(s) \in C^{l \times m}$ are proper, rational functions of s.

15.2.1 Linearisation of nonlinear systems

Many control systems problems begin with a nonlinear model in the general form

$$\dot{x} = f(x, u), \qquad y = g(x, u)$$

(15.3)

---

[2] The material on linear systems discussed in this section can be found in many texts on linear control system design; see, for example, [1]

Suppose x_0, u_0 corresponds to an equilibrium point, i.e. $f(x_0, u_0) = 0$. Then the linear perturbation equations are

$$\delta\dot{x} = \left[\frac{\partial f(x_0, u_0)}{\partial x}\right]\delta x + \left[\frac{\partial f(x_0, u_0)}{\partial u}\right]\delta u$$

$$\delta y = \left[\frac{\partial g(x_0, u_0)}{\partial x}\right]\delta x + \left[\frac{\partial g(x_0, u_0)}{\partial u}\right]\delta u$$

(15.4)

TSiControls provides a function Linearize [f,g,x,x0,u,u0] that produces the linear perturbation eqns. 15.1 around the point $x = x_0$, $u + u_0$.

Example 1. Linearisation

As a simple example of the linearisation process we consider the following example:

$$\begin{bmatrix} \dot{x}_1 \\ \dot{x}_2 \end{bmatrix} = \begin{bmatrix} x_1^2 \sin x_1 - x_2 + u_1 u_2 \\ -x_2 + x_2^3 + u_1 + 3u_2 \end{bmatrix}$$

$$\begin{bmatrix} y_1 \\ y_2 \end{bmatrix} = \begin{bmatrix} x_1 + x_2 \\ x_2 + u_2^2 \end{bmatrix}$$

(15.5)

This system has an equilibrium point at $x = x_0 = (0, 0)$, $u = u_0 = (0, 0)$; using Linearize [f,g,x,x0,u,u0] we can compute the linearisation about this point. First define the system, i.e. specify f, g, x and u:

```
f={x1^2*Sin[x1]-x2+u1*u2,-x2+x2^3+u1+3*u2};
g={x1+x2,x2+u2^2};
x={x1,x2};
u={u1,u2};
```

Now linearise it:

```
{A1,B1,C1,D1}=Linearize[f,g,x,{0,0},u,{0,0}]
```

The result is a list of the four matrices A, B, C, D in order (in this case each is 2×2):

$$\{\{\{0, -1\}, \{0, -1\}\}, \{\{0, 0\}, \{1, 3\}\}, \{\{1, 1\}, \{0, 1\}\},$$
$$\{\{0, 0\}, \{0, 0\}\}\}$$

(Note that the Mathematica list syntax for an $m \times n$ matrix is $\{\{a_{11}, a_{12}, \ldots, a_{1n}\}, \ldots, \{a_{m1}, a_{m2}, \ldots, a_{mn}\}\}$.)

15.2.2 State-space and transfer-function models

TSiControls provides functions for converting from state-space models to transfer function models and vice versa:

☐ StateSpaceToTransferFunction[A,B,C,D,s] returns the matrix

$$G(s) = C[sI - A]^{-1}B + D \tag{15.6}$$

☐ ControllableRealization [G,s] returns the A, B, C, D matrices of a controllable realisation of the function $G(s)$.
☐ ObservableRealization [G,s] returns the A, B, C, D matrices of an observable realisation of the function $G(s)$.

The realisations are the so-called 'standard controllable' and 'standard observable' realisations computed as follows. Suppose that $G(s)$ is an $l \times m$ transfer matrix and that the least common denominator of the elements of the transfer matrix, $p(s)$ has degree n:

$$p(s) = s^n + p_{n-1}s^{n-1} + \cdots + p_0 \tag{15.7}$$

Now consider the expansion in s:

$$G(s) = [Q_0 + Q_1 s + \cdots + Q_{n-1}s^{n-1}]p^{-1}(s)$$

or $$\tag{15.8}$$

$$p(s)G(s) = [Q_0 + Q_1 s + \cdots + Q_{n-1}s^{n-1}]$$

We assume that the product $p(s)G(s)$ is a matrix polynomial of degree $n - 1$ or less; that is, $G(s)$ is strictly proper. Let I_m, O_m denote, respectively, the $m \times m$ identity and zero matrices. Then the *standard controllable realisation* is given by

$$A = \begin{bmatrix} O_m & I_m & O_m & \cdots & O_m \\ O_m & O_m & I_m & \cdots & O_m \\ \vdots & \vdots & \ddots & \ddots & \vdots \\ O_m & O_m & \cdots & & I_m \\ -p_0 I_m & -p_1 I_m & \cdots & & -p_{n-1} I_m \end{bmatrix} \tag{15.9}$$

$$B = \begin{bmatrix} O_m \\ O_m \\ \vdots \\ I_m \end{bmatrix} \qquad C = [Q_0, Q_1, \ldots, Q_{n-1}] \tag{15.10}$$

This realisation has mn states and it is controllable by construction, but it may not be observable (and hence not minimal).

The *standard observable realisation* is based on the Laurent expansion $G(s)$,

$$G(s) = [L_0 s^{-1} + L_1 s^{-2} + \cdots + L_{n-1}s^{-n} + \cdots] \tag{15.11}$$

an infinite series. This expansion can be generated by polynomial division, e.g. dividing through by the least common denominator $p(s)$ yields, after $n - 1$ divisions,

$$G(s) = [L_0 s^{-1} + L_1 s^{-2} + \cdots + L_{n-2}s^{-n+1} + L_{n-1}p(s)^{-1}] \tag{15.12}$$

The standard observable realisation is

$$
A = \begin{bmatrix}
O_l & I_l & O_l & \cdots & O_l \\
O_l & O_l & I_l & \cdots & O_l \\
\vdots & \vdots & \ddots & \ddots & \vdots \\
O_l & O_l & \cdots & & I_l \\
-p_0 I_l & -p_1 I_l & \cdots & & -p_{n-1} I_l
\end{bmatrix}
\tag{15.13}
$$

$$
B = \begin{bmatrix}
L_0 \\
L_1 \\
\vdots \\
L_{n-1}
\end{bmatrix}
\qquad
C = [I_l, 0, \ldots, 0]
\tag{15.14}
$$

This realisation has ln states. It is observable by construction but may not be controllable.

These realisations are easy to construct, but they are unlikely to be minimal. The dimension of the minimal realisation can be computed from the information in the Laurent series. Let us construct the *Hankel matrix*,

$$
H_{n-1} = \begin{bmatrix}
L_0 & L_1 & \cdots & L_{n-1} \\
L_1 & L_2 & \cdots & L_n \\
\vdots & \ddots & \ddots & \vdots \\
L_{n-1} & \cdots & & L_{2(n-1)}
\end{bmatrix}
\tag{15.15}
$$

The *McMillan degree* of the minimal realisation of $G(s)$ is rank $[H_{n-1}]$, where n is the degree of the least common denominator $p(s)$ of the elements of $G(s)$. A realisation (A, B, C) is *minimal* if and only if it is both controllable and observable.

Based on this, algorithms are available to reduce one of the standard realisations to a minimal realisation.

Example 2. State space to transfer function and back again

Invoking StateSpaceToTransferFunction:
  G=StateSpaceToTransferFunction[A1,B1,C1,D1,s]
returns the 2×2 matrix expression

$$
\left\{ \left\{ \frac{-1+s}{s+s^2}, \frac{3(-1+s)}{s+s^2} \right\}, \left\{ \frac{1}{1+s}, \frac{3}{1+s} \right\} \right\}
$$

We can define the transfer function as a function of s by the statement:

  G[s_] := StateSpaceToTransferFunction[A1, B1, C1, D1, s]

The realisation functions require that $G(s)$ is defined as a function. The command

  {A2, B2, C2} = ControllableRealization[G, s]

returns the *A*, *B*, *C* matrices in the Mathematica list notation

$$\{\{\{0, 0, 1, 0\}, \quad \{0, 0, 0, 1\}, \quad \{0, 0, -1, 0\},$$
$$\{0, 0, 0, -1\}\}, \quad \{\{0, 0\}, \quad \{0, 0\}, \quad \{1, 0\}, \quad \{0, 1\}\},$$
$$\{\{-1, -3, 1, 3\}, \quad \{0, 0, 1, 3\}\}\}$$

Notice that the realisation is nothing like the original system—in fact it has four states—but it does return the same transfer matrix:

$$G2 = StateSpaceToTransferFunction[A2, B2, C2, s]$$

$$\left\{\left\{\frac{-1+s}{s+s^2}, \frac{3(-1+s)}{s+s^2}\right\}, \left\{\frac{1}{1+s}, \frac{3}{1+s}\right\}\right\}$$

We revisit this example later.

15.2.3 State-space operations

A complete list of the functions available is provided in the Reference section. We provide only a brief summary here. The available operations on state-space systems include those in Table 15.1.

Example 3. Controllability and observability

Let us again consider the system of Example 1 and test it for controllability and observability:

$$Controllable[A1, B1]$$
$$True$$

$$Observable[A1, C1]$$
$$True$$

Table 15.1   Selected functions for manipulationg linear systems

Controllable/Observable	tests for controllability and observability
ControllabilityMatrix	returns the controllability or observability
ObservabilityMatrix	matrices, respectively
PolePlace	state-feedback pole placement based on Ackermann's formula with options
Decouple	state-feedback and co-ordinate transformation that decouples input-output map
RelativeDegree	computes the vector relative degree

Now, we can also test the controllable realisation obtained in Example 2 for the same system. Of course, it is controllable by construction, so we will test observability:

$$\text{Observable[A2, C2]}$$
$$\text{False}$$

This explains the additional states as expected.

Example 4. Pole placement

The function PolePlace returns a feedback gain matrix that places the closed-loop poles at specified locations. In the system of Example 1 we choose to place both poles at -2:

$$\text{Kf} = \text{PolePlace[A1, B1, \{-2, -2\}]}$$
$$\{\{-4, 3\}, \{0, 0\}\}$$

Notice that only the first control is used. Only one control is required to do the job, and this function identifies the first 'port' as the most effective. We could also specify a desired port

$$\text{PolePlace[A1, B1, 2, \{-2, -2\}]}$$
$$\left\{\{0, 0\}, \left\{-\left(\frac{4}{3}\right), 1\right\}\right\}$$

Example 5. Lateral dynamics of an aircraft

We consider a somewhat more interesting example involving the linearised lateral dynamics of an aircraft:

$$
\begin{bmatrix} \dot{p} \\ \dot{r} \\ \dot{\beta} \\ \dot{\phi} \end{bmatrix}
=
\begin{bmatrix}
-10 & 0 & -10 & 0 \\
0 & -7 & 9 & 0 \\
0 & -1 & -7 & 0 \\
1 & 0 & 0 & 0
\end{bmatrix}
\begin{bmatrix} p \\ r \\ \beta \\ \phi \end{bmatrix}
+
\begin{bmatrix}
20 & 2.8 \\
0 & -3.13 \\
0 & 0 \\
0 & 0
\end{bmatrix}
\begin{bmatrix} \delta_a \\ \delta_r \end{bmatrix}
\tag{15.16}
$$

where the state variables p, r, β, ϕ denote the roll and yaw rates, the sideslip angle and the roll angle, respectively. The controls δ_a and δ_r are, respectively, the aileron and rudder deflections. First, let us choose as outputs the sideslip angle and roll angle, and define the system

$$\text{AA} = \{\{-10, 0, -10, 0\}, \{0, -.7, 9, 0\}, \{0, -1, -.7, 0\}, \{1, 0, 0, 0\}\}$$
$$\{\{-10, 0, -10, 0\}, \{0, -0.7, 9, 0\}, \{0, -1, -0.7, 0\},$$
$$\{1, 0, 0, 0\}\}$$
$$\text{BB} = \{\{20, 2.8\}, \{0, -3.13\}, \{0, 0\}, \{0, 0\}\}$$
$$\{\{20, 2.8\}, \{0, -3.13\}, \{0, 0\}, \{0, 0\}\}$$
$$\text{CC} = \{\{0, 0, 1, 0\}, \{0, 0, 0, 1\}\}$$
$$\{\{0, 0, 1, 0\}, \{0, 0, 0, 1\}\}$$

Now, test for controllability and observability,

$$\texttt{Controllable[AA, BB]}$$
$$\texttt{True}$$
$$\texttt{Observable[AA, CC]}$$
$$\texttt{True}$$

and compute the transfer function

$$\texttt{K[s\_] := StateSpaceToTransferFunction[AA, BB, CC, s]}$$
$$\texttt{K[s]}$$

$$\left\{ \left\{ 0, \ \frac{31.3\,s + 3.13\,s^2}{94.9\,s + 23.49\,s^2 + 11.4\,s^3 + s^4} \right\}, \right.$$

$$\left. \left\{ \frac{20(9.49 + 1.4\,s + s^2)}{94.9\,s + 23.49\,s^2 + 11.4\,s^3 + s^4}, \ \frac{-4.728 + 3.92\,s + 2.8\,s^2}{94.9\,s + 23.49\,s^2 + 11.4\,s^3 + s^4} \right\} \right\}$$

We will return to this example.

15.2.4 Transfer-function operations

In addition to the realisation constructions described above, there are other operations on transfer-function models that are implemented in **TSiControls**. A partial listing is given in Table 15.2.

Example 6. McMillan degree and Laurent series for aircraft lateral dynamics

Continuing Example 5, we compute the McMillan degree of the transfer function for the lateral dynamics of the aircraft,

$$\texttt{McMillanDegree[K, s]}$$
$$4$$

Table 15.2   Selected functions for constructing linear system realisations

LeastCommonDenominator	finds the least common denominator of the elements of a proper, rational $G(s)$
Poles	finds the roots of the least common denominator of all minors of all orders
LaurentSeries	computes the Laurent series up to specified order
HankelMatrix	computes the Hankel matrix associated with Laurent expansion of a transfer function $G(s)$
McMillanDegree	computes the degree of the minimal realisation of a transfer function $G(s)$

Next we compute the Laurent series up to order 3,

```
LaurentSeries[K, s, 3]
```

$$\left\{\left\{\left\{0, -\frac{3.13}{s^2} - \frac{4.382}{s^3} + O\left[\frac{1}{s}\right]^4\right\}, \right.\right.$$
$$\left.\left.\left\{-\frac{20}{s^2} - \frac{200}{s^3} + O\left[\frac{1}{s}\right]^4, -\frac{2.8}{s^2} - \frac{28}{s^3} + O\left[\frac{1}{s}\right]^4\right\}\right\}\right\}$$

Consider a variation on this problem in which only one control action, aileron deflection δ_a, is used to control the roll angle φ. We build the new transfer function,

```
K1[s_]:= StateSpaceToTransferFunction[AA, BB[[{1, 2, 3, 4}, {1}]], CC[[{2},
{1, 2, 3, 4}]], s]
K1[s]
```

$$\left\{\left\{\frac{20(9.49 + 1.4\ s + s^2)}{94.9\ s + 23.49\ s^2 + 11.4\ s^3 + s^4}\right\}\right\}$$

We can compute the McMillan degree

```
McMillanDegree[K1, s]
2
```

so that we determine that a minimal realisation is of dimension 2. We can compute an observable realisation as follows:

```
ObservableRealization[K1, s]
{{{0, 1}, {0, −10}}, {{0}, {20}}, {{1, 0}}}
```

The fact that the observable realisation is of order 2 (minimal) suggests that the single input, single output problem was controllable but not observable. We can confirm this as follows:

```
Controllable[AA, BB[[{1, 2, 3, 4}, {1}]]]
False
Observable[AA, CC[[{2}, {1, 2, 3, 4}]]]
True
```

15.2.5 *The linear quadratic Gaussian problem*

TSiControls includes several functions for solving the linear quadratic Gaussian design problem. The main tool is a function that solves the algebraic Riccati equation. Riccati equation solutions are used to obtain the optimal state-feedback controller and the optimal stochastic state estimator. First, we give a brief summary of the problem.

Consider a linear system described by the state and output equations

$$\dot{x} = Ax + Bu + Gw(t)$$

$$y = Cx + Du + v(t)$$

where $w(t)$ (driving noise) and $v(t)$ (the measurement noise) are uncorrelated, zero mean, Gaussian white stochastic processes, with respective covariances

$$E[w(t)w^T(\tau)] = W\delta(t-\tau), \quad E[v(t)v^T(\tau)] = V\delta(t-\tau)$$

W and V are positive definite symmetric matrices. The *Linear Quadratic Gaussian Control* (LQG) problem is to find a controller that operates on the measurements y to produce a control u such that the quadratic performance index

$$\bar{J} = \lim_{T \to \infty} E\left[\frac{1}{T}\int_0^T \{x^T Qx + u^T Ru\}dt\right]$$

is a minimum. Here Q is assumed to be a non-negative-definite symmetric matrix and R is a positive-definite symmetric matrix.

It is well known that the LQG problem can be solved by dividing it into two simpler subproblems:

1. Solve the deterministic *Linear Quadratic Regulator* (LQR) problem, i.e. find the state-feedback controller $u = Kx$ that minimises the performance index

$$\bar{J} = \lim_{T \to \infty} \frac{1}{T}\int_0^T \{x^T Qx + u^T Ru\}dt$$

subject to the differential equation constraint

$$\dot{x} = Ax + Bu$$

2. Solve the stochastic *Linear Quadratic Estimator* (LQE) problem, i.e. determine the minimum variance state estimate $\hat{x}(t)$ (minimises $E[\{\hat{x}(t) - x(t)\}\{\hat{x}(t) - x(t)\}^T]$) based on the observations $\{y(\tau), 0 < \tau < t\}$.

The control that is implemented, $u^*(t) = K\hat{x}(t)$, solves the LQG problem. This fact is called the *separation principle*. Solution of both the LQR and LQG require solution of the algebraic Riccati equation. We describe tools for solving this equation and the LQR, LQE problems in the following paragraphs.

15.2.5.1 Algebraic Riccati equation

The method of solution provided here gives infinite precision results when the data is provided in infinite precision (assuming the solution exists). Results of this type are interesting only for rather simple systems, and they are time-consuming to obtain. Numerical (finite-precision) results are produced (and much more quickly) when numerical data are provided. Consider the solution of the equation

$$PA + A^T P - PUP + Q = 0$$

with the following data:

$$A = \{\{0, 0\}, \{1, -\mathrm{Sqrt}[2]\}\}; U = \{\{1, 0\}, \{0, 0\}\}; Q = \{\{0, 0\}, \{0, 1\}\};$$

The result is obtained with

```
{P, Eigs} = AlgRiccatiEq[A, Q, U]
Hamiltonian has nonsimple divisors.
```

$$\left\{\left\{\left\{\frac{2+\mathrm{Sqrt}[2]}{3+2\mathrm{Sqrt}[2]}, \frac{1}{3+2\mathrm{Sqrt}[2]}\right\}, \left\{\frac{1}{3+2\mathrm{Sqrt}[2]}, \frac{2}{3+2\mathrm{Sqrt}[2]}\right\}\right\}, \{-1, -1\}\right\}$$

The matrix P is the solution of the Riccati equations, and the list Eigs contains the eigenvalues of the associated Hamiltonian matrix that have negative real parts (the 'closed-loop' eigenvalues). Notice that the results are indeed given in infinite precision. Also, a message is received indicating that the Hamiltonian matrix has an incomplete set of eigenvectors so that a more time-consuming method must be used. Consider another set of data,

$$A2 = \{\{0, 1, 0\}, \{0, 0, 1\}, \{0, 0, 0\}\};$$
$$Q2 = \{\{1, 0, 0\}, \{0, 0, 0\}, \{0, 0, 0\}\};$$
$$U2 = \{\{0, 0, 0\}, \{0, 0, 0\}, \{0, 0, 1\}\};$$
$$\{P, Eigs\} = AlgRiccatiEq[A2, Q2, U2]$$
$$\{\{\{2, 2, 1\}, \{2, 3, 2\}, \{1, 2, 2\}\}, \{-1, -(-1)^{1/3}, (-1)^{2/3}\}\}$$

In this case, a complete set of eigenvectors does exist so the preferred method is used and no messages appear. Even in this simple case the increase in speed obtained when using finite precision is quite evident. For example, reformulating the problem as

$$\{P, Eigs\} = AlgRiccatiEq[N[A2], Q2, U2]$$
$$\{\{\{2, 2, 1\}, \{2, 3, 2\}, \{1, 2, 2\}\},$$
$$\{-1, -0.5 + 0.866025I, -0.5 - 0.866025I\}\}$$

leads to much more rapid computation.

15.2.5.2 Optimal state feedback

Consider the linear control system with (A, B) parameter matrices

$$A = \{\{0, 0\}, \{1, -\mathrm{Sqrt}[2]\}\}; B = \{\{1\}, \{0\}\};$$

We wish to design a state-feedback controller that minimises a quadratic performance index with control weighting matrix R and state weighting matrix Q:

$$R = \{\{1\}\}; Q = \{\{0, 0\}, \{0, 1\}\};$$

This is accomplished with the LQR function:

```
{K, P, Eigs} = LQR[A, B, Q, R]
Hamiltonian has nonsimple divisors.
```

$$\left\{\left\{\left\{-\left(\frac{2+\mathrm{Sqrt}[2]}{3+2\mathrm{Sqrt}[2]}\right), -\left(\frac{1}{3+2\mathrm{Sqrt}[2]}\right)\right\}\right\},\right.$$

$$\left.\left\{\left\{\frac{2+\mathrm{Sqrt}[2]}{3+2\mathrm{Sqrt}[2]}, \frac{1}{3+2\mathrm{Sqrt}[2]}\right\}, \left\{\frac{1}{3+2\mathrm{Sqrt}[2]}, \frac{2}{3+2\mathrm{Sqrt}[2]}\right\}\right\}, \{-1, -1\}\right\}$$

LQR returns the state feedback gain K, the Riccati solution P and the closed-loop eigenvalues, i.e. the eigenvalues of the matrix $(A + BK)$.

15.2.5.3 Optimal stochastic state estimator

When the states are not directly measured, it is necessary to use a state estimator. The function LQE computes the optimal least-squares estimator parameters for a system with (white) driving noise and measurement noise. With the dynamical process defined above, the estimator is of the form

$$\dot{\hat{x}} = A\hat{x} + Bu + L(C\hat{x} + Du - y)$$

and the matrix parameter L is obtained from the function LQE. For example, consider the system used above with additional parameters:

$$CC = \{\{0, 1\}\}; G = \{\{1\}, \{1\}\}; W = \{\{1\}\}; V = \{\{1\}\};$$

LQE is used to compute L:

$$\{L, S, Eigs\} = LQE[A, G, CC, W, V];$$

While these computations are carried out with infinite precision, we can get a more compact output of the results with

```
N[%]
{{{-1}, {-0.821854}}, {{1.23607, 1}, {1, 0.821854}}, {-1.61803, -0.618034}}
```

As does LQR, LQE returns three objects, the observer parameter matrix L, the associated Riccati solution S (which is the error covariance matrix), and the estimator eigenvalues Eigs, the eigenvalues of $(A + LC)$.

Here again, finite-precision computation is substantially faster, i.e. entering the problem in the form

```
{L, S, Eigs} = LQE[N[A], G, CC, W, V]
{{{-1}, {-0.821854}}, {{1.23607, 1}, {1, 0.821854}}, {-1.61803, -0.618034}}
```

initiates a finite precision calculation.

15.3 Design of nonlinear control laws

Given the capability to generate models with embedded (control) forces and torques, the natural complement is a system for the computation of effective control laws. Since we are interested in designing the architecture of the control system as well as in crafting specific algorithms, it is important to use symbolic computing methods in the design process. As noted earlier, while there has also been a large body of work on software for the design and analysis of linear control systems, there has been much less work on tools for the design of nonlinear control systems. In this section we describe one approach to the synthesis of such tools starting from the geometric formulation of nonlinear control theory.

In 1987, O. Akhrif developed the first computational tools for the design of nonlinear control systems using symbolic computing (MACSYMA)[2]. This work was inspired by the work of J.P. Quadrat and his colleagues on the use of MACSYMA (and Prolog) in the treatment of optimal stochastic control problems [3]. The work here builds on the tools developed by Akhrif and Ghanadan. It employs techniques of nonlinear adaptive control [2,4–6] and performance evaluation by simulation.

The **TSiControls** package includes several functions for computation of mathematical objects frequently encountered in control system analysis, such as Lie derivatives, Lie brackets and controllability distributions, along with functions for synthesis (e.g. the dynamic extension algorithm, decoupling control algorithms of Hirschorn and Singh, adaptive and approximate linear-isation algorithms of Ghanadan and Blankenship, and Kokotovic and Kanellopoulos, Krener, Sastry, Slotine, and many others), as well as functions for automatic C and Matlab code generation.

The tools presented here have been applied to realistic nonlinear problems for which hand calculation is not feasible and for which conventional tools (e.g. Matlab, MatrixX, etc.) are not well suited. Earlier versions of this package were used to design controllers for an active automotive suspension and a magnetic levitation system [4,7]. The power of the tools can be seen from our work on designing an adaptive tracking controller for conical magnetic bearings, an 18-dimensional system with complicated nonlinear dynamics [8]. We also consider a vehicle steering problem and show how to compute the (local) zero dynamics of the model produced by the **TSiDynamics** package.

15.3.1 Controls package description

The **TSiControls** package deals with MIMO nonlinear systems in the following form:

$$\dot{x} = f(x, \theta) + \sum_{k=1}^{m} g_k(x)u_k(t), \qquad y(t) = h(x)$$

where θ is a vector of (unknown) parameters. The tools in the **TSiControls** package fall into four general categories: (i) basic analysis tools; (ii) model representation; (iii) controller design and (iv) simulation.

15.3.2 Basic analysis tools

There are several mathematical operations that occur frequently in nonlinear control systems design. Although these operations involve straightforward mathematics, actual computation is tedious and time consuming, especially for large ($n > 5$ state) systems. The most common of these mathematical tools are Lie derivatives and Lie brackets. The Lie derivative of a function h relative to a function f is defined by

$$L_f^0(h) = h$$

$$L_f^k(h) = \frac{\partial L_f^{k-1}(h)}{\partial x} \cdot f$$

This algorithm may be expressed in Mathematica as the following sequence of 'rules':

```
LieDerivative[f_, h_, x_, 0] := h
LieDerivative[f_, h_, x_] :=
Dot[Jacob[h, x], f]
LieDerivative[f_, h_, x_, 1] := LieDerivative[f, h, x]
LieDerivative[f_, h_, x_, n_] :=
LieDerivative[f, LieDerivative[f, h, x], x, n − 1]
```

Here `Jacob[h,x]` is the Jacobian matrix of h with respect to x and `Dot` is a Mathematica function for multiplying arrays (matrices). The definitions of `LieDerivative` make use of Mathematica's pattern checking and conditional definition capabilities to ensure that both the arguments and the answers make sense.

Mathematica has the capability to use 'pure' functions in rules. This is a particularly convenient construction for creation and maintenance of control system models. The power of this feature can be seen in the definition of the Jacobian in Mathematica.

The first line in the following is a usage statement associated with the help system in Mathematica. The second line is the computation of the gradient:

```
Grad::usage="Grad[f,varlist] computes the Grad of the
function f with respect to the list of variables varist."
Grad[f_, var_List] := D[f, #]&/@var
```

In the definition of `Grad`, the expression `D[f,#]&` is a pure (un-named) function. The symbol D stands for derivative; so `D[f,x]` is the derivative of f with respect to a (single) variable x. To compute the gradient of a scalar function of a vector, we must compute its derivative with respect to each element of the vector. This is accomplished by 'mapping' the operation 'take the deriva-

tive of f with respect to a variable' (this is the meaning of the expression
D[f,#]&). The symbol & stands for a 'name' that one might assign to the
function 'take the derivative'. However, since we will only use the pure function
once, we do not need to name it. Similarly, we do not need to name the variable
that is its argument, so the symbol # is used as a place marker.

Arguments to function definitions in Mathematica are of the form
h[x_]:=x^2, which means that any symbol substituted for the place holder x_
is raised to the second power. The form var_List means that the argument
must be a list, a form of data validation built into Mathematica.

The symbol /@ stands for the Mathematica operation Map; so we could have
written the definition as

$$\text{Grad}[f\_, \text{var}\_\text{List}] := \text{Map}[D[f, \#]\&, \text{var}]$$

The use of pure functions and the capability to map functions over sets of
arguments are powerful constructions which increase the expressive power of
Mathematica programs. Map is especially useful in avoiding procedural programming constructions. The use of Map in the definition of Grad illustrates the
capability of Mathematica to treat functions as objects like symbols or numbers
and use them as arguments to other functions.

We use two lines (rules) to define the Jacobian of a function with respect to a
vector. The first handles the case when the function is a vector function of a
vector argument. The second handles the case of scalar functions (of vector
arguments). These may be regarded as rules for the computation. Mathematica
uses a kind of pattern matching to find the case that applies.

$$\text{Jacob}[f\_\text{List}, \text{var}\_\text{List}] := \text{Outer}[D, f, \text{var}] /; \text{VectorQ}[f]$$
$$\text{Jacob}[f\_, \text{var}\_\text{List}] := \text{Grad}[f, \text{var}]$$

Outer is a built-in Mathematica function which provides a generalised outer
product. The test VectorQ[f] defined by the condition symbol '/;' checks that
f is a vector. If the test succeeds, this rule is used. If not, the next one is used.

The next function illustrates the use of condition checking in Mathematica
in more detail. The symbol '&&' is logical 'and'. In the first rule, we check that
the functions are vector valued, that their lengths are identical (==), and that
the lengths equal the length of the vector of variables. If this compound test
succeeds, the rule is used.

```
LieBracket[f_List, g_List, var_List] := (Jacob[g, var].f – Jacob[f, var].g/;
VectorQ[f]&&VectorQ[g]&&Length[f] == Length[g] == Length[var])
(* Test the data *)
LieBracket[f_, g_, var_List] :=
Jacob[g, var]f – Jacob[f, var]g
```

The next sequence illustrates the recursive power of the language to define the Ad operator (we omit the vector cases):

```
Ad :: usage = "Ad[f, g, varlist, n] computes the nth Adjoint of the
functions f, g with respect to the variables varlist."
Ad[f, g, var, 0] = g
Ad[f, g, var, n] = LieBracket[f, Ad[f, g, var, n - 1], var]
Ad[f, g, var] = Ad[f, g, var, 1]
```

Using these functions, we can express the Hunt-Su-Meyer conditions [9,10] in Mathematica functions:

```
ControllabilityDistribution[f_, g_, var_List] := Module[{k},
Table[Ad[f, g, var, k], {k, 0, Length[var] - 1}]]
Controllable[f_, g_, var_List] :=
If[Rank[ControllabilityDistribution[f, g, var]] == Length[var],
True, False];
FeedbackLinearizable[f_, g_, var_List] := Module[{cm, cm1, k},
cm = Table[Ad[f, g, var, k], {k, 0, Length[var] - 1}]
cm1 = Drop[cm, -1]; (* drop last element *)
If[Rank[cm] == Length[var]
(* system is controllable *)
&&Involutive[cm1, var], True, False]];
```

The `Module` construction permits the use of local variables in the definition of functions. We use the Mathematica `Table` function to construct a set of derived vector fields. The function `Involutive` checks that a set of vector fields is *involutive*, that is, closed under the Lie Bracket:

```
Involutive[f_List, var_List] :=
Module[{k, h, vec},
k = Length[f];
h = Table[LieBracket[f[[i]], f[[j]], var], {i, 1, k}, {j, i + 1, k}];
vec = Union[Flatten[h, 1], f];
If[Rank[vec] > Rank[f], False, True]]
```

In this expression the notation `f[[i]]` takes the element of the list (vector) `f`. `Union` and `Flatten` are Mathematica functions for manipulating lists.

With these simple operations we can define several useful functions for the analysis of nonlinear control systems, including

```
RelativeDegree[f,g,h,x],   VectorRelativeDegree[f,g,h,x],
DynamicInverse[f,g,h,x], ZeroDynamics[f,g,h,x],
```

etc. These functions were implemented by translating into Mathematica functional syntax the definitions found in standard nonlinear control texts [9,33].

---

[3]The System construct discussed here is not in fact used in the package TSiControls. We present it to illustrate the interesting concept of representing a 'system' as an object, abstracted to a fairly high level

15.3.3 Model representation as system objects

While it is natural to work with conventional function definitions for the vector fields that occur in nonlinear control problems, it is interesting to create a 'data structure' for maintaining models.[3] The pure function construction in Mathematica is an effective means for accomplishing this.

15.3.3.1 System

The data defining the controlled nonlinear system are stored as a Mathematica data object with 'head' System and the associated structure

System[f, g, h, x, y, u, theta, analysisdata]

where f, g and h are Mathematica functions, x, y and u are lists containing labels of the states, the outputs and the inputs, respectively, and theta is a list of uncertain parameters found in f, g or h. As various functions are applied to the System, their results are appended in the list analysisdata. The System object provides an economical and efficient organisation for often bulky and unenlightening expressions.

15.3.3.2 MakeSystem

Constructing a System object is made relatively easy by the MakeSystem function, which has the following syntax:

MakeFunction[f, g, h, x, u, y, u, theta]

Although generally the components of the system model (f,g,h) are stored as pure functions, the first three arguments of MakeSystem can also be given as ordinary Mathematica expressions. MakeSystem, if necessary, automatically converts the f,g,h to functions and makes sure the dimensions agree before returning a valid System object. For example, the data

```
var = {x1[t], x2[t]};
f := {#[[2]], 2 omega xi(1 – mu#[[1]]^2)#[[2]] – omega^2#[[1]]}&;
g := {{0}, {1}}&;
h := {#[[1]]}&;
sys = MakeSystem[f, g, h, var];
```

construct the equations of the controlled Van der Pol Oscillator with output x1.

15.3.3.3 ShowSystem and GetResults

In order to examine the contents of the System object and extract the results that were appended to it by previous analyses, two useful functions are provided: ShowSystem and GetResults.

ShowSystem[sys], where sys is a valid System, will display the data of the system, f,g,h, etc., as well as a list of any functions which have been applied and whose results are contained in the data portion of this System.

`GetResults[sys, ''analysis'']` will return the results of a function called 'analysis' which has been applied to `sys` earlier. For example, to extract results of `Singh` from `demosys` one would use

$$\texttt{GetResults[demosys,"Singh"]}$$

If the results are not contained within the `System` a string 'Not found' is returned.

15.3.4 Design functions

In this section we describe functions for the design of nonlinear control laws, including adaptive and approximate methods.

15.3.4.1 Hirschorn and Singh

Predecessors to the **TSiControls** package implemented two functions for partial (input-output) feedback linearisation via construction of right inverse systems using the algorithm of Hirschorn [13,14] and its extension by Singh [15,8]. Since Singh's algorithm is applicable to a wider class of systems, we discuss its implementation. (Hirschorn's algorithm is implemented in a function called Hirschorn with syntax identical to Singh.) **TSiControls** includes a version of Singh's algorithm.

The recursive nature of Singh's algorithms is well suited for implementation in Mathematica. The command to apply Singh's algorithm is

$$\texttt{Singh[sys, opts]}$$

where `sys` is a valid System and `opts` are options described below. Singh will append the following to the System:

$$\texttt{SinghResults[D, c, K, z]}$$

where $z = c + Du$ and K defines the relationship between z and y and its derivatives. Several options are available for `Singh`. `ScreenOutput-> False` will disable almost all screen output. `ReturnObject-> List`, instead of returning the original System with the results appended, will simply return a list of the results.

15.3.4.2 AdaptiveTracking

The function `Singh` forms the foundation for `AdaptiveTracking` which implements the adaptive algorithm of Ghanadan and Blankenship [4–6], basically an adaptive observer. Given a System object with a list of uncertain parameters, `AdaptiveTracking` computes the control law and the parameter update law to track a desired trajectory.

The syntax for `AdaptiveTracking` is

$$\texttt{AdaptiveTracking[sys, poles, adgain, opts]}$$

where sys is a valid System object, poles is a list of lists and adgain is the adaptive gain used in constructing the parameter update law. adgain can be supplied in two forms: a constant which sets the same gain for all parameters or a vector in which the ith element sets the gain for the ith parameter.

As in Singh, options for AdaptiveTracking include ScreenOutput and ReturnObject. In addition, Simulate->Matlab option prepares the output to be simulated using MATLAB as described below.

15.3.4.3 ApproximateAdaptiveTracking

Results of approximate feedback linearisation theory [5,11,17] are useful design alternatives to the more restrictive schemes based on exact (partial) feedback linearisation. This scheme assumes milder involutivity and invertibility restrictions and can be applied to slightly non-minimum-phase nonlinear systems as well. In Reference 5 an adaptive approximate tracking and regulation scheme was presented for nonlinear systems with uncertain parameters. The function ApproximateAdaptiveTracking implements this scheme as a Mathematica function with syntax

```
ApproximateAdaptiveTracking[sys,poles,observerpoles,
UpdateLawGain]
```

where observerpoles is a list of desired observer poles for the adaptive scheme of [5]. The tracking function searches for linear functions of unknown/uncertain parameters theta specified in the dynamics. The regulation version of this algorithm can handle parameters that do not necessarily appear linearly in the system.

15.3.5 Simulation

15.3.5.1 C, MATLAB and Simulink code generation

Included in the package are two functions for automatic code generation in C and Fortran. These functions automatically write a subroutine compatible with the Numerical Recipes [34] integrator, odeint, compile the program, execute it and return the results to Mathematica. The following Mathematica command line will execute the operations listed above:

```
SimulateC[sys,rules,ic,tfin,"AdaptiveTracking",tol]
```

where sys contains results of AdaptiveTracking, rules is a list of substitutions which are made before simulation is executed, ic are the initial conditions and tol is an optional tolerance specification.

15.3.5.2 Simulate

Functions called Simulate and MatlabSimulate are included in the package to provide simulation capabilities in MATLAB. This is important for systems, like the magnetic bearing described in the following section. Due to memory limitations it is not possible to analyse such a large model using Mathematica alone. For example, in computing a control law for the conical magnetic bearings, the function Singh found the 5×8 decoupling matrix, D, which occupied 1.6 KB (ASCII) and its pseudoinverse D^∇ which occupied 3.87 MB (ASCII). Thus, D^∇ was too large to be manipulated, and the control law, when saved as ASCII text, was approximately 16 MB. Consequently, straightforward inclusion of the control and parameter update laws into C or Fortran code was impractical for this application.

MatlabSimulate writes a Matlab function which at each time instant evaluates the components of the control law, numerically computes the pseudoinverse of D using the Matlab function pinv and then performs the necessary matrix multiplications and additions to find the control. Besides allowing simulation for large systems, linking to Matlab in this way provides extra flexibility in selecting time limits, tolerance and initial conditions without the need to recompile every time a change is made. The disadvantage of this method is slower computation time.

If the simulation is to be performed using MatlabSimulate, the option Simulate -> Matlab must be used when performing Adaptive-Tracking, e.g.

AdaptiveTracking[sys, poles, adgain, MatlabSimulate ->True]

Next, we need to form the substitution rules for the desired output trajectory and its derivatives as well as the actual output and its derivatives. The latter can be accomplished using the function BuildSubRules with the following syntax:

BuildSubRules[sys, vectorrelativedegree]

The vector relative degree is displayed in the course of running Singh or it can be computed using VectorRelativeDegree. The substitution rules for the desired output and its derivatives must be provided by the user.

Finally, we can automatically write a Matlab function for simulation using

Simulate[sys,"Matlab","dir","filename", rules]

The Matlab function will be stored in the file called filename.m in directory dir and can be integrated using standard Matlab integrators, e.g. ode45. In fact, two options are available to use Matlab to simulate systems. If the option Matlab is selected, then Simulate generates a Matlab function that will simulate the system using the Matlab ODE solvers. If the option Simulink is selected, then Simulate generates a file that generates a Simulink block diagram, and the simulation can be run from the Simulink environment.

15.3.6 Application: adaptive control of a conical magnetic bearing

The design of control laws for conical magnetic bearings provides a nontrivial test for linear and nonlinear design methodologies. For the bearing configuration shown in Fig. 15.2 we use the model derived by Mohamed and Emad [8], which has 18 states, eight controls, eight outputs and several disturbances. We include an uncertain parameter representing rotor angular velocity. Using the functions in the **TSiControls** package we first model the magnetic bearings as a System object and then design and simulate a nonlinear adaptive control that achieves asymptotic tracking.

15.3.6.1 Model

The following Mathematica script defines the model based on the analysis in [19]. First we define the right-hand side of the model,

```
f11 := k x11[t]*x11[t]*(1 + 2(D0 + x1[t])/(pi*h))
f12 := k x12[t]*x12[t]*(1 + 2(D0 − x1[t])/(pi*h))fr1 := k x13[t]*x13[t]*
(1 + 2(D0 + x2[t])/(pi*h))fr2 := k x14[t]*x14[t]*(1 + 2(D0 − x2[t])/(pi*h))
f13 := k x15[t]*x15[t]*(1 + 2(D0 + x3[t])/(pi*h))f14 := k x16[t]*x16[t]*
(1 + 2(D0 − x3[t])/(pi*h))fr3 := k x17[t]*x17[t]*(1 + 2(D0 + x4[t])/(pi*h))fr4 :=
k x18[t]*x18[t]*(1 + 2(D0 − x4[t])/(pi*h))
ran = 2R/(mu0*A*N); cos = Cos[sigma]; mg = m*g/2; H1 = ((l*l/Jy) + 1/m)
cos; H2 = ((l*l/Jy) − 1/m)cos;
rhs1 := x6[t];
rhs2 := x7[t];
rhs3 := x8[t]
rhs4 := x9[t];
rhs5 := x10[t];
rhs6 := alpha/(2m)(x1[t] + x2[t]) − pJx/(2Jy)(x8[t] − x9[t]) − H1((f11 − f12)
cos − mg) + H2((fr1 − fr2)cos − mg);
rhs7 := alpha/(2m)(x1[t] + x2[t]) + pJx/(2Jy)(x8[t] − x9[t]) − H1((fr1 − fr2)
cos − mg) + H2((f11 − f12)cos − mg);
rhs8 := alpha/(2m)(x3[t] + x4[t]) + pJx/(2Jy)(x6[t] − x7[t]) − H1(f13 − f14)
cos + H2(fr3 − fr4)cos;
```

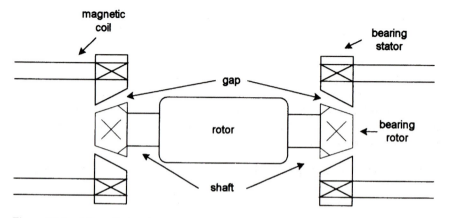

Figure 15.2   Magnetic bearing system

```
rhs9 := alpha/(2m)(x3[t] + x4[t]) - pJx/(2Jy)(x6[t] - x7[t]) - H1(fr3 - fr4)
cos + H2(fl3 - fl4)cos;
rhs10 := -beta/mx5[t] - 2gamma/mx10[t] + Sin[sigma]/m((fl1 + fl2 + fl3
+ fl4) - (fr1 + fr2 + fr3 + fr4));
rhs11 := ((43.284433 + u1) - ran(D0 + x1[t])x11[t])/N; rhs12 := ((44.208506
+ u2) - ran(D0 - x1[t])x12[t])/N;


rhs13 := ((43.284433 + u3) - ran(D0 + x2[t])x13[t])/N;
rhs14 := ((44.208506 + u4) - ran(D0 - x2[t])x14[t])/N;
rhs15 := ((44.208506 + u5) - ran(D0 + x3[t])x15[t])/N;
rhs16 := ((44.208506 + u6) - ran(D0 - x3[t])x16[t])/N;
rhs17 := ((44.208506 + u7) - ran(D0 + x4[t])x17[t])/N;
rhs18 := ((44.208506 + u8) - ran(D0 - x4[t])x18[t])/N;
rhs := {rhs1, rhs2, rhs3, rhs4, rhs5, rhs6, rhs7, rhs8, rhs9, rhs10,
rhs11, rhs12, rhs13, rhs14, rhs15, rhs16, rhs17, rhs18};
Erhs := Expand[rhs];
```

Then we identify $f(x, \theta)$, $g(x, \theta)$ from $f(x, u, t, \theta)$ using the fact that u appears linearly.

```
g1 := Coefficient[Erhs, u1, 1];
g2 := Coefficient[Erhs, u2, 1];
g3 := Coefficient[Erhs, u3, 1];
g4 := Coefficient[Erhs, u4, 1];
g5 := Coefficient[Erhs, u5, 1];
g6 := Coefficient[Erhs, u6, 1];
g7 := Coefficient[Erhs, u7, 1];
g8 := Coefficient[Erhs, u8, 1];
g := Transpose[{g1, g2, g3, g4, g5, g6, g7, g8}];
f := Erhs - g.u;
x := {x1[t], x2[t], x3[t], x4[t], x5[t], x6[t], x7[t], x8[t], x9[t], x10[t],
x11[t], x12[t], x13[t], x14[t], x15[t], x16[t], x17[t], x18[t]}; y := {y1[t], y2[t],
y3[t], y4[t], y5[t]}; u := {u1[t], u2[t], u3[t], u4[t], u5[t], u6[t], u7[t], u8[t]};
```

To speed up the calculation, wherever possible we substitute numerical values for any parameters which are known with certainty. In this case, Values is the list of substitution rules. The Mathematica symbol /. is a short-hand notation for the function Substitute. Chop removes values below a given precision level (due to numerical errors).

```
nf = Chop[f/.Values]; ng = Chop[g/.Values]; nh = Chop[h/.Values];
```

Using MakeSystem we create a valid System object:

```
magbear = MakeSystem[nf, ng, nh, x, y, u, {p}];
```

15.3.6.2 Control system design

Given the model, we apply AdaptiveTracking with options that prepare the output for Matlab simulation:

```
poles = Table[-10^3, {5}, {3}]; adgain = 10^4; magbear =
AdaptiveTracking[magbear, adgain, poles, Simulate->
Matlab, ScreenOutput ->False];
```

Throughout the above manipulations the rotor angular velocity was allowed to remain in its symbolic form, p. In general, the angular velocity may not be known with certainty and we treat p as uncertain parameters. The parameter update law computed by AdaptiveTracking will allow us to track the desired output.

Since the outputs are the deviation from gap equilibrium value, our goal is to track zero. Therefore, we define the desired output and its derivatives

```
ydes[1] = 0;
ydes[2] = 0;
ydes[3] = 0;
ydes[4] = 0;
ydes[5] = 0; outdrule = Table[outd[i][y]-> D[ydes[i], {t, j}], {i, 5}, {j, 0, 3}];
```

Next, we use BuildSubRules to write $y(t)$ and its derivatives as functions of x,

```
        thetarule = Thread[p, p^2]- > thetabar 1[t],
        thetabar 2[t]]
        yrule = BuildSubRules[magbear, {3, 3, 3, 3, 3}, thetarule]
```

Finally, we combine the above rules with the substitution rule for the actual value of p,

```
rules = Join[outdrule, yrule, {p- > 10^4}];
```

and write a Matlab function to simulate the controlled system

```
MatlabSimulate[magbear"~/MagBear/","magsim", rules];
```

The following Matlab code simulates the magnetic bearings:

```
ic =[zeros(1, 10),
0.002061, 0.002105, 0.002061, 0.002105, 0.002105, 0.002105, 0.002105,
0.002105, 0, 0];
tfin=1; tstart=0; [t, state]=ode45('magsim', tstart, tfin, ic, 1.e-7, 1
```

The simulation was performed for values of p ranging from 10 to 105. The nonlinear control law stabilised the system with initial errors of the order of the equilibrium gap length (0.5 mm)—a substantial improvement (factor of 5) over the results obtained using linear controls in [8]. Additional details of the analysis of this system can be found in Reference 20.

15.3.7 *Computing the local zero dynamics*

To illustrate the integrated use of the **TSiDynamics** [20,32] and **TSiControls** packages, we consider the problem of steering the car. The equations for the car following a prescribed trajectory can be easily generated by the **TSiDynamics** package [21]. We use the **TSiControls** package to compute the zero dynamics for the case of motion along a straight path. The vector relative degree is found to be [1, 1]. Therefore, the zero dynamics involve three first-order differential equations in the zero dynamics 'state' variables. The zero dynamics are computed using the following program:

```
≪TSiControls`(* Load Control Package *)
(* compute relative degree *)
ro = VectorRelativeOrder[f, g, h, var];
(* compute feedback linearizing/decoupling control *)
{R1, R2, R3, R4, u} = IOLinearize[f, g, h, var];
(* compute linearizable coordinates *)
z = NormalCoordinates[f, g, h, var, ro]/.{wd- > 0};
(* shift origin to point of interest *)
{f, g, h, u, z} = {f, g, h, u, z}/.{x2- > x2 + Vd};
(* compute zero dynamics *)
u0 = u/.{v1- > 0, v2- > 0};
f0 = LocalZeroDynamics[f, g, h, var, u0, z];
(* linearize zero dynamics and determine stability of origin *)
Anu = Jacob[f0, {w1, w2, w3}]/.{w1- > 0, w2- > 0, w3- > 0, b- > a + nu};
Eigenvalues[Anu/.{nu- > 0, s- > 0}]
```

Up to fourth order terms the zero dynamics are defined by the equations w1dot = {w2}

w2dot = {(kappa*(2*a + R*s)*w1)/(2*Izz) – (kappa*(a + R*s)*w1^3)/
(2*Izz) + (kappa*(–2*a + 2*b – R*s)*w3)/(2*Izz*Vd) +
(kappa*(–2*a + 2*b – R*s)*w3^3)/(12*Izz*Vd^3) + w1^2*((kappa*(a + R*s)*
w3)/(2*Izz*Vd)) + w2*((a*kappa*R*s)/(2*Izz*Vd) +
(a*kappa*R*s*w1*w3)/(2*Izz*Vd^2) – (a*kappa*R*s*w3^2)/
(4*Izz*Vd^3) + w1^2*(–(a*kappa*R*s)/(2*Izz*Vd)))}
w3dot = {(kappa*w1)/m1 – (kappa*w1^3)/(2*m1)(2*kappa*w3)/(m1*Vd)
 – (kappa*w3^3)/(3*m1*Vd^3) + w1^2*((kappa*w3)/(2*m1*Vd)) +
w2*((kappa*R*s)/(2*m1*Vd) + (kappa*R*s*w1*w3)/(2*m1*Vd^2) –
(kappa*R*s*w3^2)/(4*m1*Vd^3)(kappa*R*s*w3^4)/(16*m1*Vd^5) +
w1^2*(–(kappa*R*s)/(2*m1*Vd)))}

We can test the stability of the equilibrium point w=0 by examining the linearised zero dynamics computed at the end of the program.

Table 15.3   Nonlinear systems: geometric control

Function Name	Operation
VectorRelativeOrder	computes the relative degree vector
DecouplingMatrix	computes the decoupling matrix
IOLinearize	computes the linearising control
NormalCoordinates	computes the partial state transformation
LocalZeroDynamics	computes the local form of the zero dynamics
StructureAlgorithm	computes the parameters of an inverse system
DynamicExtension	applies dynamic extension as a remedy for singular decoupling matrix

The eigenvalues are readily obtained, but they are lengthy functions of the parameters. Some insight is obtained, however, by examining the special case, a=b and s=0, in which case the eigenvalues are

$$\lambda_1 = -\frac{2\kappa}{m_1 V_d}, \qquad \lambda_{2,3} = \pm \frac{\sqrt{a\kappa}}{\sqrt{I_{zz}}}$$

From this we see that the zero dynamics are unstable. Because the eigenvalues vary smoothly as a function of the parameters, this situation will be true for a-b and s small, but not necessarily zero. Furthermore, since Izz is small, $\lambda_{2,3}$ are a pair of 'parasitic' zeros, one of which is far into the right half-plane, and the other to the left. These locations may or may not make the vehicle difficult to control.

The technique for (numerically) computing the (local) zeros of a nonlinear control system is described in [21].

Some of the tools we have developed for the design of nonlinear control systems are listed in Table 15.3.

15.4 Design of variable structure systems

In this section we describe tools in **TSiControls** that support the design and implementation of variable-structure control systems. The tools enable the efficient design of sliding surfaces and reaching controllers including the inclusion of 'smoothing' and 'moderating' functions and the assembly of C source code for simulation and real-time implementation of the resulting control systems.

The specific methods we have implemented are described in detail in [21–26]. These references deal with variable structure control system design for smooth affine systems that are feedback linearisable in the input-output sense. Such systems are of the form:

$$\dot{x} = f(x) + G(x)u$$
$$y = h(x)$$

where f, G and h are sufficiently smooth and satisfy certain feedback linearisability conditions. All of the basic functions needed for design including reduction to regular form (as described in [22]) and computation of the zero dynamics (as previously reported in [21]), as well as functions for designing sliding surfaces as switching controllers, have been integrated into **TSiControls**. To do this has required extending Mathematica's facilities for working with nondifferentiable nonlinear functions. Ongoing work is focused on extending these techniques to plants containing hard nonlinearities such as dead zone, backlash, hysteresis and coulomb friction. We describe applications of this type below.

In this we summarise the methods and computations that we have implemented. We include some preliminary remarks concerning nonsmooth plant dynamics and we briefly discuss chattering reduction techniques. A very simple system with nonsmooth friction is used for illustrative purposes. The effects of control smoothing and moderation are illustrated.

15.4.1 Variable-structure control design

15.4.1.1 Sliding

Using feedback linearising constructions [22], we reduce the affino system to a regular form [19]. However, we do not feedback-linearise. Instead, we choose a variable-structure control law with switching surface of the form $s(x) = Kz(x)$, where K is chosen to stabilise $A + EK$ and $Z(x)$ are the linearisable co-ordinates. We can prove that, during sliding, the equivalent control is Kz, so that we achieve feedback linearised behaviour in the sliding phase.

15.4.1.2 Reaching

The second step in variable-structure control system design is the specification of the control functions $u_i^{\pm}$ such that the manifold $s(x) = 0$ contains a stable submanifold which ensures that sliding occurs. There are many ways of approaching the reaching design problem [27]. We consider only one. Consider the positive-definite quadratic form in s,

$$V(x) = s^T Q s$$

A sliding mode exists on a submanifold of $s(x) = 0$ which lies in a region of the state space on which the time rate of change V is negative. On differentiation we obtain

$$\frac{d}{dt} V = 2\dot{s}^T Q s = 2[KAz + \alpha]^T QKz + 2u^T \rho^T QKz$$

If the controls are bounded, $|u|_i \bar{U}_i > 0 \, (0 > U_{min,i} \le u_i \le U_{max,i} > 0)$, then obviously, to minimise the time rate of change of V, we should choose

$$u_i = U_{min,i} \text{ step}(s_i^* + U_{max,i} \text{ step}(-s_i^*), \qquad i = 1, \dots, m, \; s^*(x) = \rho^T(x)QKz(x)$$

Notice that if $U_{min,i} = -U_{max,i}$, the control reduces to

$$u_i = -U_{max,i} \text{ sign}(s_i^*)$$

In this case it follows that $\dot{V}$ is negative, provided

$$|U_{max}^T \rho^T QKz| > |[KAz + \alpha]^T QKz| \tag{15.17}$$

A useful sufficient condition is that

$$|(\rho(x)U_{max})_i| > |(KAz(x) + \alpha(x))_i| \tag{15.18}$$

Expr. 15.17 and 15.18 may be used to ensure that the control bounds are of sufficient magnitude to guarantee sliding and to provide adequate reaching dynamics. This rather simple approach to reaching design is satisfactory when a 'bang-bang' control is acceptable.

15.4.1.3 Chattering reduction

The state trajectories of ideal sliding motions are continuous functions of time contained entirely within the sliding manifold. These trajectories correspond to the equivalent control $u_{eq}(t)$. However, the actual control signal, $u(t)$—definable only for nonideal trajectories—is discontinuous as a consequence of the switching mechanism which generates it. Persistent switching is sometimes called chattering. In some applications this is undesirable. Several techniques have been proposed to reduce or eliminate chattering. One approach is 'regularisation' of the switch by replacing it with a continuous approximation. A second is 'extension' of the dynamics by using additional integrators to separate an applied discontinuous pseudocontrol from the actual plant inputs. A third is 'moderation' of the reaching control magnitude as errors become small.

Switch regularisation entails replacing the ideal switching function, $\text{sign}(s(x))$, with a continuous function such as

$$\text{sat}\left(\frac{1}{\varepsilon}s(x)\right) \quad \text{or} \quad \frac{s(x)}{\varepsilon + |s(x)|} \quad \text{or} \quad \tanh\left(\frac{s(x)}{\varepsilon}\right)$$

This intuitive approach is employed by Young and Kwatny [28] and Slotine and Sastry [16,29], and there are probably historical precedents. Regularisation induces a boundary layer around the switching manifold of size $O(\varepsilon)$. Within this layer the control behaves as a high-gain controller. The reaching behaviour is altered significantly because the approach to the manifold is now exponential and the manifold is not reached in finite time as is the case with ideal switching. On the other hand, within the boundary layer the trajectories are $O(\varepsilon)$ approximations to the sliding trajectories as established by Young et al. [30] for linear dynamics with linear switching surfaces. The justification for this approach for linear systems is provided by the results in [30]. Some of those results have been extended to single-input/single-output nonlinear systems by Marino [12]. Switch regularisation for nonlinear systems has been extensively discussed by Slotine and co-workers, e.g. [16,29]. With nonlinear systems there are subtleties, and regularisation can result in an unstable system.

Dynamic extension is another obvious and effective approach to control input smoothing. Our characterisation of it differs very little from that of Emelyanov *et al.* [31]. A sliding mode is said to be of *p*th order relative to an output y if the time derivatives $\dot{y}, \ddot{y}, \ldots, y^{(p-1)}$ are continuous in t but y^p is not. The following observation is a straightforward consequence of the regular form theorem. Suppose eqn. 15.8 is input-output linearisable with respect to the output $y = h(x)$ with vector relative degree $(r_r, \ldots, r_m)$. Then the sliding mode corresponding to the control law (eqn. 15.23) is of order $p = \min(r_1, \ldots, r_m)$ relative to the output y. We may modify the relative degree by augmenting the system with input dynamics as described. Hence, we can directly control the smoothness of the output vector y.

Control moderation involves design of the reaching control functions $u_i(x)$ such that $|u_i(x)| \to$ small as $|e(x)| \to 0$. For example,

$$u_i(x) = |e(x)|\text{sign}(s_i(x))$$

Control moderation was used by Young and Kwatny [28], and the significance of this approach for chattering reduction in the presence of parasitic dynamics was discussed by Kwatny and Siu [25].

15.4.1.4 *Example 7. A rotor with friction*

The following is a simple rotor with friction and input torque:

$$\dot{x}_1 = x_2$$
$$\dot{x}_2 = -\phi_{fr}(x_2) + u$$

Suppose the input torque u is bounded, say, $u \in [-U, U]$. We can easily show that the controller $u = -U \, \text{sgn}(cx_1 + x_2)$, $c > 0$, and U sufficiently large stabilises the origin for all piecewise-smooth friction functions with a discontinuity at the origin such that $\phi_{fr}(0)$ is bounded.

Consider a variable structure controller with

$$u(x) = \begin{cases} u^+(x) & s(x) > 0 \\ u^-(x) & s(x) < 0 \end{cases}, \qquad s(x) = cx_1 + x_2, c > 0$$

Imposing the sliding condition $s(x) \equiv 0$ leads to

$$\dot{x}_1 = -cx_1, \quad u_{eq} = -cx_2 + \phi_{fr}(x_2)$$

Now, we need to design the reaching control. Choose $V(x) = s^2(x)$ and compute

$$\dot{V} = 2(cx_1 + x_2)(cx_2 - \phi_{fr}(x_2) + u)$$

If u is bounded, say, $u \in [-U, U]$, choose $u = -U \, \mathrm{sgn}(cx_1 + x_2)$. Then

$$\dot{V} = 2 \, \mathrm{abs}(cx_1 + x_2)\{\mathrm{sgn}(cx_1 + x_2)(cx_2 - \phi_{fr}(x_2)) - U\}$$

Certainly, $\dot{V} < 0$ if $\mathrm{abs}(cx_2 - \phi_{fr}(x_2)) < U$. It follows that so long as $U > \sup \mathrm{abs}(\phi_{fr}(0))$ there is a neighbourhood of the origin $\mathcal{N}$ such that each trajectory beginning in $\mathcal{N}$ converges to the origin.

15.4.2 Computing tools

It should be clear that the calculations described above are most effectively carried out in a symbolic computing environment. Some of the functions we have implemented for the design of variable-structure control systems are listed in Table 15.4.

15.4.2.1 Sliding surface computations

There are several methods for determining the sliding surface, $s(x) = Kz(x)$, once the system has been reduced to normal form. We have included a function SlidingSurface that implements two alternatives depending on the arguments provided. The function may be called via

$$\{\mathrm{rho}, \mathrm{s}\} = \mathtt{SlidingSurface[f, g, h, x, lam]}$$

or

$$\mathrm{s} = \mathtt{SlidingSurface[rho, vro, z, lam]}$$

In the first case the data provided are the nonlinear system definition f, g, h, x and an m-vector lam which contains a list of desired exponential decay rates, one for each channel. The function returns the decoupling matrix rho and the switching surfaces s as functions of the state x. The matrix K is obtained by solving the appropriate Ricatti equation.

Table 15.4   Nonlinear systems: variable-structure control

Function Name	Operation
SlidingSurface	generates the sliding (switching) surface for feedback linearisable nonlinear systems
SwitchingControl	computes the switching functions – allows the inclusion of smoothing and moderating functions
SmoothingFunctions	an option for SwitchingControl that introduces specified smoothing functions
ModeratingFunctions	an option for SwitchingControl that introduces specified moderating functions

The second use of the function assumes that the input-output linearisation has already been performed so that the decoupling matrix rho, the vector relative degree and the normal co-ordinate (partial) transformation $z(x)$ are known. In this case the dimension of each of the m switching surfaces is known so that it is possible to specify a complete set of eigenvalues for each surface. Thus, lam is a list of m-sublists containing the specified eigenvalues. Only the switching surfaces are returned. In this case K is obtained via pole placement.

15.4.2.2 Switching control

The function is SwitchingControl[rho,s,bounds,Q,opts], where rho is the decoupling matrix, s is the vector of switching surfaces, 'bounds' is a list of controller bounds each in the form {lower bound, upper bound}, Q is an $m \times m$ positive-definite matrix (a design parameter), and 'opts' are options which allow the inclusion of smoothing and/or moderating functions in the control.

Smoothing functions are specified by a rule of the form

SmoothingFunctions[x_]-> {function1[x],...,functionm[x]}

where m is the number of controls. Moderating functions are similarly specified by a rule

ModeratingFunctions-> {function1[z],...,functionm[z]}

The smoothing function option replaces the pure switch sign (s) by a smooth switch as specified. The moderating function option multiplies the switch by a specified function. We give an example below.

15.4.2.3 Example 7 (continued)

We apply some of the above computations to Example 7. For illustrative purposes the friction function is taken to be

$$\phi_{fr} = \text{sign } \omega$$

and the plant equations are repeated as:

$$\dot{x}_1 = x_2$$
$$\dot{x}_2 = -\phi fr(x_2) + u$$

```
{rho2, s2} = SlidingSurface[f, g, h, {theta, omega}, {2}]
Computing Decoupling Matrix
Computing linearising/decoupling control
{{{1}}, {8.03066 omega + 16.1844 theta}}
```

Now, we compute the switching control using various combinations of smoothing and moderating functions:

1. no smoothing or moderating
2. smoothing
3. moderating
4. smoothing and moderating.

The particular functions chosen for this example are shown in Fig. 15.3.

We specify the control bounds as ± 5 and $Q = 1$. The following computation yields the four controls:

In[29] := vsc1 = SwitchingControl[rho2, s2, ctrlbnds, Q]
    vsc2 = SwitchingControl[rho2, s2,
        ctrlbnds, Q, SmoothingFunctions[x_]-> {(1 − Exp[−Abs[x/.1]])}]
    vsc3 = SwitchingControl[rho2, s2, ctrlbnds, Q,
        ModeratingFunctions->
        {(Abs[theta] + Abs[omega]/10)/(0.002 + Abs[theta] + Abs[omega]/
    vsc4 = SwitchingControl[rho2, s2, ctrlbnds, Q,
        ModeratingFunctions->
        {(Abs[theta] + Abs[omega]/10)/(0.002 + Abs[theta] + Abs[omega]/
    SmoothingFunctions[x_]- > {(1 − Exp[−Abs[x/.1]])}]

Out[29] = {5Sign[−8.03066omega − 16.1844theta]}

Out[30] = {5Sign[−8.03066omega − 16.1844theta]−
        $5E^{-10.\text{Abs}[-8.03066\text{omega}-16.1844\text{theta}]}$
        Sign[−8.03066omega − 16.1844theta]}

Out[31] = $\left\{ \dfrac{\text{Abs[omega]Sign}[-8.03066\text{omega}-16.1844\text{theta}]}{2(0.002 + \dfrac{\text{Abs[omega]}}{10} + \text{Abs[theta]})} + \right.$
        $\left. \dfrac{5\text{Abs[theta]Sign}[-8.03066\text{omega}-16.1844\text{theta}]}{(0.002 + \dfrac{\text{Abs[omega]}}{10} + \text{Abs[theta]})} \right\}$

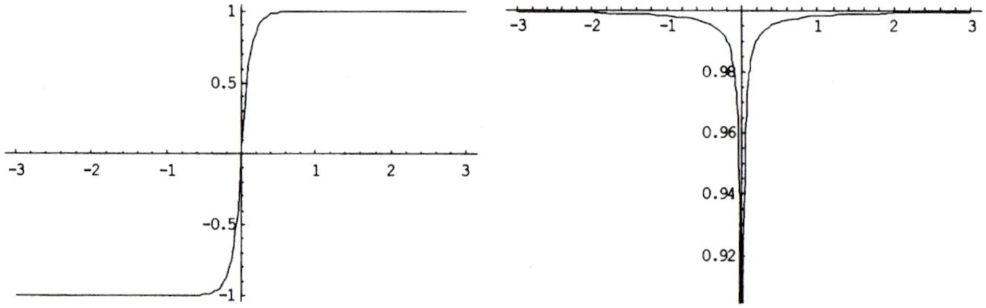

$$1 - e^{-Abs(x)/.1}$$

Smoothing Function

$$\frac{Abs(x)}{.005 + Abs(x)}$$
$$Abs(x) \leftarrow Abs(\theta) + Abs(\omega)/10$$

Moderating Function

Figure 15.3   Smoothing and moderating functions used in the example

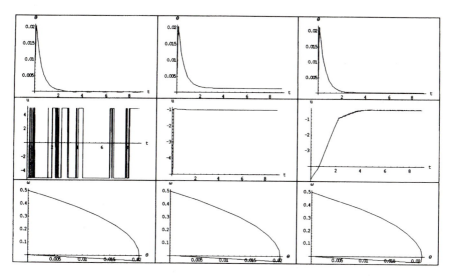

Figure 15.4   *This figure compares, left to right, pure switching control, switching with smoothing and switching with moderating. From top to bottom: q position response, control and the state trajectory*

$$\text{Out}[32] = \left\{ \begin{array}{l} \dfrac{\text{Abs[omega]Sign}[-8.03066\,\text{omega} - 16.1844\,\text{theta}]}{2\left(0.002 + \dfrac{\text{Abs[omega]}}{10} + \text{Abs[theta]}\right)} - \\[2em] \dfrac{5E^{-10.\text{Abs}[-8.03066\,\text{omega}-16.1844\,\text{theta}]}\text{Abs[omega]}\ \text{Sign}[-8.03066\,\text{omega} - 16.1844\,\text{theta}]}{2\left(0.002 + \dfrac{\text{Abs[omega]}}{10} + \text{Abs[theta]}\right)} + \\[2em] \dfrac{5\text{Abs[theta]Sign}[-8.03066\,\text{omega} - 16.1844\,\text{theta}]}{0.002 + \dfrac{\text{Abs[omega]}}{10} + \text{Abs[theta]}} - \\[2em] \dfrac{5E^{-10.\text{Abs}[-8.03066\,\text{omega}-16.1844\,\text{theta}]}\text{Abs[theta]}\ \text{Sign}[-8.03066\,\text{omega} - 16.1844\,\text{theta}]}{0.002 + \dfrac{\text{Abs[omega]}}{10} + \text{Abs[theta]}} \end{array} \right.$$

Figure 15.4 compares the closed-loop performance of the first three controllers.

15.5 Conclusions

Integrated design of systems and their controls requires tools for symbolic modelling and manipulation. Our approach involves the integration of symbolic and numerical computing. Successful modelling of complex vehicle dynamics and design of adaptive tracking controls for a detailed model of a magnetic bearing demonstrate that this methodology can solve realistic problems. The commercial packages **TSiControls** and **TSiDynamics** are now in use at several hundred installations.

15.6 References

1 CHEN, C. T.: 'Linear system theory and design' (Holtz, Rinkhart & Winston, NY)
2 AKHRIF, O.: 'Using computer algebra in the design of nonlinear control systems'. MS thesis, University of Maryland at College Park, 1987
3 CHANCELIER, J.P., GOMEZ, C., QUADRAT J.P., and SULEM, A.: 'Pandore' (Proc. NATO Advanced Study Institute on CAD of Control Systems, Il Ciocco, September 1987), in 'Advanced computing concepts and techniques in control engineering,' M. Denham and A. Laub (Eds.) (Springer-Verlag, New York, 1988) pp. 81–125
4 GHANADAN, R.: 'Adaptive control of nonlinear systems with applications to flight control systems and suspension dynamics.' PhD thesis, Institute for Systems Research, University of Maryland, College Park, 1993
5 GHANADAN, R., and BLANKENSHIP, G.L.: 'Adaptive approximate tracking and regulation of nonlinear systems.' Proc. of 32nd IEEE CDC, San Antonio, December 1993; also *IEEE Trans. Autom. Control*, to be published
6 GHANADAN, R., and BLANKENSHIP, G.L.: 'Adaptive output tracking of invertible MIMO nonlinear systems.' Proc. of 26th Conf. on Inf. Sciences and Systems, Princeton, 1992, pp. 767–772
7 BLANKENSHIP, G.L., GHANADN R., and POLYAKOV, V.: 'Nonlinear adaptive control of active vehicle suspension.' Proc. 1993 ACC, San Francisco, June 1993
8 MOHAMMED, A.M., and EMAD, F.P.: 'Conical magnetic bearings with radial and thrust control.' *IEEE Trans. Autom. Control*, 1992, **37**, pp. 1859–1868
9 ISIDORI, A.: 'Nonlinear control systems'. (Springer-Verlag, London, 1995, 3rd edn.)
10 POLYAKOV, V., GHANADAN, R., and BLANKENSHIP, G.L.: 'Nonlinear adaptive control of conical magnetic bearings'
11 KRENER, A.J.: 'Approximate linearisation by state feedback and coordinate changes.' *Syst Control Lett*, 1984, **5**, pp. 181–185
12 MARINO, R.: 'High gain feedback non-linear control systems.' *Int. J. Control*, 1985, **42**, (6), pp. 1369–1385
13 HIRSCHORN, R.M.: 'Invertibility of multivariable nonlinear systems.' *SIAM J. Optim. Control*, 1979, **17**, pp. 289–297
14 HIRSCHORN, R.M.: 'Invertibility of multivariable nonlinear control systems.' *IEEE Trans. Autom. Control*, 1979, **AC-24**, (6), pp. 855–865
15 SINGH, S.N.: 'A modified algorithm for invertibility in nonlinear systems.' *IEEE Trans. Autom. Control*, 1981, **25**, pp. 595–598
16 SLOTINE, J.J. and SASTRY, S.S.: 'Tracking control of non-linear systems using sliding surfaces, with application to robot manipulator.' *Int. J. Control*, 1983, **38**, (2), pp. 465–492
17 HAUSER, J., SASTRY, S., and KOKOTOVIC, P.: 'Nonlinear control via approximate input-output linearisation: the ball and beam example.' *IEEE Trans. Autom. Control*, 1992, **37**, pp. 392–398
18 SINGH, S.N.: 'Decoupling of invertible nonlinear systems with state feedback and precompensation.' *IEEE Trans. Autom. Control*, 1980, **AC-25**,(6), pp. 1237–1239
19 LUK'YANOV, A.G. and UTKIN, V.I.: 'Methods of reducing equations of dynamic systems to regular form.' *Avtom Telemekh*, 1981, (4), pp. 5–13

20 BLANKENSHIP, G.L., GHANADAN, R., KWATNY, H.G., LAVIGNA, C., and POLYAKOV, V.: 'Tools for integrated modeling, design, and nonlinear control.' *IEEE Control Syst. Mag.*, 1995, **15**,(2), pp. 65–79

21 KWATNY, H.G., and BLANKENSHIP, G.L.: 'Symbolic tools for variable structure control system design: the zero dynamics.' IFAC Symposium on Robust Control via Variable Structure and Lyapunov Techniques, Benevento, Italy, 1994

22 KWATNY, H.G., and KIM, H.: 'Variable structure regulation of partially linearisable dynamics.' *Syst. Control Lett.*, 1990, **15**, pp. 67–80

23 KWATNY, H.G., and BERG, J.: 'Variable structure regulation of power plant drum level,' in J. Chow, R.J. Thomas and P.V. Kokotovic (Eds.), 'Systems and control theory for power systems' (Springer-Verlag, New York, 1995), pp. 205–234

24 KWATNY, H.G., and BERG, J.: 'Variable·structure regulation of power plant drum level', in J. Chow, R.J. Thomas and P.V. Kokotovic (Eds.) 'Systems and control theory for power systems,' (Springer-Verlag, New York, 1993)

25 KWATNY, H.G., and SIU, T.L.: 'Chattering in variable structure feedback systems.' 10th IFAC World Congress, 1987, Munich

26 KWATNY, H.G.: 'Variable structure control of AC drives', in K.D. Young (Ed.),'Variable structure control for robotics and aerospace applications.' (Elsevier, Amsterdam, 1993)

27 UTKIN, V.I.: 'Sliding modes and their application' (MIR, Moscow, 1974, in Russian; 1978 in English)

28 YOUNG, K.D., and KWATNY, H.G.: 'Variable structure servomechanism and its application to overspeed protection control.' *Automatica*, 1982, **18**, (4), pp. 385–400

29 SLOTINE, J.J.E.: 'Sliding controller design for non-linear control systems.' *Int. J. Control*, 1984, **40**, (2), pp. 421–434

30 YOUNG, K.D., KOKOTOVIC, P.V., and UTKIN, V.I.: 'Singular perturbation analysis of high gain feedback systems.' *IEEE Trans. Autom. Control*, 1977, **AC-22**, (6), p. 931

31 EMELYANOV, S.V., KOROVIN, S.K., and LEVANTOVSKY, L.V.: 'A drift algorithm in control of uncertain processes.' *Prob Control Inf. Theory*, 1986, **15**, (6), pp. 425–438

32 KWATNY, H.G., and BLANKENSHIP, G. L.: 'Symbolic construction of models for multibody dynamics.' *IEEE Trans. Robot. Autom.* 1995, **11**, pp. 271–281

33 NIJMEIJER, H., and VAN DER SCHAFT, H.J.: 'Nonlinear dynamical control systems.' (Springer-Verlag, New York, 1990)

34 PRESS, W.H. *et al.*: 'Numerical recipes in C: the art of scientific computing' (University Press, New York, 1992)

Index